*Mammal-like Reptiles and the Origin of Mammals*

# *Mammal-like Reptiles and the Origin of Mammals*

*T. S. KEMP*
*University Museum and Department of Zoology*
*Oxford, England*

1982

ACADEMIC PRESS

*A Subsidiary of Harcourt Brace Jovanovich, Publishers*

LONDON NEW YORK
PARIS SAN DIEGO SAN FRANCISCO SÃO PAOLO
SYDNEY TOKYO TORONTO

ACADEMIC PRESS INC. (LONDON) LTD.
24/28 Oval Road,
London NW1

*United States Edition published by*
ACADEMIC PRESS INC.
111 Fifth Avenue
New York, New York 10003



*British Library Cataloguing in Publication Data*

Kemp, T.S.
Mammal-like reptiles and the origin of mammals.
1. Vertebrates, Fossil 2. Vertebrates—Anatomy
3. Vertebrates—Evolution
566 QE841

ISBN 0-12-404120-5

LCCCN 81-67894

Phototypesetting by Oxford Publishing Services, Oxford
Printed in Great Britain by Whitstable Litho Ltd.,
Whitstable, Kent

*In memory of*
*Dr F. R. Parrington, FRS—*
*a great teacher*

# Preface

This book is an attempt to summarise our present knowledge about the nature and evolution of the mammal-like reptiles and their early mammal descendants. The subject has not been comprehensively reviewed before, the information is widely scattered throughout the specialised scientific literature, and the treatment of the group in standard vertebrate text-books is often hopelessly out of date and biased towards views not by any means universally accepted by research workers in the field. This has always seemed to me a pity, because here is perhaps the most revealing part of the entire terrestrial fossil record, complete enough and extensive enough to illustrate many aspects of the patterns and processes of large-scale evolution.

Several developments of recent years have influenced my approach. The most obvious is the increase in basic anatomical knowledge of the group, derived partly from the discovery of important new finds, and partly from greatly improved methods of laboratory preparation of specimens. A second is the rise of cladistic analysis of evolutionary relationships, which has greatly increased the clarity and precision with which phylogenetic relationships are established. A third has been the spread of interest into the functional interpretation of fossil structure and the presumed evolutionary modifications to anatomy. A fourth is the revolution in palaeontology of the last decade that has so stimulated attempts to use fossils for the elucidation of the mechanisms underlying macroevolution, and the causes behind long-term evolutionary change.

The book has developed from a series of lectures I give to Oxford University Zoology undergraduates, and as such I suppose it is intended primarily for this level. I hope, however, that it may contain some interest for all who are involved in vertebrate palaeontology and in evolution generally, both at the research level and more widely.

I should like to take this opportunity of thanking all those who have helped me in the production of this work. Dr Gordon Chancellor, Dr Arthur Cruickshank and Dr Gillian King have all read parts of the manuscript and offered many helpful comments. I am particularly grateful to Dr Adrian Friday for reading the whole draft. Of course, I retain complete responsibility for any omissions, inaccuracies, inconsistencies and want of logic that emerge.

Most of the figures have been taken from the literature, and I am grateful to the numerous authors and publishers who gave me permission to make use of published illustrations. Individual acknowledgements have been given where appropriate. I thank David Nicholls for the original Figure 41 A–C and Figure 42 A–C, and similarly Gillian King for Figure 52 E–G. Denise Blagden patiently and sympathetically undertook all the photography associated with the book, and Mrs O. Hebden skilfully deciphered my scrawl to produce the final typescript. I thank both of them. I should also like to express my gratitude to Mrs Dorothy Sharp of Academic Press for the efficient, friendly way that she got this book into a presentable form.

Finally, and most of all, my thanks to my wife Sue for a great deal of patience and tolerance over the last three years.

*October 1981* TOM KEMP

# Contents

# *List of Figures*

# 1 Introduction

THE MAMMAL-LIKE REPTILES or synapsids are one of the great Orders of reptiles and they are of particular interest because it was from them that the mammals evolved. Although the synapsids have been extinct since about the end of the Triassic period some 190 million years ago, a remarkably detailed fossil record of them remains. The first of the primitive members of the group dates from the early part of the Upper Carboniferous (Lower Pennsylvanian in North American terminology), over 300 million years ago, and in the course of their history mammal-like reptiles radiated widely to become for a time the dominant members of the terrestrial fauna. Indeed, the popularly termed "Age of Reptiles" consisted of two distinct phases. The first phase was very largely the mammal-like reptiles and it lasted throughout the later part of the Carboniferous and the Permian. During the succeeding Triassic Period, they were gradually replaced by the more familiar dinosaur reptiles of the Mesozoic, leaving only their mammalian descendents. Throughout the long Jurassic and Cretaceous Periods the mammals remained very small animals, mostly insectivorous, whilst the dinosaurs occupied all the niches for large terrestrial forms. It was not until the eventual extinction of the dinosaurs at the close of the Cretaceous that the mammals could radiate and flourish, evolving in the course of time into the different kinds of mammals familiar today.

Due to a series of fortunate geological coincidences the fossil record of the mammal-like reptiles is more complete than that of any other group of terrestrial vertebrates, with the exception of the Tertiary mammals. Moreover, their evolution spanned a huge morphological progression, from early forms of a very primitive reptilian grade through to others which are technically to be regarded as mammals. Therefore this is the one example known where the evolution of one class of vertebrates from another class is well documented by the fossil record. Even amongst the invertebrates there is no example of a comparable morphological or

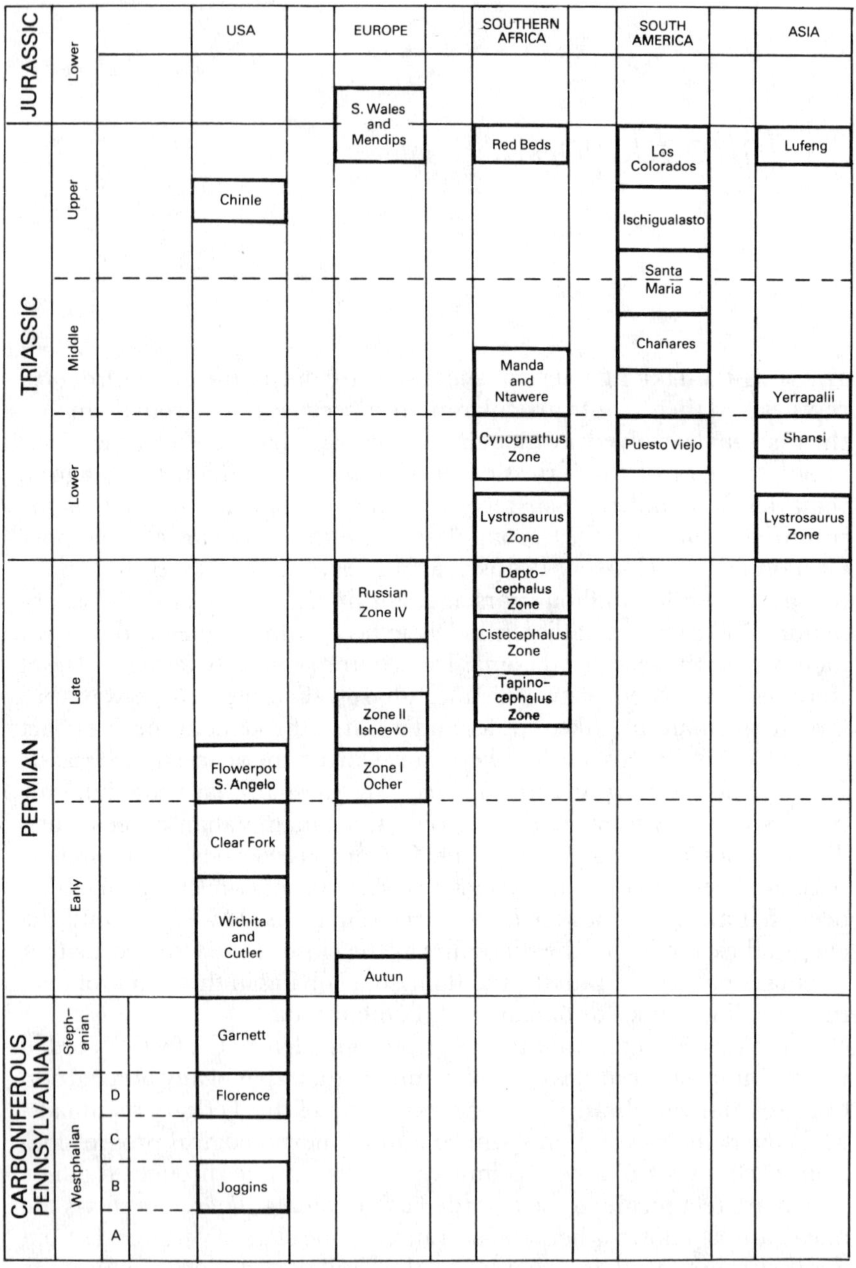

| Period | Epoch | Stage | USA | EUROPE | SOUTHERN AFRICA | SOUTH AMERICA | ASIA |
|---|---|---|---|---|---|---|---|
| JURASSIC | Lower | | | S. Wales and Mendips | | | |
| TRIASSIC | Upper | | Chinle | S. Wales and Mendips | Red Beds | Los Colorados<br>Ischigualasto<br>Santa Maria | Lufeng |
| TRIASSIC | Middle | | | | Manda and Ntawere | Santa Maria<br>Chañares | Yerrapalii |
| TRIASSIC | Lower | | | | Cynognathus Zone<br>Lystrosaurus Zone | Puesto Viejo | Shansi<br>Lystrosaurus Zone |
| PERMIAN | Late | | Flowerpot S. Angelo | Russian Zone IV<br>Zone II Isheevo<br>Zone I Ocher | Dapto-cephalus Zone<br>Cistecephalus Zone<br>Tapino-cephalus Zone | | |
| PERMIAN | Early | | Clear Fork<br>Wichita and Cutler | Autun | | | |
| CARBONIFEROUS PENNSYLVANIAN | Steph-anian | | Garnett | | | | |
| CARBONIFEROUS PENNSYLVANIAN | Westphalian | D | Florence | | | | |
| CARBONIFEROUS PENNSYLVANIAN | Westphalian | C | | | | | |
| CARBONIFEROUS PENNSYLVANIAN | Westphalian | B | Joggins | | | | |
| CARBONIFEROUS PENNSYLVANIAN | Westphalian | A | | | | | |

FIG. 1.

taxonomic distance being illustrated by a good, continuous fossil record. Of course there are many gaps in the synapsid fossil record, with intermediate forms between the various known groups almost invariably unknown. However, the known groups have enough features in common that it is possible to reconstruct hypothetical intermediate stages. The record is also geographically patchy, no locality yielding more than a relatively short segment of the history of the mammal-like reptiles, and in many cases a region contains fossils of a single age. Similarly, no single taxonomic group of synapsid occurs world-wide even though there is little doubt that at least some of them had a very extensive distribution in life.

Figure 1 charts the major localities of mammal-like reptiles in time and space. The vast majority of the early synapsids occur in Upper Carboniferous and Early Permian rocks of North America, along with a few isolated localities in various parts of Europe. The American record ceases at the beginning of Late Permian times, but more or less contemporaneous and slightly younger deposits in Russia allow the story to be continued. The focus then shifts to the great Karroo System of southern Africa, where Late Permian through to Lower Triassic forms occur in abundance. Middle and Upper Triassic Synapsids are also present in parts of the Karroo, but are far better represented in certain regions of South America. The mammals evolved some time in the later part of the Triassic and the earliest members are most abundantly found in deposits in South Wales, although similar animals have been discovered in South Africa and China, as well as mainland western Europe. The Mesozoic mammals of the Jurassic and Cretaceous are still poorly known, as much on account of their small size as anything, and occur in several localities around the world. The intercontinental shifts that are necessary in order to follow the story of the mammal-like reptiles and their mammalian descendants is of far less significance than might be supposed, since throughout the period of time spanned by their history, the present continents were joined together as the supercontinental land mass Pangaea. The various synapsid groups probably had wide distributions and therefore knowledge of them from one particular area may well reflect their nature in other areas.

The quality of the individual fossils varies enormously. At one extreme there are specimens so incomplete, poorly preserved and distorted that they give no more information than the fact of their presence. At the

Fig. 1. Chart of the main localities yielding mammal-like reptiles. (Data from Chudinov 1965; Reisz, 1972; Anderson and Cruickshank, 1978.)

other extreme, however, there are complete skeletons so well preserved that the animals could almost have died only a short time ago. One of the chief attractions of studying vertebrate fossils is that the skeleton closely reflects many of the biological features of the animal. A well-preserved specimen allows the investigator to consider such attributes as the mechanism of feeding and diet, the mode of locomotion, the nature of certain of the sense organs and so on, to a degree quite impossible for most of the invertebrate groups. This high biological information content of the individual fossils, coupled with the extensive record of the evolution of the group puts the mammal-like reptiles in an exceedingly good position for studying several fundamental aspects of the evolutionary process.

There are several reasons for studying mammal-like reptiles in depth. At a superficial level they have an intrinsic interest simply as a group of extinct animals unlike any that occur today, and which include remote ancestors of Man. Although lacking some of the drama of the dinosaurs, they nevertheless included a number of bizarre forms, and produced certain interesting solutions to various of the problems of life on land. Synapsids are also of some use to geologists as stratigraphic indicators of the relative ages of the continental rocks in which they are found, and they throw some light on the palaeoenvironmental conditions of their times.

However, by far the greatest importance of this group of animals concerns the way in which it can be used to investigate the processes and patterns of evolution over long periods of time, and involving major morphological transitions. This is the one important area of biology in which fossils alone provide the information to test hypotheses, and should indeed be seen as the ultimate justification for the science of palaeontology. The broad, long-lasting and highly complex radiation of the mammal-like reptiles, coupled with the considerable knowledge of the functional anatomy and biology of the members can be used to illustrate what may well be a fairly general kind of pattern of evolution of at least terrestrial organisms. It may therefore form the basis of a model of evolution, and be used to generate questions about the causes of phylogenetic divergence in general. A more specific macro-evolutionary question that arises from the fossil record of the mammal-like reptiles concerns the way in which a major new taxon arises from its ancestors. The details of the structural, functional and environmental features involved in the origin of the mammals may again form the basis of generalisations, this time about large evolutionary advances in organisms.

Following a brief consideration of the scientific methods available to palaeontology, the bulk of this book is devoted to a documentation of the geological occurrence, nature and relationships of the various groups of mammal-like reptiles and early mammals, treated in systematic order. For each main group an attempt is made to interpret the functional anatomy and biology, in the light of current knowledge. Inevitably the systematic approach has tended to obscure the equally interesting comparative approach, but it is hoped that the layout is such that it is possible to follow the evolution of any particular system, for example jaw musculature, from group to group with reasonable ease.

The final two chapters contain attempts to interpret the fossil record of the mammal-like reptiles in broad macroevolutionary terms.

# 2 Methods

PROGRESS IN PALAEONTOLOGY, as in all science, depends on the association of observed facts as hypotheses. These suggest relationships between the observations, for example as causes and effects. The hypothesis must also be expressed in such a way as to predict that certain other things will be true and it can therefore be tested by checking whether this is so. If the predictions are not verified, the hypothesis is said to have been refuted, and must be modified or discarded.

Understanding the methods of study which are available to any particular branch of science is essential because it is these that define the tests that can be applied to the hypotheses which observations generate. Thus the methodology defines the limits of knowledge of the particular field of enquiry. Palaeontology, certainly as much as any other branch of biology, and perhaps more than most, is prone to speculation. This consists of ideas that cannot be falsified, because suitable methods for testing them are simply not available. The most obvious constraint to our gaining knowledge of fossil animals is a result of the limited extent to which it is possible to conduct experiments on them or on their evolution. Nevertheless, one of the striking developments in the last few years has been the continuing attempts to define and refine those methods which are available to the palaeontologist.

## Practical methods

### *Finding and collecting fossils*

The initial discovery of most of the localities yielding mammal-like reptiles and early mammals was a fortuitous by-product of geological work throughout the world, often conducted in association with prospecting for coal and minerals. Mention of the remains of bones would lead to a more intensive search by palaeontologists and the beginning of serious collecting. It is an unfortunate fact that knowledge of the nature of the fossilisation process is very limited, and few successful predictions beforehand about likely sites have ever been made. Therefore palaeontological hypotheses which may predict that certain types of fossil must have occurred at certain times and in certain places cannot generally be tested by simply looking for such specimens at appropriate localities. The failure to find the expected fossils does not necessarily falsify the hypothesis, since it may be a result of failure for some unknown reason for the fossils to be preserved there.

Methods of actually collecting fossils from the localities have not changed much over the years, except perhaps for an increasing carefulness to pick up all the pieces. Before the advent of modern transportation, understandable but regrettable examples occurred of the postcranial skeletons of large individuals being left behind because they were difficult to move. Thus in certain older collections, the presence of a skull alone cannot be taken as a guarantee that the rest of the skeleton had not been preserved. Another collecting bias often found is the failure to collect very poor specimens of common forms in rich localities. Measurements of the relative numbers of different forms may therefore be quite misleading, and of no great value in ecological assessments. The only important innovation in collecting seen in recent years is the widespread application of screening techniques for accumulating microfossils such as mammal teeth from a bulk matrix. The matrix is broken down and washed through sieves, and the concentrate retained for laboratory investigation. Although greatly increasing the potential yield, one limitation of such methods is the tendency to lose associations between fragments of the same individual specimens.

The extent to which details of the location and lithology of the fossils will prove important in subsequent evolutionary interpretations cannot always be predicted at the time of collection. Many collections in the past have not been accompanied by what is now regarded as adequate data, but a growing awareness of the possible significance of such extraneous information has led to much better field documentation.

## *Preparation*

With rare exceptions, fossils of mammal-like reptiles come to the laboratory in consolidated nodules of rock of one sort or another. Removal of the rock is frequently the most time consuming part of the palaeontological exercise, and has often not been carried out as far as possible. Most early descriptions concerned only superficial aspects of the skeleton, such as the external views of the suture pattern of the skull; even much of the more recent work has been barely more detailed. For purposes of classification and to some extent assessment of phylogenetic relationship, such studies are adequate. For a more comprehensive understanding of the nature and evolution of particular forms however, a more complex degree of preparation may be required. The development of the acetic acid technique, whereby the matrix can be removed by dissolving in acetic acid, while the bone remains, has produced a small revolution in preparation of vertebrate fossils. It only works if the matrix is high in calcium carbonate, while the bone is still fairly pure phosphate, and even then is a rather slow process. However, in the cases where it does work well, it is possible to remove the matrix entirely and without damage to the surfaces of the bone, revealing a great deal more anatomical detail.

Even in cases where acetic acid techniques do not work, they have nevertheless had an oblique influence on other methods of preparation. Mechanical methods, mainly the use of a reciprocating needle in the dental mallet or engraver tend to be taken further than in the past in order to try to at least approach the degree of matrix removal of the acid method.

As the desire for ever more detailed structural information grows, well-known specimens are frequently restudied using more detailed methods of preparation than were previously applied, and it is probably true to say that furtherance of the study of mammal-like reptiles still depends to a greater extent on the detailed reinvestigation of known material than the discovery of new specimens.

### *Anatomical reconstruction*

Almost without exception, mammal-like reptiles and early mammals have suffered damage during the processes of fossilisation. Breakage and loss of certain parts of the skeletons before they were finally entombed is frequent, while a plastic distortion of the bones, particularly the skull, is practically universal. Presumably it results from the compression of the sediments as rock accumulates above them. As most specimens are discovered because they have become partially exposed on the surface of the ground, a degree of damage due to weathering is also common.

The extent to which these various kinds of damage can be corrected for, in the course of study, is very much a matter of the skill and experience of the investigator. If several specimens happen to be available, then together they will normally permit a more confident reconstruction of the anatomy because each will have suffered damage and distortion in different ways. The assumption that the right and left sides of the animal were identical is also helpful of course. Otherwise, however, it is a question of trying to assess the degree and nature of the damage, and then making a subjective attempt to restore the structure to a reasonably life-like conformation. The conventions of vertebrate palaeontology require the investigator himself to make a restoration of the fossil, complete with illustrations, in its supposed original form. This is on the grounds that such a person is likely to be in the best position to perform the process satisfactorily, being intimately familiar with the material. However, because of the inevitable subjective element of the reconstruction that ensues, there is always the possibility of error. The reported "facts" of the anatomy may not always be as reliable as supposed and careful reading of the accompanying text may be necessary in order to discover the areas and regions of doubt. Regrettably not all the papers on the subject are as explicit as they ought to be about the level of confidence of the reconstructed anatomy.

## Theoretical methods

### *Functional hypotheses*

The first level of interpretation of the anatomy of fossils concerns the purposes of the various parts of the skeleton. The underlying assumption is that the structure is adapted to perform particular functions or groups of functions. In the study of living organisms, hypotheses about function can be tested by direct observations, noting the actions of the structure in question, and by experimental manipulations of one sort or another. In the case of fossils neither of these kinds of tests are possible and more indirect methods have to be used. The first method is comparison of the fossil structure with anatomically similar structures of known function in living animals. By analogy the same function can then be ascribed to the fossil form. The limitations of this particular approach are evident. The fossil structure is unlikely to resemble the living structure in any great detail, and yet the very differences may relate to important functional differences. Indeed in many cases concerning the mammal-like reptiles, structures are present which simply have no reasonable analogy amongst living tetrapods.

The second and much more important method of testing a functional hypothesis is based on the details of the design of the structure in question. The hypothesis predicts that in order to function as proposed, the structure will possess certain mechanical properties, an appropriate arrangement of muscles as indicated by the bone surfaces, and possibly other features such as suitable foramina for nerves or blood vessels, or evidence of say, glands. In so far as these can be demonstrated on the fossil, the hypothesis is supported.

There will be complications in functional analyses in many cases, because a particular structure may be involved in more than one function. It will therefore be a compromise between the idealised design appropriate for each separate function. The role of the vertebral column in supporting the animal against gravity, in transmitting the locomotory forces of the limbs, and protecting the nerve cord, is an example. Structure of the vertebrae cannot be interpreted in the light of only one of these functions. There may also be cases where the precise design of the structure is not particularly critical for its functioning, for example social signalling structures.

A third possible method of testing a functional hypothesis arises when something is known of the evolutionary changes that occurred to a particular structure. Interpretation of the later, more highly evolved version may be assisted by the interpretation of the earlier version, and vice versa. Such reasoning does however devolve into a circular argument at times, and at least one version of the structure needs to be analysed without reference to the other.

### *Phylogenetic hypotheses*

Of all the methods and techniques of palaeontology, none have been more intensively scrutinized in recent years than those whose purpose is the interpretation of the evolutionary relationships between organisms.

The traditional or classical method of reconstructing the phylogenetic relationships of organisms is based on a comparison of homologous characters, those forms with the most similarities being regarded as most closely related to one another. The main drawback to this approach is that the rates of morphological evolution along different lineages appear to vary enormously. Therefore forms which have a relatively recent common ancestor, crocodiles and birds for example, may be very dissimilar organisms superficially. Forms whose last common ancestor was relatively remote, such as crocodiles and lizards, may have a great many points of similarity, simply because neither type has changed to any great extent from that common ancestor. A conflict arises because the general degree of similarity does not necessarily reflect the closeness of phylogenetic relationship.

The response to this problem has been the rise of the cladistic method of phylogenetic interpretation, which incorporates much more clearly defined rules and is based essentially on two simple tenets. The first is the assertion that new taxa generally arise initially by the splitting of an ancestral species into two daughter species. A complete phylogenetic diagram showing the relationships of several forms should consist therefore of a hierarchical arrangement of dichotomous pairs of what are called sister groups. The possibility that a single species might simultaneously split into more than two species, a polychotomy, exists in theory, but in practice is regarded as unlikely. In any case, it is held to be impossible to distinguish from the resulting taxa whether the evolutionary event was, say a trichotomous split or two successive dichotomous splits. The latter is therefore normally assumed.

The second tenet of cladistics is that all the descendants of a single hypothetical common ancestor should possess certain characters which had evolved uniquely in the common ancestor (Fig. 2). These are the uniquely derived, or synapomorphic characters of the descendants. Put

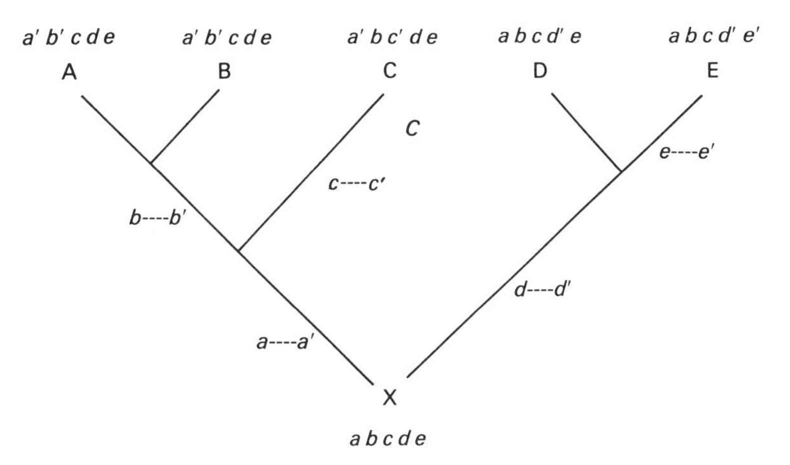


FIG. 2. Cladistic analysis of the relationships of five taxa A–E. All are descended from a hypothetical common ancestor X, which is assumed to have had the five characters *a–e*, all in the ancestral or plesiomorphic state. Evolution of the respective characters from the ancestral to a new, derived state are indicated by *a----a'*, *b----b'* etc. Assuming that these changes each occurred once only, then the only possible interrelationships of A–E are those indicated. Those taxa which share a common ancestor more recent than X possess derived characters, or synapomorphies in common. A, B and C all possess *a'*, and A and B share *b'* for example. It will be seen that the common possession of ancestral characters, such as *e* in A, B, C and D, or *b* in C, D and E, has no bearing on the relationships.

the other way round, the presence of synapomorphic characters in certain forms demonstrates that those forms constitute a monophyletic group, having evolved from a common ancestor that did not also give rise to other forms which lack the synapomorphies. It follows that only the synapomorphic characters and their distribution amongst the taxa in question can be used to work out phylogenetic relationships. Primitive, or plesiomorphic, characters which may be present not only in some of the descendants and their common ancestor, but also in other forms which branched off before the evolution of the common ancestor, have no value in phylogenetic determination.

The logical basis for relying solely on synapomorphy to determine phylogenetic relationships is indisputable, but of course the conclusions can be no more reliable than the reliability with which particular characters are judged to be synapomorphic. By definition, a synapomorphic character is a newly evolved rather than primitive feature, and also one that has not arisen more than once by parallel evolution; and there are several criteria for assessing both these aspects of the characters. The commonest test of whether a character is derived rather than primitive is to look at a wide range of relatively (and presumed) unrelated forms, such as reptiles in general. Primitive characters will be expected to occur at least sporadically amongst them, and the alleged derived character not at all. The geological age of appearance of the character may also indicate whether it is derived or primitive, since the latter state must have occurred before the former. This test is clearly somewhat ambiguous, because the fossil record being incomplete, it is possible for a form with the derived character to be actually known from an earlier date than forms which have retained the ancestral character. Nevertheless, with a good fossil record this remains an important criterion for certain groups, including the mammal-like reptiles. Some indication of whether a particular character present in modern organisms is derived may be gained from a knowledge of its embryological development. Generally, although by no means always, the more primitive characters appear earlier and the more evolved characters later. None of these tests is completely reliable, but taken together they are believed to give a strong indication of the state of a particular character.

Parallel evolution of a certain derived character independently in more than one lineage can be tested for in two ways. The first way concerns the degree of similarity of the particular character in the different forms. Frequently parallel evolution has been detected because the character in question has been shown to differ unexpectedly in different groups. This may indicate that the exact evolutionary pathway by which the character was achieved differed in the different lines.

The second possible indication is when the character which is suspected to have evolved in parallel is not associated with other synapomorphies, and therefore leads to a phylogeny manifestly different from that based on several other characters.

In many ways, the most important test of synapomorphy is whether several apparent synapomorphies occur independently in the same taxon. The greater the number of such characters found, the greater becomes the probability that the taxon is a monophyletic group, and therefore that the supposed synapomorphies are indeed just that. Nevertheless, there are many cases where one cannot, with any confidence, recognise suitably distributed synapomorphic characters amongst the taxa under study. The usual reason for this is probably that the actual taxa to hand have evolved considerably since their original differentation from one another, and the original synapomorphic characters have subsequently evolved further in several different ways, and can no longer be recognised. These cases are referred to as unresolved dichotomies, and their resolution depends either on better fossil specimens showing hitherto undescribed characters, or else the discovery of more primitive members of the groups in which the original synapomorphies are still present. Meanwhile, the situation must be recognised for what it is and the degree of ignorance about their phylogenetic interrelationships admitted.

A secondary tenet of the cladistic approach to phylogenetics is that ancestral forms cannot be recognised amongst known fossil specimens. They are always regarded as hypothetical. The common ancestor of a monophyletic group must have possessed those synapomorphies that are present in all its descendants, but none of those which define subgroups among the descendants. More importantly it could not have possessed any synapomorphic characters where the homologous character of the descendants is in the primitive condition, which is to say that the common ancestor could not have had any specialisations of its own. Since actual ancestors must of necessity be extinct they could only occur as fossil forms, which will have relatively few characters preserved compared to living organisms. It may be argued therefore that one can never assert than an alleged actual ancestor did not have specialisations of certain of the characters which are not preserved. The possibility always remains that such a specimen or taxon had branched off the evolutionary line and was not actually ancestral to any other known form. The cladistic point of view over the matter of ancestors has caused a certain amount of controversy, since more classically minded taxonomists are generally prepared to recognise ancestry. In fact, the difference between the two positions is semantic. Few organisms or taxa in the fossil record qualify for actual ancestral status on anyone's reckoning, since they almost always have some manifest specialisation. And when a classical taxonomist does claim ancestral status for a particular taxon compared to another, he naturally implies that the taxon could be ancestral in all the features we know about, and therefore is a suitable model for the actual ancestral taxon. The possibility that the alleged ancestor merely resembled the actual ancestor extremely closely, but had actually diverged slightly away from it can never be discounted.

The phylogenetic discussions in the present work are based on the cladistic methods of using apparent synapomorphies, reconstructing phylogenies as dichotomous hierarchies where possible, and generally avoiding the use of known taxa as direct ancestors of other taxa.

The construction of formal classifications based upon phylogenetic reconstruction has led to a great deal more controvery and misunderstanding than the method of phylogenetic reconstruction itself. A classification is held to serve two functions: an expression of the relationships of the organisms concerned, and a device for aiding communication amongst biologists by offering generalised statements that apply to certain sets of similar organisms. The lack of correlation between the closeness of phylogenetic relationship on the one hand and the overall degree of similarity in structure on the other leads to a certain level of conflict between the two desired properties of a classification. A strictly cladistic classification arbitrarily accepts only the criterion of relationship. The taxa are defined only in terms of the hierarchical branching pattern of the phylogeny, and every time a new branching point is reached a new pair of sister taxa are defined at the next lowest rank. Whilst this approach leads to an unambiguous expression of phylogenetic beliefs, it involves certain difficulties in the field of communication. Taxa at the same rank may contain sets of species with widely different morphological ranges. A species for example could well prove to be the sister group of a whole class, and therefore could finally be classified at the same rank. Equally, very similar forms can become classified separately at quite a high taxonomic level. There is also necessarily a complex multiplication of taxa when a large group of organisms is to be classified. The result is therefore a very unfamiliar classification, which does not correlate easily with the widely accepted classifications hitherto used. There are also problems associated with how to deal with unresolved sets of dichotomies.

In contrast, the more familiar forms of classification of classical taxonomists suffer from ambiguity because subjective decisions must be made on exactly where to draw lines between arbitrarily defined taxa.

They do, however, permit the association of broadly similar organisms in the same taxa, and the erection of taxa at the same rank for broadly comparable sets of species. The approach adopted in the present work is that phylogenetic relationships are best expressed in the form of diagrams, with no particular regard to the ranking of the various groups of animals contained. For purposes of communication, the classifications used are more classical and therefore more familiar, easily understood, and comparable with previous classifications of this particular group of organisms.

## *Macroevolutionary hypotheses*

The kinds of hypothesis discussed so far concern the course of evolution of the organisms studied and the functional significance of the changes which occurred. Macroevolutionary hypotheses may be defined as higher-order hypotheses which attempt to explain how and why a particular evolutionary pattern took place.

The starting point of a macroevolutionary hypothesis is the evolutionary pattern itself, as deduced from the fossil record and incorporating both the known fossil forms and the hypothetical intermediate forms in a phylogeny. Deductions already made concerning the functional and biological importance of the particular evolutionary steps are added. The factors which caused and controlled the evolution along the particular path it took must now be considered. A direct knowledge of these is not possible and reliance is placed on analogy with modern organisms, assuming that the general biological properties of the extinct animals were identical to those of the living, and therefore that the kinds of evolutionary mechanisms available now were equally available in the past. To the best of present beliefs, organisms evolve by the appearance of random genetic changes which cause modifications to the pattern of development of the individual. This leads to changes in the phenotype which become established in the breeding population because an appropriate environment happens to be available (Fig. 3). At each of the four points along this chain of cause and effect one can speak of constraining mechanisms, which permit certain changes but prevent others, and it is the nature of these four "filters" which actually determines the course of evolution. In order to create a macroevolutionary hypothesis, as much as possible needs to be known about the "filters". The constraint at the genetic level is that the entire genome must remain a single, integrated whole, which therefore affects the possible size and nature of a single genetic change, and also perhaps the rate at which gene changes can accumulate. There is, however, much current debate about the quantitative values of these limits. At the developmental level, it is fairly obvious that one part of an organism cannot develop in a radically different way unless appropriate adjustments to the development of related parts also occur. There is a volume of evidence that many kinds of feedback between the development of different parts, nerves, muscles, bones, etc., can occur to maintain integration of the overall development, despite alterations to one particular part. However the details of such mechanisms are not understood at all well yet. At the functional level of the phenotype, it is similarly clear that each function depends on the other functions of the organism, and therefore a substantial change in any one must be accompanied by suitable adjustments to the rest. Certain possible changes to the organism cannot therefore occur unless and until suitable evolutionary modifications to other parts have been incorporated. The final constraint is environmental, for unless an opportunity exists for the new organism to remain viable in the environment it cannot give rise to a breeding population. Traditionally natural selection has been regarded as the main form of constraint at this level, based primarily on the observations of population geneticists. But there are certain other theoretical possibilities, such as the virtual absence of selective pressure for brief periods in a tiny, isolated population. The random accumulation of selectively neutral features could also arguably occur. As well as the evidence from modern ecological studies, a certain amount of palaeontological evidence must also be used at this level. These are the observations bearing upon the palaeo-ecology of the known fossils, such as the climatic conditions of the time and the nature of the associated fauna and flora.

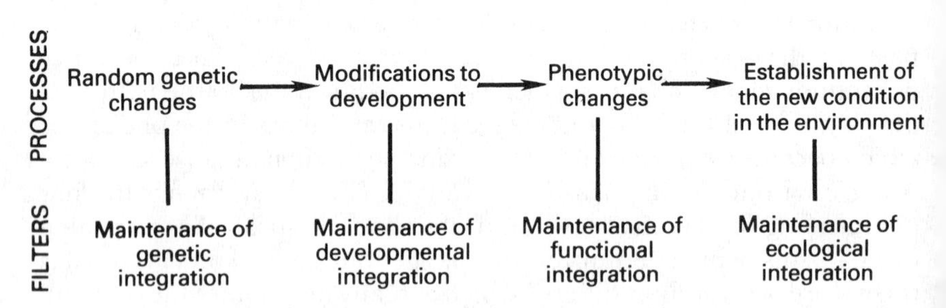


FIG. 3. The processes of evolution and their associated filters.

A macroevolutionary hypothesis results from the combination of the palaeontological observations with the data about evolutionary mechanisms derived from modern animals. It takes the form of a network interconnecting the various observations and factors as causes, effects and correlations—that is, a sort of flow-diagram. Testing of the hypothesis is done by seeing whether it is internally self-consistent. If any fact of the fossil record or property of living organisms in general fails to fit harmoniously into the scheme, then the hypothesis is refuted. Equally, new facts and new information that can be fitted in satisfactorily will strengthen it.

Macroevolutionary hypotheses of this kind are not yet very respectable, notwithstanding the classical work of G. G. Simpson (1949; 1953) and the recent review of Stanley (1979). Indeed the mildly scathing epithet of "story" is to be found applied to such efforts. While it has to be admitted that little confidence can yet be placed in currently available high-level interpretations of the fossil record, nevertheless such work represents attempts to answer questions of fundamental importance to biology. Present weakness in this field results as much from the absence of adequate knowledge about living organisms as from poor interpretation of the fossil record, and progress will depend as much on advances in genetics and developmental biology as in palaeontology. None the less, the ultimate aim of palaeontology should be towards the fullest elucidation of the fossil record in terms of biological evolution, both for specific groups and in general terms. In this spirit, the last two chapters in this book are devoted to a rudimentary attempt to define macroevolutionary questions and offer tentative answers with respect to the mammal-like reptiles and the origin of the mammals.

# 3 | The Origin of Synapsids

DURING THE UPPER CARBONIFEROUS PERIOD, some 300 million years ago, the climate of what is now Nova Scotia was warm and moist, and great forests dominated by giant lycopods occurred. These lycopods were related to the modern club mosses but grew to 100 feet in height. The hard, woody part lay at the periphery of the trunk, and after the death of the plant they remained as hollow stumps. Numerous small tetrapods appear to have sheltered within the stumps and in due course died in them, eventually to be covered by sediments and fossilised. Today a number of such trees have been discovered containing the fossil remains of the animals. Various kinds of amphibians are present, along with the very earliest known reptiles. One particular form, *Archaeothyris* (Fig. 4) comes from lycopod stumps of Florence (Reisz, 1972). There are several specimens, none of them complete, and the bones are disarticulated and scattered. Nevertheless, much of the skull and postcranial skeleton is represented and a reasonably comprehensive picture of the animal can be built up. It was a lizard-like reptile, with a skull about 9 cm long and a total body length of perhaps 50 cm including the tail. The limbs were short, but well built and *Archaeothyris* was clearly an agile, active animal. The most significant feature, however, is the presence of a temporal fenestra or space in the side of the skull behind the orbit. The fenestra is bounded above by the postorbital and squamosal bones, and below by the jugal, and is therefore of the synapsid type. Other features of the skull which identify *Archaeothyris* as a primitive synapsid include the long snout, the relatively short postorbital region, the posterior position of the jaw articulation so that the back of the skull is sloping, and the slightly enlarged teeth in the canine region of the maxilla. Various points of the postcranial skeleton also show the synapsid nature, such as the well-developed neural spines and transverse processes of the vertebrae.

*Archaeothyris* is in fact the earliest adequately known member of the

Pelycosauria, the most primitive of the mammal-like reptiles, and it belongs to the family Ophiacodontidae which is much better known from later forms (p. 28). Other pelycosaurs are also found in the tree stumps of Nova Scotia, which are much less complete than *Archaeothyris*, but appear to represent at least three other genera, one of which is a member of the more advanced pelycosaur group, the Sphenacodontia. It is therefore clear that even at this early stage adaptive radiation of the pelycosaurs had already commenced. The Florence locality is dated as Westphalian D, the youngest part of the Westphalian series of the Pennsylvanian or Upper Carboniferous Period. Contemporary pelycosaurs have been found in other deposits, in the coal swamps of Linton in Ohio, and Nýřany in Czechoslovakia, but these latter specimens are extremely rare and very poorly preserved (Reisz, 1975).

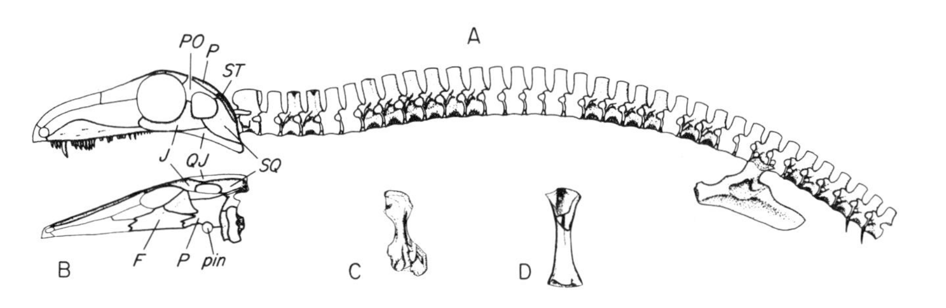


FIG. 4. *Archaeothyris florensis*. A, partial reconstruction of the skull and skeleton in side view. B, dorsal view of the skull. C, dorsal view of the humerus. D, dorsal view of the femur. (From Reisz, 1972.)

*F*, frontal; *J*, jugal; *P*, parietal; *pin*, pineal foramen; *PO*, postorbital; *QJ*, quadratojugal; *SQ*, squamosal; *ST*, supratemporal.

Magnification *c*. × 0.30.

An even earlier pelycosaur occurs in another locality in Nova Scotia. This is at Joggins, where again fossil tetrapods are found in lycopod stumps. Carroll (1964) described a humerus and Reisz (1972) a few vertebrae of a reptile named *Protoclepsydrops*. In the absence of a skull, there is a lingering doubt about whether *Protoclepsydrops* is truly a mammal-like reptile. If it is, then it is the earliest known pelycosaur, since the Joggins deposits are dated as Westphalian B, although until better material is found little of interest can be said about it.

The origin of the mammal-like reptiles was believed for a long time to have been from a primitive stem-reptile of the captorhinomorph family

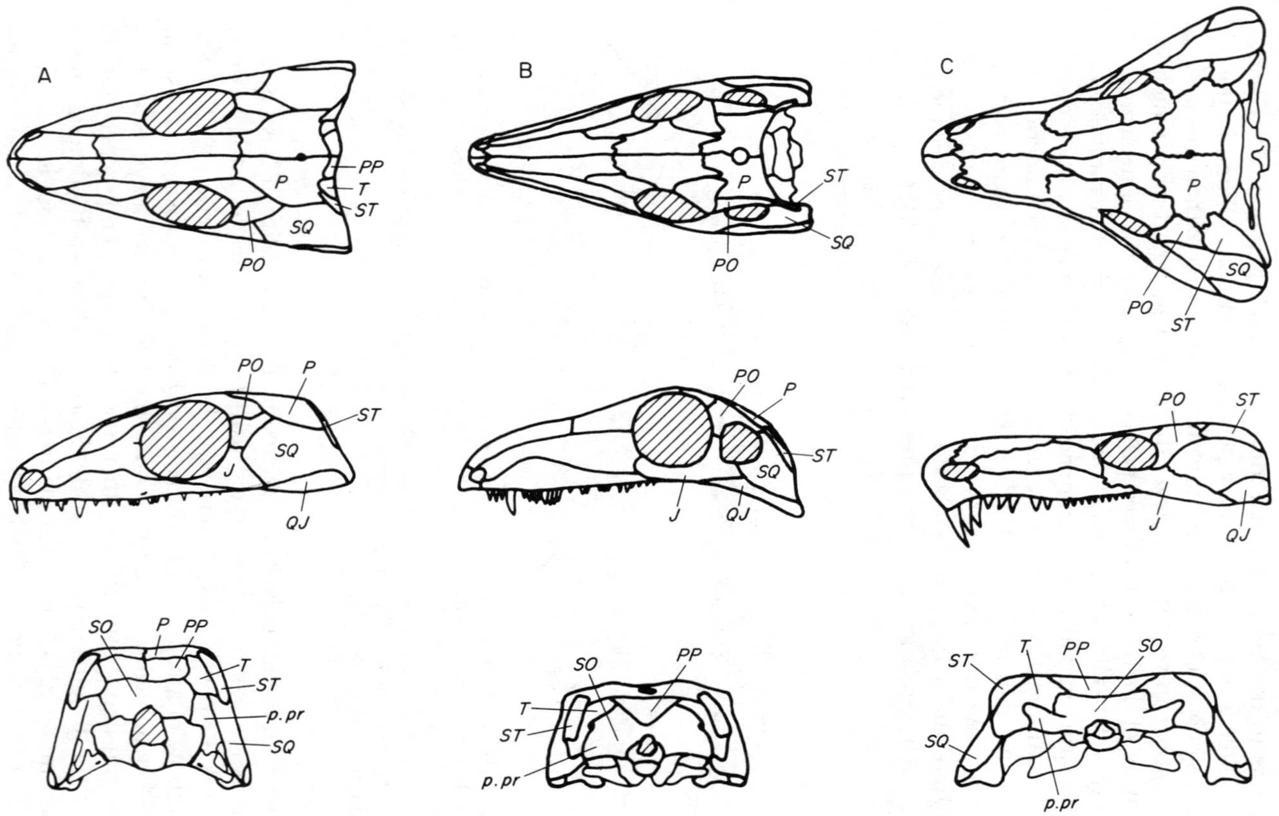

Fig. 5.

Romeriidae (Romer and Price, 1940; Carroll, 1969a). These were small, unspecialised reptiles which lacked fenestration of the skull altogether (Fig. 5A). They appear in the fossil record at about the same time as the pelycosaurs, the earliest form, *Hylonomus*, being from the Westphalian B deposits of Joggins. Certainly the general structure of these two groups of reptiles is similar and they must be quite closely related. There are, however, two apparent specialisations of the romeriid skull not found in the pelycosaur skull which indicate that the common ancestor of the two was more primitive than a romeriid. The first is that the supratemporal bone and postorbital bone are both reduced and do not contact one another. Instead, the parietal bone of the skull roof sutures with the squamosal bone of the cheek. In the pelycosaurs (Fig. 5B), the supratemporal bone and postorbital bone do meet, excluding the parietal from contact with the squamosal. This is probably the more primitive condition. The second romeriid specialisation is a reduction of the size of the tabular bone on the occipital surface of the skull, so that it is no longer closely attached to the paroccipital process.

A more recent view of the origin of synapsids compares them to a different group of archaic reptiles, the limnoscelids (Kemp, 1980a). These also occur in the Upper Carboniferous, and a possible member of the group is the very poorly preserved *Romeriscus* (Baird and Carroll, 1967), which dates from Westphalian A times, and, if it is correctly interpreted, is the earliest known of all reptiles. In limnoscelids (Fig. 5C), the supratemporal is still a relatively large bone and makes a broad contact with the postorbital. No temporal fenestra is present, but there is an open suture, or line of weakness between the skull roof and cheek region, bounded above by the supratemporal and postorbital, and below by the squamosal. It is believed (Panchen, 1972) that the line of weakness is a remnant of the kinetic hinge of the crossopterygian fish ancestor of all tetrapods, and a similar feature is still present in several lines of early amphibians. The arrangement of the bones of the temporal region of the limnoscelid skull is much more likely to have been ancestral to the synapsid skull than is that of the romeriids. Enlargement of the "slit" between the roof and the cheek

FIG. 5. Early tetrapod skulls. A, dorsal, lateral and posterior views of the captorhinomorph *Paleothyris*. B, dorsal and lateral views of the pelycosaur *Archaeothyris*, and posterior view of the pelycosaur *Varanops*. C, dorsal, lateral and posterior views of *Limnoscelis*. (From Kemp, 1980a. Reprinted by permission from *Nature*, ©1980 Macmillan Journals Limited.)

*J*, jugal; *P*, parietal; *PO*, postorbital; *PP*, postparietal; *p.pr*, paroccipital process; *QJ*, quadratojugal; *SO*, supraoccipital; *SQ*, squamosal; *ST*, supratemporal; *T*, tabular.

would produce the temporal fenestra, still bounded above by the postorbital and supratemporal bones and not involving the parietal. Secondary extension of the postorbital backwards below the supratemporal must then have occurred, strengthening the posterior part of the fenestra, and ultimately the postorbital contacted the squamosal. Limnoscelids, like all pelycosaurs, also retain a large tabular bone which is buttressed strongly by the paroccipital process of the occipital surface of the skull, a point which further supports the idea of the origin of the synapsid skull from a limnoscelid-like form, rather than a romeriid-like form.

The functional significance of the evolution of the synapsid fenestra is also clearer on the basis of a limnoscelid-like origin. An early view of temporal fenestrae in general was that they provided windows, out of which the jaw-closing muscles could bulge when they contracted, thus permitting the animal to have larger muscles. Unfortunately this does not account very well for the incipient stages in the evolution of the fenestra, and also does not apparently apply to modern fenestrated reptiles, where the muscles are not observed to bulge outwards in the predicted manner. An alternative view proposed by Frazzetta (1969) is that the edges of a temporal fenestra provide a more secure anchorage for jaw muscles than does a flat surface of bone, a view that accords well with the actual arrangement of the muscle fibres in modern reptiles. A muscle may therefore contain more fibres, or alternatively attach to less bone, without the muscle tearing away from its origin. Applied specifically to the synapsid fenestra (Fig. 6), what may have occurred is that some of the muscle fibres of the main adductor jaw muscle of a limnoscelid-like skull actually attached to the connective tissue between the skull roof and cheek. These particular fibres would have been particularly favourably anchored. The next stage would have been some enlargement of the gap between roof and cheek, giving a larger connective tissue sheet into which a greater number of muscle fibres could attach. Thus the "line of weakness' became transformed into the temporal fenestra, covered by the aponeurotic sheet of connective tissue for attachment of muscle fibres. At the same time, the potentially weakest part of the fenestra, the section bounded by the small supratemporal bone, became strengthened by a posterior extension of the postorbital bone. Originally this merely supported the supratemporal, but eventually it made a new contact with the squamosal at the back of the fenestra, producing the definitive synapsid arrangement.

The synapsids were one of the first of the distinctive reptilian lines to diverge from the basic stem reptile which probably existed in pre-Westphalian times. The origin of the reptiles as a whole is a matter of

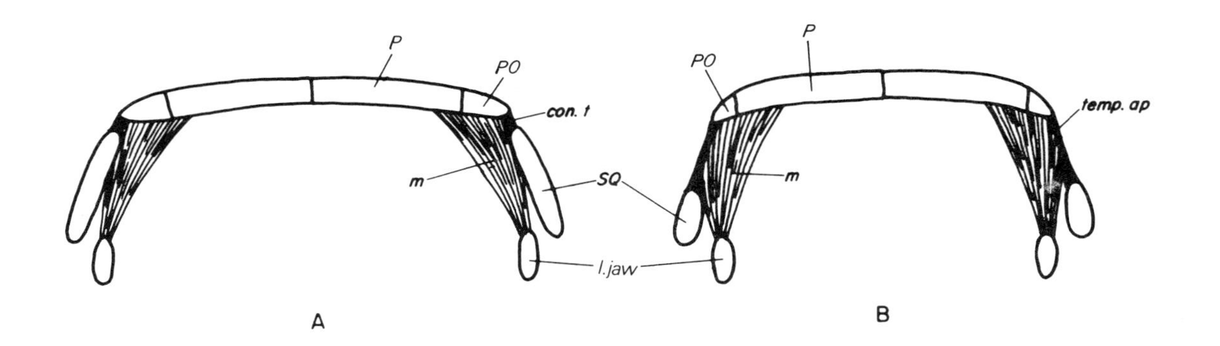


FIG. 6. Origin of the synapsid temporal fenestra. A, transverse section through the temporal region of a *Limnoscelis*-like skull. B, the same of a pelycosaur skull.

*con.t*, connective tissue; *l.jaw*, lower jaw; *m*, adductor muscle; *P*, parietal; *PO*, postorbital; *SQ*, squamosal; *temp.ap*, temporal aponeurosis.

some debate. There are two current major views. Carroll, in a series of papers (e.g. 1970) has argued that among the labyrinthodont amphibians of the Carboniferous, forms called gephyrostegids share a great deal of features with early reptiles, particularly in the structure of the postcranial skeleton. The vertebrae, for example, are constructed similarly, with a dominant, posterior pleurocentrum and a small, ventral, wedge-shaped intercentrum in front, and the limbs were well developed and generally reptile like. On the other hand, Panchen (1972) has stressed certain rather profound differences in the skull structure of gephyrostegids compared with reptiles, and particularly the structure of the middle ear region and the temporal part of the skull roof. Panchen's view is that gephyrostegids and indeed all the labyrinthodonts on the one hand, and the reptiles on the other, represent two distinct lines that evolved from an unknown, hypothetical "prototetrapod".

The ecological conditions under which the reptiles arose, radiated and gave rise to the synapsids may be assessed rather tentatively from the conditions of deposition of the rocks in which the earliest members are fossilised, and from the structure of the animals themselves. Both the romeriids and the pelycosaurs are relatively abundant in the tree stumps of Nova Scotia. In contrast they are extremely scarce in the contemporary coal swamp deposits such as at Linton and Nýřany. Yet other tetrapods, labyrinthodont and lepospondyl amphibians are common in the latter. At least during Westphalian times, therefore, it seems likely that the reptiles lived in forest regions under relatively dry land conditions and only rarely, perhaps accidentally, did they find themselves in the swampier regions. This is in contrast to the amphibians, which were much more at home in the semi-aquatic conditions of the swamps. Further support for the view that the reptiles were more or less fully terrestrial animals comes from the observations of Milner and Panchen (1973) that there is a greater faunal similarity between the reptiles of respective European and North American Upper Carboniferous localities than between the undoubtedly aquatic members of the Amphibia. At this time, the two continents were connected, but the Appalachian–Caledonian mountain range intervened between the two areas. Such a mountain barrier would be expected to constitute a major barrier to migration of aquatic animals, but not to terrestrial ones (Panchen, 1977), and therefore only the reptiles were capable of extensive migration between the two areas.

The structure of primitive reptiles also indicates that they were essentially adapted to dry land conditions. The most obvious feature is the postcranial skeleton, with its fully ossified, relatively large and robust limbs, in contrast to many of the contemporary amphibians. The vertebrae too, with massive, cylindrical pleurocentra and strong interlocking zygapophyses indicate that reptiles are primarily designed for terrestrial rather than aquatic locomotion. The reptilian skull is very strongly built, for the cheeks are firmly sutured to the skull table, at least in those forms more advanced than the limnoscelids. The paroccipital processes, lateral outgrowths from the otic capsule region of the braincase, are massive and firmly support the jaw articulation area of the skull. These mechanical features combine with an apparently large posterior jaw-closing muscle, the capiti-mandibularis, which attached to the internal surfaces of the posterior skull bones and ran almost vertically to insert over the hind part of the lower jaw (Carroll, 1969b). The manner in which such a jaw apparatus was used is described as a static pressure system, whereby the jaw muscle produced a large bite force when the jaws were more or less closed. This sytem is characteristic of animals feeding on land, eating for example small invertebrates. The question of whether the earliest reptiles had evolved that most fundamental terrestrial adaptation of modern reptiles, the cleidoic egg, remains open. The first fossil shelled, and therefore presumably cleidoic egg, does not occur until the Permian period. However, eggs do not generally fossilise well and therefore the absence of eggs in the Upper Carboniferous is not particularly significant. Carroll (1969b) has made the interesting suggestion that the small size of the earliest reptiles is correlated with the absence of proper cleidoic eggs with extraembryonic membranes. The eggs must necessarily have been small to keep the surface area to volume ratio high for surface respiration, assuming they were laid on land. Only later, after the evolution of the membranes, could large reptiles laying large eggs exist.

The origin of the synapsids from stem reptiles must also be assumed to have been a dry land event, since all the relevant known forms were evidently terrestrial. Reisz (1972) argued that the single key change underlying the development of synapsid features was simply an increase in size of the animals. *Archaeothyris* and, to judge by the size of its humerus, *Protoclepsydrops* were about twice as long as the contemporary romeriids. The exact niche occupied by the new, larger synapsids is not clear, but it might be supposed that they ate larger insects. At any rate, a larger animal may be expected to improve its feeding apparatus, partly because its food is liable to be tougher, and partly because the bulk of its body will increase disproportionately to the area of its mouth. The early synapsids adapted to the requirements of increased size, therefore, by developing a relatively larger head and longer jaws and presumably by evolving the temporal fenestra.

The postcranial skeleton of the early pelycosaurs does not differ in great detail from that of the romeriids, apart from being a little more robust, with relatively larger limbs and large processes on the vertebrae. Such changes as occurred no doubt reflected the increased strength and musculature required by a larger animal.

# 4 Pelycosaurs

THE PELYCOSAURS were the earliest and most primitive of the mammal-like reptiles, forming the first of the several phases of adaptive radiation of synapsids. Having arisen some time in the middle of the Carboniferous Period, all the principal groups had evolved before the close of that period, and the pelycosaurs continued to dominate the known terrestrial faunas throughout the first half of the succeeding Permian. They are particularly common in the Early Permian Redbeds of Texas and the surrounding States, and occasional specimens have also been found in parts of Europe, and possibly but very dubiously in South Africa.

## Systematics

The structural similarity between all the pelycosaurs is so great that there is no doubt that they all shared a common ancestor which had already evolved the essential features of the group, a form probably rather like *Archaeothyris*. In 1940, Romer and Price reviewed the whole of the Pelycosauria in a monograph that remains seminal to the present understanding of the group. They recognised three suborders of pelycosaurs, the Ophiacodontia which were the least modified members, the Sphenacodontia which were specialised carnivores, and the Edaphosauria which were specialised for a herbivorous habit. As Romer and Price themselves acknowledged, this is probably a rather over-simplified view of pelycosaur classification.

### *Ophiacodonts*

The ophiacodonts are those pelycosaurs that are generally primitive in nature, lacking the characteristic modifications of the other two, more specialised groups. The occipital surface of the skull slopes forwards and upwards and its upper margin is markedly concave as seen from above, which reflects the smallness of the temporal fenestra and the posterior position of the jaw articulation. The lower jaw is simple, lacking the dorsal development of a coronoid eminence and the jaw articulation is at about the same level as the tooth-row. The teeth themselves are conical, slender and a little recurved.

The postcranial skeleton (Fig. 7A) is also primitive, with short, simple vertebrae and well-developed intercentral bones between adjacent centra. The transverse processes are short and the neural spines broad and rather heavy structures. A particularly marked character is the holocephalous nature of the ribs of the dorsal region, the two heads of each rib, tuberculum and capitulum, being confluent. Only two sacral ribs are present, compared to three or more in the advanced pelycosaurs. In the shoulder girdle, the scapula blade is short and rather broad, and the humerus has a strong supinator process on its shaft for extensor muscles of the lower limb (Fig. 19B). The pelvis also shows a primitive construction, the ilium is not expanded either anteriorly or posteriorly, and bears a dorsal trough for the attachment of axial muscles. The limbs as a whole are relatively short, and the feet broad.

Since the Suborder Ophiacodontia has to be defined solely by ancestral pelycosaur characters, it should be seen as a horizontal grade of organisation rather than a monophyletic taxon, an interpretation that becomes apparent when the two constituent families are considered (Fig. 17). The best known of these is the family Ophiacodontidae, to which *Archaeothyris* (p. 18) belongs. Another early member is *Clepsydrops*, from the Carboniferous (Stephanian) deposits at Danville, Illinois, which was a small form, some 2–3 feet in length. Only a little larger was *Varanosaurus* (Fig. 7A) from the Early Permian Wichita and Clear Fork localities of Texas. Much the commonest genus was *Ophiacodon*, (Fig. 8) which first appeared in the Stephanian, but persisted until well into the Early Permian in North America, and some species of which achieved a considerable size. *Ophiacodon major*, for example, was about 3 metres long and must have weighed over 200 kg. The ophiacodontids are a true monophyletic family, as they all have certain specialisations in common for a semi-aquatic, fish-eating habit. The skull is large, with a noticeably narrow, elongated facial region, and there are a great many marginal teeth, around forty, in each jaw. These teeth are small and sharp, and there is little tendency to enlarge the canine teeth. Such a dentition would not have been suitable for eating terrestrial prey or vegetation, but, as in modern crocodiles, would have been well adapted for catch-

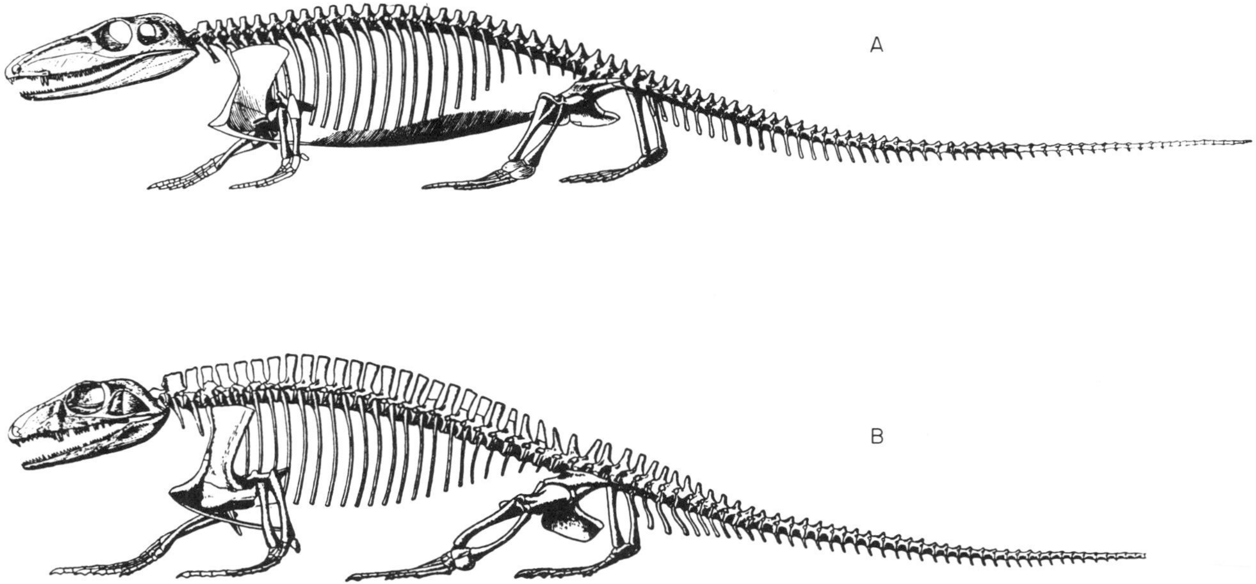

A

B

Fig. 7. A, skeleton of the ophiacodontid *Varanosaurus*. B, skeleton of the primitive sphenacodontid *Varanops*. (From R
1940.)
Magnification *c*. × 0.16.

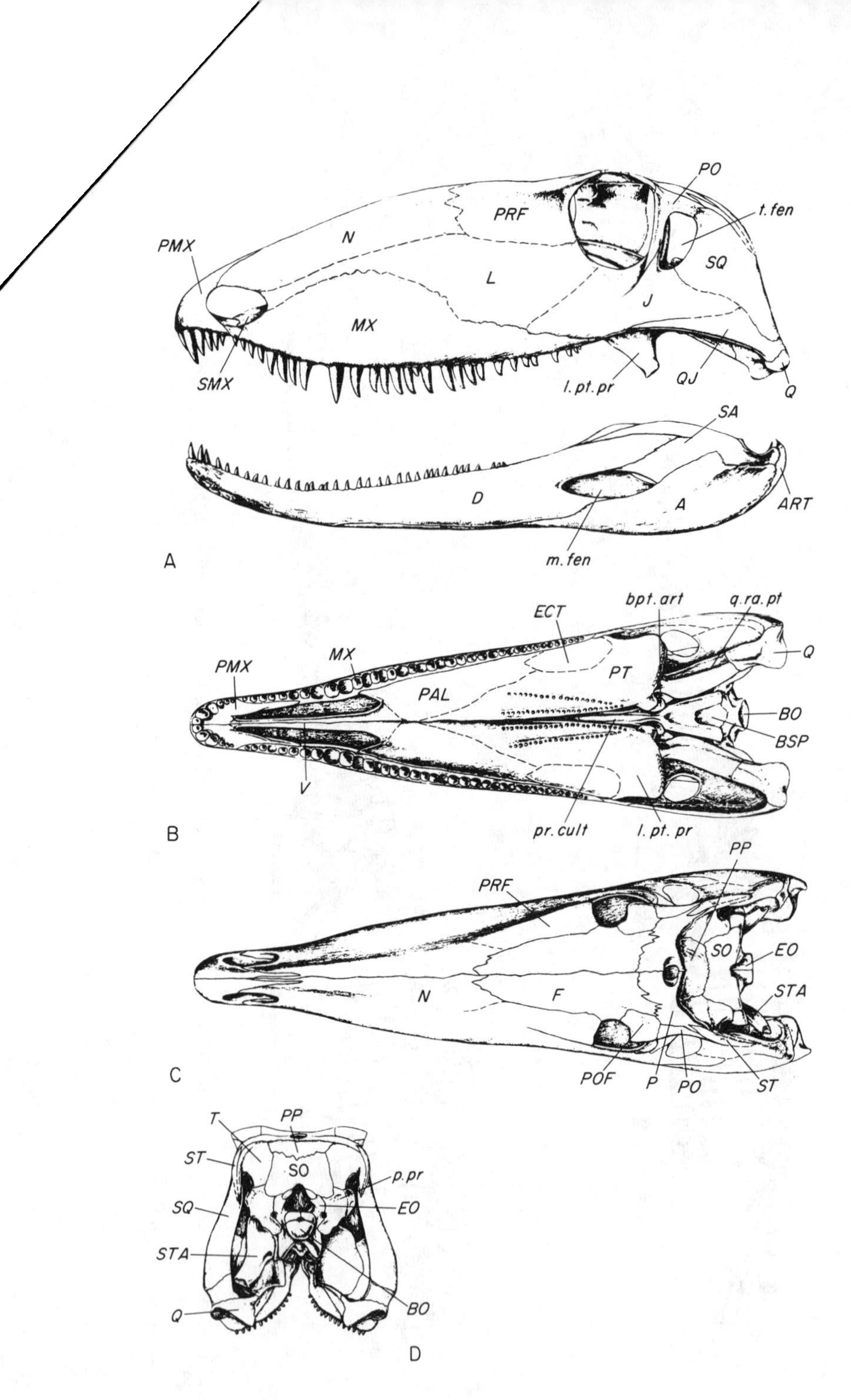
PO
t.fen
PRF
N
PMX
SQ
L
J
MX
SMX
l.pt.pr
QJ
Q
SA
D
A
ART
A
m.fen
ECT
bpt.art
q.ra.pt
MX
Q
PMX
PT
PAL
BO
BSP
V
B
pr.cult
l.pt.pr
PRF
PP
SO
EO
N
F
STA
C
POF
P
PO
ST
T
PP
ST
SO
p.pr
SQ
EO
STA
BO
Q
D

FIG. 8.

ing fish in the water. The postcranial skeleton also shows aquatic adaptations, in the form of hindlegs markedly longer than the forelegs.

The second ophiacodont family is the Eothyrididae, of which the single skull of the little *Eothyris* (Fig. 9) is the only well-known specimen. It occurred in the uppermost part of the Wichita Formation and is only about 60 mm long. Quite unlike the more typical ophiacodonts, the skull is broad and the facial region very short. The prefrontal bone and postfrontal bone meet above the orbit, and the supratemporal bone is very large—features that are primitive compared to the condition in all other pelycosaurs. However, the most striking feature of *Eothyris* is the great enlargement of two of the anterior maxillary teeth to form canines,

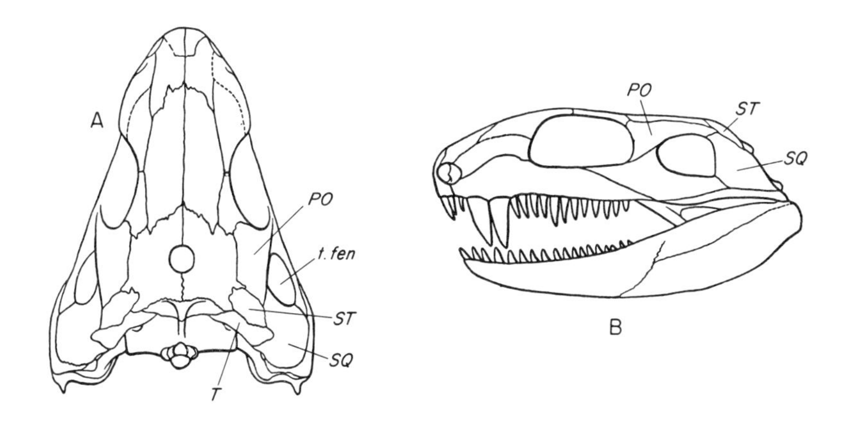


FIG. 9. The skull of *Eothyris*. A, dorsal view. B, lateral view. (From Romer and Price, 1940.)
*PO*, postorbital; *SQ*, squamosal; *ST*, supratemporal; *T*, tabular; *t.fen*, temporal fenestra.
Magnification *c.* × 0.8.

FIG. 8. The skull of *Ophiacodon*. A, lateral view. B, ventral view. C, dorsal view. D, posterior view. (From Romer and Price, 1940.)
*A*, angular; *ART*, articular; *BO*, basioccipital; *bpt.art*, basipterygoid articulation; *BSP*, basisphenoid; *D*, dentary; *ECT*, ectopterygoid; *EO*, exoccipital; *F*, frontal; *J*, jugal; *L*, lachrymal; *l.pt.pr*, lateral pterygoid process; *m.fen*, mandibular fenestra; *MX*, maxilla; *N*, nasal; *P*, parietal; *PAL*, palatine; *PMX*, premaxilla; *PO*, postorbital; *POF*, postfrontal; *PP*, postparietal; *p.pr*, paroccipital process; *pr. cult*, processus cultriformis; *PRF*, prefrontal; *PT*, pterygoid; *Q*, quadrate; *QJ*, quadratojugal; *q.ra.pt*, quadrate ramus of the pterygoid; *SA*, surangular; *SMX*, septomaxilla; *SO*, supraoccipital; *SQ*, squamosal; *ST*, supratemporal; *STA*, stapes; *T*, tabular; *t.fen*, temporal fenestra; *V*, vomer.
Magnification *c.* × 0.31.

paralleling the more advanced sphenacodonts in this respect. Several other possible eothyridids are known from fragments of much larger pelycosaurs which, so far as it is known, show a similar combination of extremely primitive skull and skeleton with enlarged canine teeth. All are from the Upper Carboniferous (Stephanian) and are therefore among the earliest pelycosaurs, and include *Stereophallodon* from Texas, *Baldwinosaurus* from New Mexico and *Stereorhachis* from France. It seems that towards the end of the Carboniferous Period there was a brief radiation of predaceous pelycosaurs descended directly from the basal pelycosaur stock. This radiation paralleled that of the sphenacodonts, and was largely replaced by members of the latter group. *Eothyris* itself must be a persistent relic, lasting well into the Early Permian, and perhaps owing its survival to a specialised habit reflected by its small size.

### *Sphenacodonts*

Both the skull and the postcranial skeleton of sphenacodonts evolved towards a highly predaceous mode of life. The most advanced form is the well-known, much-studied *Dimetrodon*, but other sphenacodonts of more primitive structure are also known. However, a number of derived characters shared by all forms, primitive and advanced alike, indicate that the group is monophyletic. The skull of sphenacodonts shows signs of an increase in jaw musculature, for the posterior region is broad, and the horizontal dorsal surface is sharply demarcated from the almost vertical cheeks. A retroarticular process is present, extending behind the jaw articulation for the insertion of jaw-opening muscles. The occipital plate, which takes much of the stress arising from the action of the jaw muscles, is strongly built as the supraoccipital bone has well-developed lateral processes, and the paroccipital processes are elongated and curve backwards and downwards to buttress the cheeks. All the teeth are sharp and recurved, and enlarged canine teeth occur in the maxilla. Two minor but characteristic features of all sphenacodont skulls are the prominence of the postorbital bone in the postero-dorsal corner of the orbit, and the reduction of the dorsal process of the premaxilla, which in the ophiacodonts extends well back on the skull roof between the nasal bones.

The sphenacodont skeleton shows many modifications from the primitive, ophiacodont condition, designed to improve the locomotory ability. The centra of both the neck and dorsal vertebrae are compressed from side to side and have a mid-ventral keel, and there is some reduction in the size of the intercentra. The neural arches are characteristically excavated on either side, near to the base. Thus, without increasing the weight of the vertebral column, it was strengthened, particularly in the sagittal plane. The ribs have become dolichocephalus, the tuberculum and capitulum being sharply distinct from one another. The shape of the body, as indicated by the form of the rib cage, was rather narrow from side to side, with relatively deep flanks, a shape typical of active predaceous animals of relatively large size. The limbs were elongated and relatively slender, and the limb musculature had increased as indicated by the larger surfaces for its origin from the limb girdles. In the shoulder girdle, the scapula is tall, and the clavicles expanded medially. The ilium of the pelvic girdle expands backwards, and the primitive dorsal trough for axial muscle attachment has merged with the internal surface. One very characteristic feature is the development of a pronounced rugosity marking the fourth trochanter of the femur, for insertion of the caudi femoralis muscle (Fig. 20B). The feet of sphenacodonts have become longer and more slender.

The sphenacodonts must have arisen from an ancestral pelycosaur of the ophiacodont grade. More particularly, they are related to the family Ophiacodontidae because, for example, in both groups the supratemporal is reduced and the frontal bone enters the margin of the orbit, in contrast to the presumed primitive condition of eothyridids. The most primitive sphenacodonts constitute the family Varanopidae, represented in the late Carboniferous of Illinois by *Milosaurus*, a rather poorly known form (DeMar, 1970). However, far the best known varanopid is *Varanops* itself, (Fig. 7B) which occurs at a single locality in the Arroyo Formation of the Texas Clear Fork. It is only about a metre in length and is exceedingly primitive in structure. Indeed, it is almost ideally intermediate between the ophiacodont and typical sphenacodont grades. The skull is short and low and, as in ophiacodonts, the jaw articulation is level with the tooth row and lies far posterior. The notch in the ventral border of the angular bone of the jaw, so characteristic of all other sphenacodonts, is absent, and there is little development of canine teeth. The marginal teeth of the jaws are not compressed from side to side, but are round in cross-section. Several primitive features are found in the postcranial skeleton as well. The vertebrae are short, the centra but little compressed, and the ventral keel not at all well developed, while the transverse processes are relatively short, although still somewhat longer than in ophiacodonts. There are only two sacral vertebrae instead of the three of later forms, and the ilium is low. (See note 1 added in proof, p. 351.)

Despite the relatively early appearance and primitive nature of the varanopids, there is evidence that they were among the longest surviving of all pelycosaurs. One form, *Varanodon*, has been found in the Chickasha formation of Oklahoma (Olson, 1965), which dates from the lowest part of the Late Permian (Guadalupian). From even later deposits, the *Tapinocephalus*-zone of South Africa, come two poorly preserved skulls called *Anningia* and *Elliotsmithia* respectively (Broom, 1932). That they are small, extremely primitive mammal-like reptiles is certain, although their assignment to the family Varanopidae or indeed to the pelycosaurs in general is tentative.

All the rest of the sphenacodonts constitute the family Sphenacodontidae, and have continued the trend towards a highly developed predaceous existence. The most primitive form is *Haptodus* (Currie, 1977, 1979), best known from the Early Permian (Autunian) of several places in Europe but also present in the late Upper Carboniferous (Stephanian) of North America. Compared to most forms, it is a fairly small sphenacodontid, about 1.5 metres long, and has a deep, narrow, generally powerfully built skull (Fig. 10) compared to the varanopids. The most significant advances were in the lower jaw which has developed a large coronoid eminence rising above the level of the tooth row for increasing the area of insertion of the adductor musculature. The jaw hinge is depressed to a level just below the tooth row by a ventral extension of the squamosal coupled with a down-turning of the articular region of the jaw. A consequence of the latter process was that a space developed between the large ventral keel of the angular bone laterally and the articular region medially. This space, or angular notch, is only just present in *Haptodus* although it became more prominent in the advanced sphenacodontids and has an important bearing on the subsequent evolution of the mammal-like reptiles. The postcranial skeleton of *Haptodus* has also progressed beyond the varanopid level by developing larger, more laterally compressed vertebral centra with more prominent mid-ventral keels, and larger neural spines and transverse processes. A third vertebra is added to the sacrum.

The origin of haptodontines must have been from a very *Varanops*-like ancestor, and they in turn are structurally close to the presumed ancestor of the more advanced forms, the sphenacodontines such as *Dimetrodon*. Apart from a tendency to increase in size, up to about 3 metres in the case of *D. grandis*, the modifications of sphenacodontines mostly relate to an increase in the size of the canine teeth (Fig. 11). Both the upper and lower canines are substantially larger than the adjacent teeth, and by reduction in the number of incisors and maxillary precanines in the upper tooth-row a gap is created into which the lower canine fits when the jaws are closed. The snout is long, high and massively built, and the lachrymal bone is shortened to permit expansion

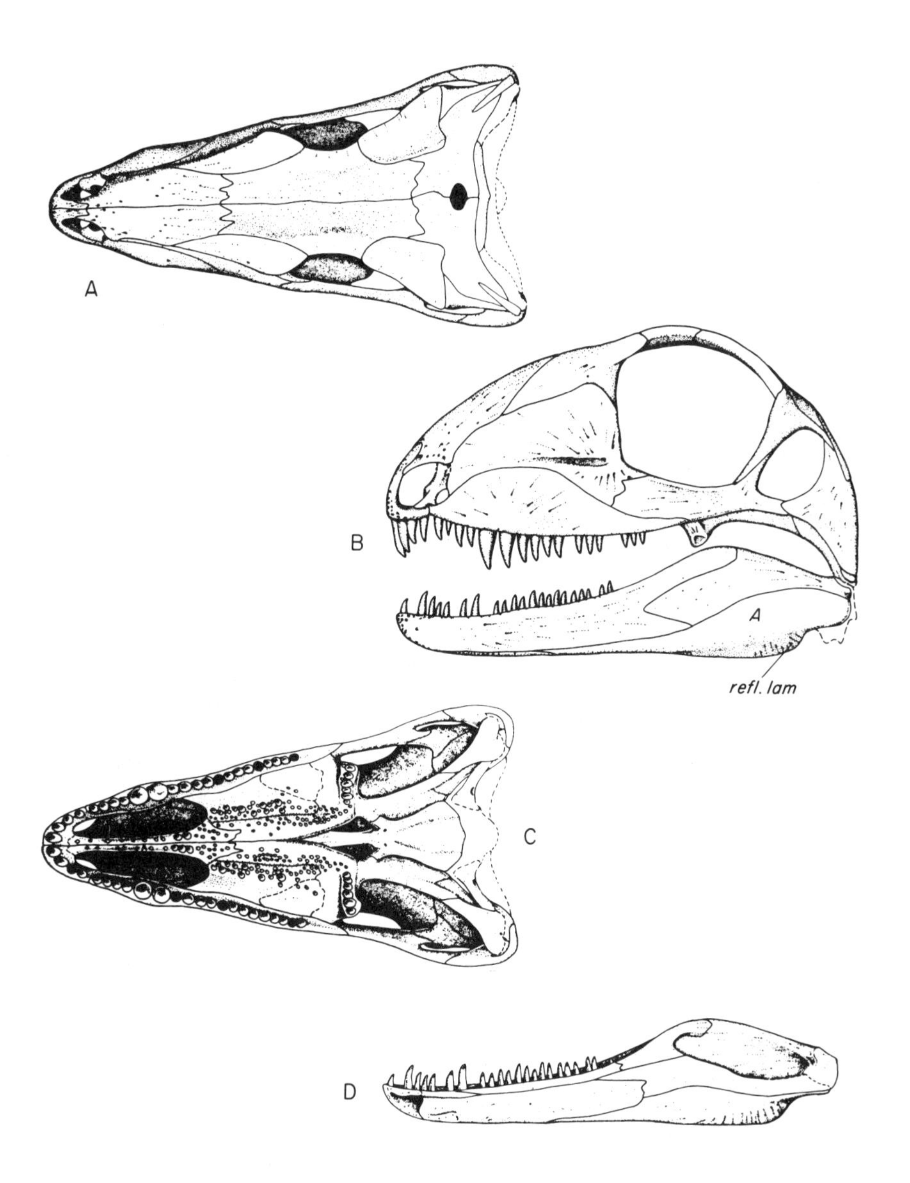


Fig. 10. The skull of *Haptodus*. A, dorsal view. B, lateral view. C, ventral view. D, internal view of the lower jaw. (From Currie, 1979.)
*A*, angular; *refl.lam*, reflected lamina of the angular.

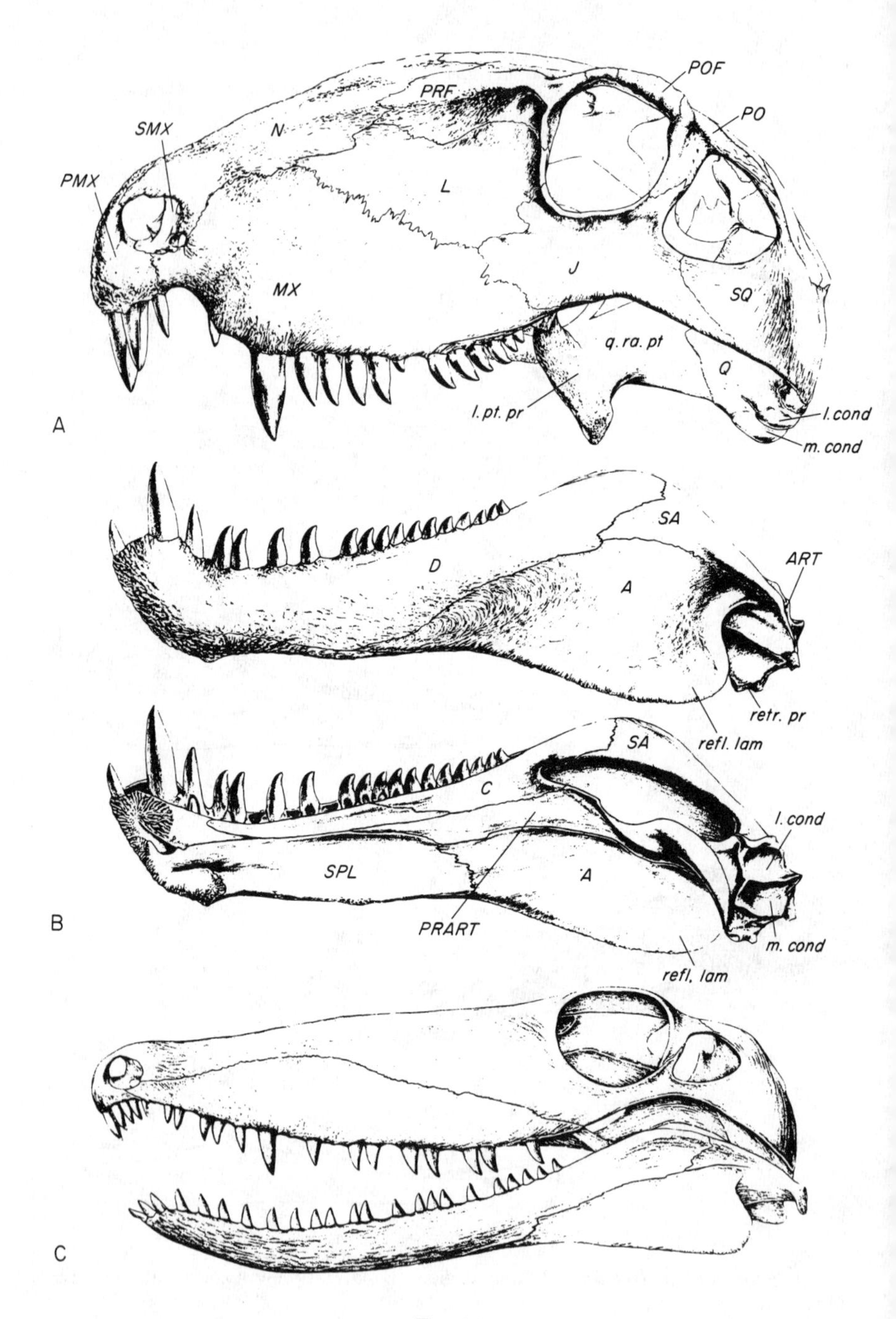
POF
PRF
PO
SMX
N
PMX
L
J
MX
SQ
q. ra. pt
Q
l. pt. pr
l. cond
A
m. cond
SA
ART
D
A
retr. pr
refl. lam
SA
C
l. cond
SPL
A
B
PRART
m. cond
refl. lam
C

FIG. 11.

of the maxilla, carrying the root of the upper canine tooth. The lower jaw is deepened, and the two rami meet anteriorly in a powerful symphysis. The alterations to the muscle-bearing region of the lower jaw initiated in *Haptodus* are carried further, by increase in the size of the coronoid eminence and an even more ventrally reflected jaw articulation. The angular notch therefore became much more prominent. As is appropriate for a large predator, the lateral pterygoid processes of the palate against which the lower jaws bear are particularly large.

*Dimetrodon* (Fig. 12) is the commonest of all the pelycosaurs in the Wichita and Clear Fork localities of the Early Permian of Texas and Oklahoma, and is one of the few forms which survived into the Late Permian, being found in the San Angelo Formation of Texas. The genus is characterised by the huge elongation of the neural spines of the dorsal vertebrae, which supported a presumably thermo-regulatory sail along the back of the animal. *Sphenacodon* is very similar to *Dimetrodon* except that it lacked the sail. It occurs only in the Abo Formation of New Mexico.

*Secodontosaurus* (Fig. 11C) is a specialised offshoot of the main sphenacodontine line. It was a semi-aquatic, fish-eating form, with a long, low skull, slender snout and poorly developed canines.

## *Edaphosaurs*

The edaphosaurs constitute the second advanced suborder recognised by Romer and Price (1940), and include forms adapted to a herbivorous mode of life. As such, they are of particular interest as the first members of the mammal-like reptiles, indeed of any tetrapods, to invade this particular adaptive zone in any abundance. There are two main families, the edaphosaurids and the caseids, which are superficially similar to one another, but which actually differ so markedly in details that they

FIG. 11. The skull of *Dimetrodon*. A, lateral view, with the lower jaw; B, internal view of the lower jaw. The skull of *Secodontosaurus*: C, lateral view. (From Romer and Price, 1940.)

*A*, angular; *ART*, articular; *C*, coronoid; *D*, dentary; *J*, jugal; *L*, lachrymal; *l.cond*, lateral condyle; *l.pr.pt*, lateral process of the pterygoid; *m.cond*, medial condyle; *MX*, maxilla; *N*, nasal; *PMX*, premaxilla; *PO*, postorbital; *POF*, postfrontal; *PRART*, prearticular; *PRF*, prefrontal; *Q*, quadrate; *q.ra.pt*, quadrate ramus of the pterygoid; *refl.lam*, reflected lamina of the angular; *retr.pr*, retroarticular process; *SA*, surangular; *SMX*, septomaxilla; *SPL*, splenial; *SQ*, squamosal.

Magnification *c.* × 0.32.

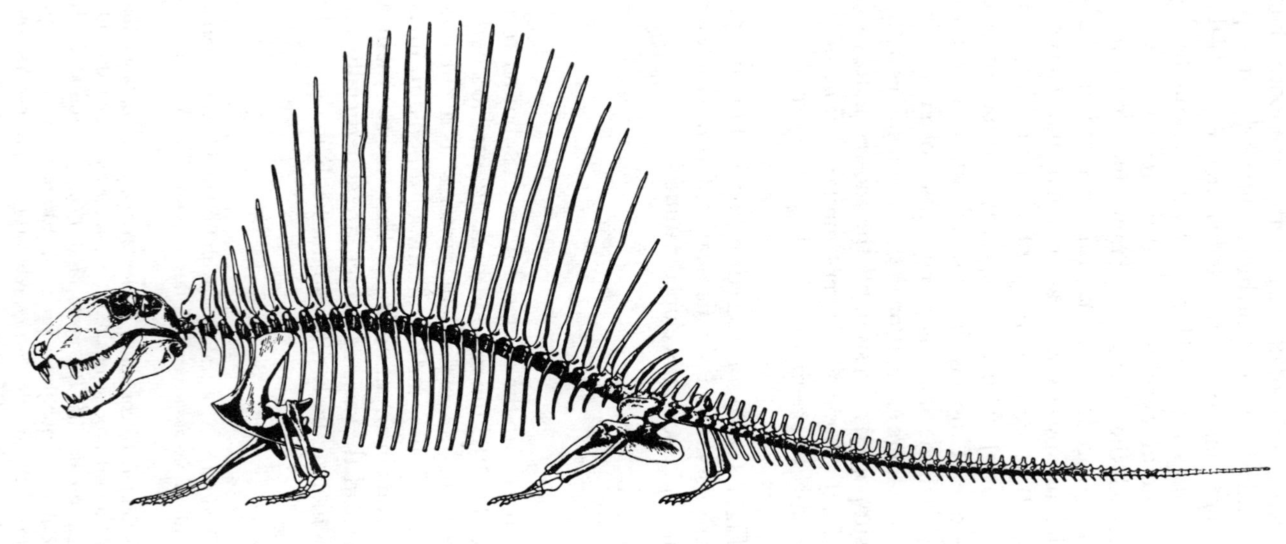

FIG. 12. Skeleton of *Dimetrodon*. (From Romer and Price, 1940.) Magnification *c*. × 0.06.

must have evolved their herbivorous adaptations independently of one another. There is, in fact, a controversy about just how closely related to one another they are.

In both families, the skull is short and low, the occiput vertical, and the jaw articulation depressed below the level of the tooth-row (Fig. 16). The jaws tilt inwards ventrally, and are united at the front by a strong symphysis. There is no differentiation of canines, while the individual marginal teeth are stout, blunt and bulbous. The teeth on the palate are well developed to increase the masticatory surface. These, however, are fairly basic modifications to the skull found commonly in herbivores. In other respects, the caseid skull differs markedly from the edaphosaurid skull. That of caseids (Fig. 16A) is relatively much smaller and has a characteristic anterior extension of the snout, overhanging the huge external nostril. A large supratemporal bone is retained, and on the occipital surface the paroccipital processes are massively developed and make powerful buttressing contacts with the squamosals. The equally characteristic edaphosaurid skull is widened between the orbits, with the supratemporal bone reduced, and large opposing tooth plates developed on the palate and inner sides of the lower jaws.

The postcranial skeletons of these two respective groups of herbivorous pelycosaurs show a number of points of similarity. The dorsal vertebrae tend to be elongated, while the neck vertebrae are reduced, perhaps in correlation with the relatively small skull. The individual centra are spool-shaped with no trace of a mid-ventral keel. As in the sphenacodonts, the transverse processes of the vertebrae are elongated, and the articulating surfaces of the zygapophyses set at an angle to the horizontal. One very marked characteristic is the reduction of the tuberculum of the rib head to a patch of roughness on the surface of the bone. The overall shape of the body, as indicated by the form of the ribs, was broad and rounded, which is typical of herbivores with their large digestive tract. Three sacral vertebrae are present, as in the more advanced sphenacodonts. Edaphosaur features of the shoulder girdle are the broad coracoid and narrow, short scapula, while the ilium of the pelvic girdle tends to expand anteriorly. The limbs are stoutly built, with somewhat elongated humerus and femur but relatively short lower legs.

The family Edaphosauridae is much the best known from the genus *Edaphosaurus* (Figs 15 and 16A), which occurs from Late Carboniferous (e.g. Peabody, 1957) to Early Permian times in North America and Europe. The most familiar specialisation is the enormous elongation of the neural spines of the vertebrae, superficially similar to those of *Dimetrodon*, but differing in the presence of cross-pieces, or at least tubercules. In size, different species of *Edaphosaurus* vary from under 1 metre to over 3 metres in length.

In several respects, the most important herbivorous family is the caseids (Figs 13B and 16A). They did not appear in the fossil record until relatively late, as *Casea broilii* in the Arroyo Formation of the Texan Early Permian, and as *C. rutena* from France (Sigogneau-Russell and Russell, 1974), but they survived into the early part of the Late Permian in both North America and Russia (Olson, 1968) where they overlapped the first therapsids. Within the group, a modest adaptive radiation occurred, producing a number of different species and genera, which presumably reflects subdivision of the herbivorous adaptive zone into several niches. One of the trends was towards a marked increase in size, at least in some lines. *Cotylorhynchus hancocki* from the San Angelo Formation at the base of the Late Permian achieved some 3–4 metres in length and is the largest known pelycosaur (Fig. 13).

Romer and Price (1940) recognised a third group of edaphosaurs, the Nitosauridae, as very primitive and close to the ancestry of the two advanced families. The best known nitosaurid is *Mycterosaurus* (Figs 13A and 14A) of the Early Permian Clear Fork of Texas. It is only about 0.6 metres in length, and the skull is very primitive compared to other edaphosaurs. The snout is long and the canine teeth slightly enlarged. Based on the skull alone, there is no reason to relate *Mycterosaurus* to the edaphosaurs. On the other hand, its postcranial skeleton is claimed to possess certain of the diagnostic features of the edaphosaur skeleton. The vertebral centra are spool-shaped and have no trace of a mid-ventral keel, while the limbs and girdles resemble those of the advanced herbivores. Romer and Price (1940) therefore claim that *Mycterosaurus* and other nitosaurids strengthen the view that edaphosaurids and caseids are closely related, because of the common possession of these postcranial characters in all three families.

Other authors, however, have denied any such particular relationship between edaphosaurids and caseids. Langston (1965) has argued that caseids were derived from primitive eothyridid pelycosaurs (p. 31), basing his view on the interpretation of a small pelycosaur *Oedaleops* (Fig. 14B). The skull of this form resembles that of *Eothyris*, being short, wide and low, and possessing for example a large supratemporal bone. On the other hand, *Oedaleops* has certain characteristics of the caseid skull at least incipiently present. The snout tends to be extended anteriorly, above the large external nostril, the pineal foramen is very large, and the maxilla forms part of the ventral edge of the orbit. None of these are found in edaphosaurids, and therefore the conclusion must be that caseids and edaphosaurids separated at a pre-eothyridid

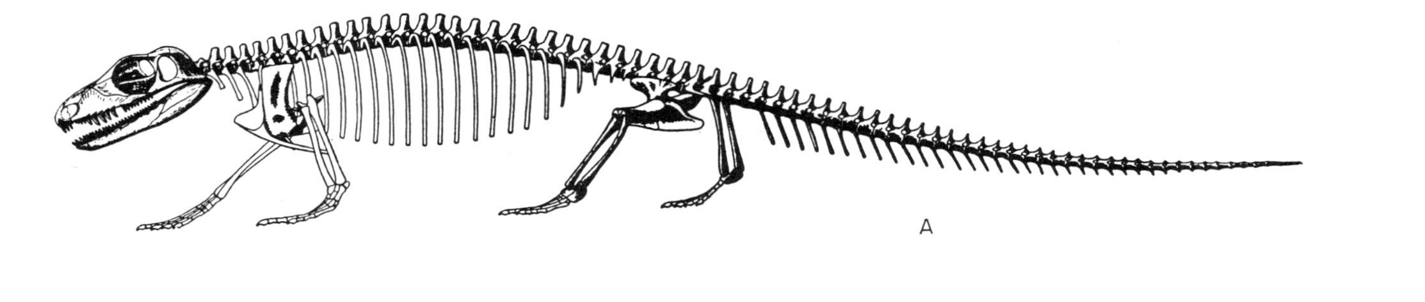


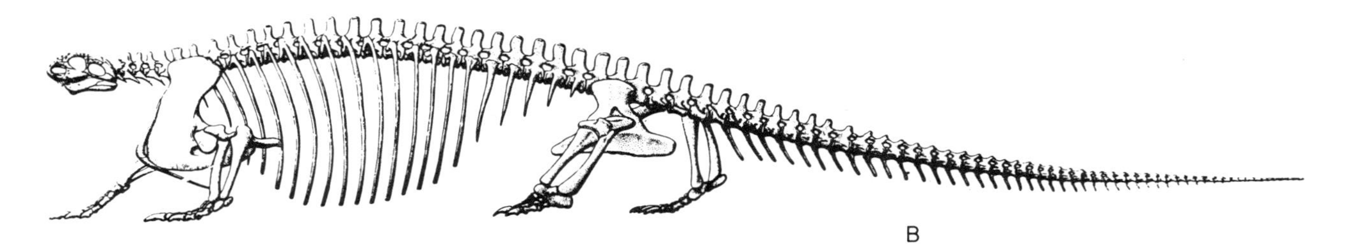


FIG. 13. A, skeleton of the primitive edaphosaur *Mycterosaurus*. B, skeleton of the caseid *Cotylorhynchus*. (A from Romer and Price, 1940; B from Stovall, *et al.*, 1966.)
Magnifications: A, *c.* × 0.24; B, *c.* × 0.06.

level of pelycosaur evolution (Fig. 17). Reisz (1981) has accepted this view, and has challenged the validity of Romer and Price's family Nitosauridae. *Mycterosaurus* itself may be a varanopid sphenacodont skull, incorrectly associated with postcranial bones of other animals including amphibians. Other supposed nitosaurids seem likely to be poorly preserved remains of a variety of primitive pelycosaurs. Reisz also regards the similarities between the postcranial skeletons of the herbivorous pelycosaurs as due to the retention of ancestral pelycosaur features, rather than as evidence for relationship.

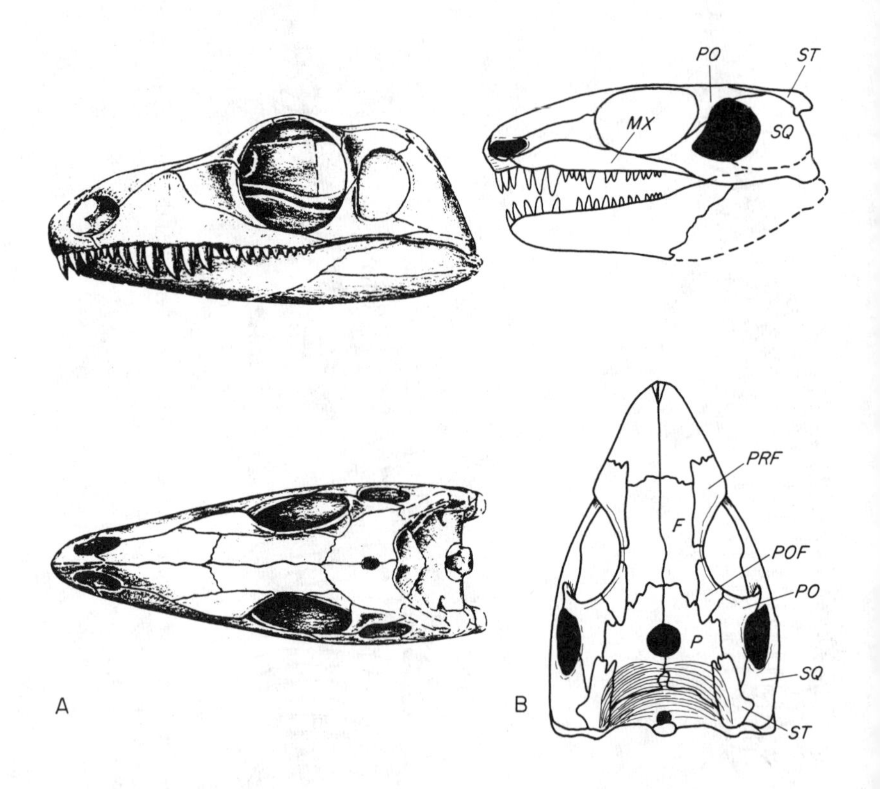


Fig. 14. A, skull of the primitive edaphosaur *Mycterosaurus*, in lateral and dorsal views. B, skull of *Oedaleops* in lateral and dorsal views. (A from Romer and Price, 1940; B, redrawn after Langston, 1965.)

*P*, parietal; *PO*, postorbital; *POF*, postfrontal; *PRF*, prefrontal; *SQ*, squamosal; *ST*, supratemporal.

Magnifications: A, *c.* × 0.66; B, *c.* × 0.68.

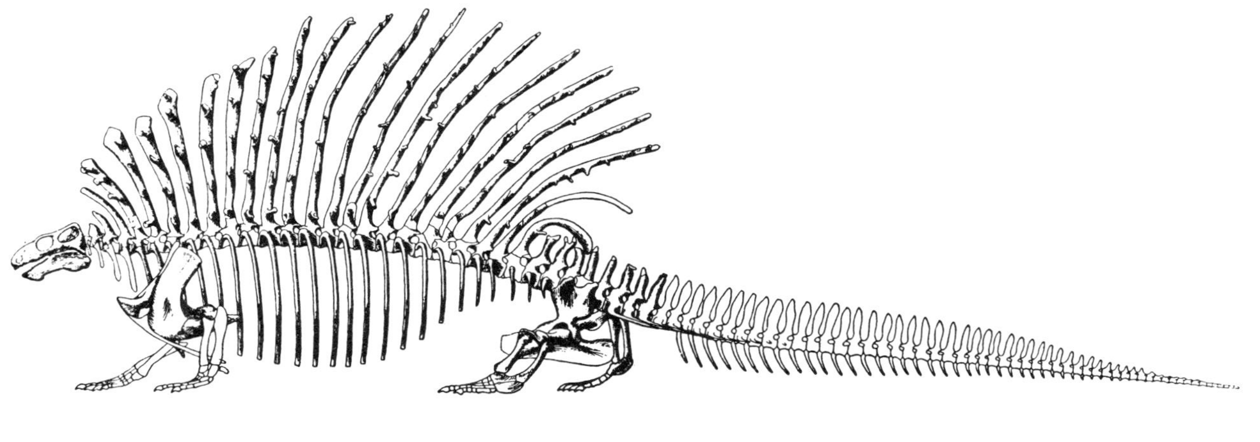

FIG. 15. Skeleton of *Edaphosaurus*. (From Romer and Price, 1940.) Magnification *c.* × 0.05.

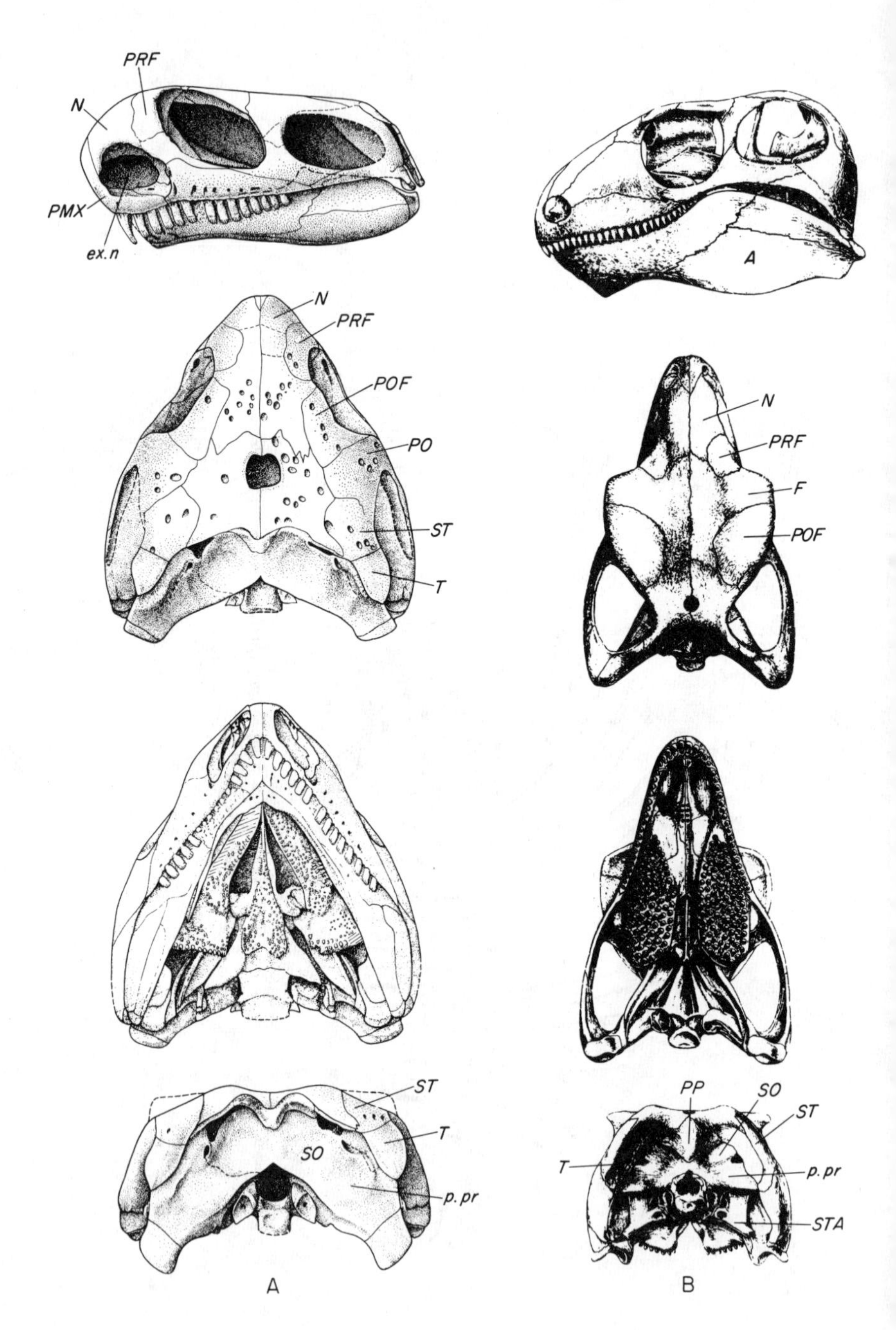
PRF
N
PMX
ex.n
A
N
PRF
POF
PO
ST
T
N
PRF
F
POF
ST
T
SO
p.pr
PP
SO
ST
T
p.pr
STA
A
B

FIG. 16.

If it is true, as seems likely at the present time, that caseids are related to eothyridids, then the origin of the edaphosaurids remains unknown. They may yet prove to have descended from eothyridid-like forms as well, or they may possibly have a closer relationship to ophiacodontids and sphenacodontids (Reisz, 1981). The reduced supratemporal bone and generally rather higher build of the skull is suggestive of the second possibility.

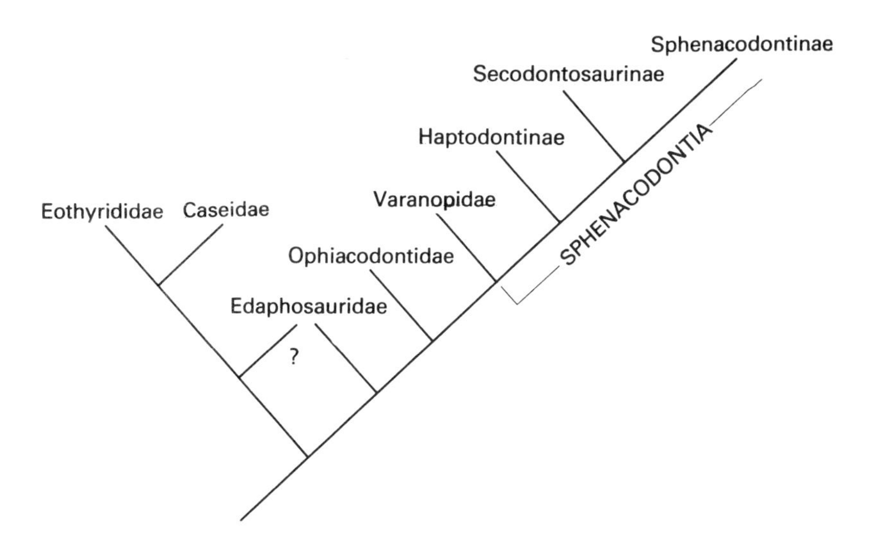


Fig. 17. Phylogeny of the main groups of pelycosaurs.

## Functional anatomy

The differences in structure amongst the various kinds of pelycosaurs means of course that there must have been corresponding differences in function. However, this diversity was relatively small, especially when

Fig. 16. A, skull of *Casea* in lateral, dorsal, ventral and posterior views. B, skull of *Edaphosaurus* in lateral, dorsal, ventral and posterior views. (A from Sigogneau-Russell and Russell, 1974; B from Romer and Price, 1940.)

*A*, angular; *D*, dentary; *ex.n*, external naris; *F*, frontal; *N*, nasal; *P*, parietal; *PMX*, premaxilla; *PO*, postorbital; *POF*, postfrontal; *PP*, postparietal; *p.pr*, paroccipital process; *PRF*, prefrontal; *SA*, surangular; *SO*, supraoccipital; *ST*, supratemporal; *STA*, stapes; *T*, tabular.

Magnifications: A, *c.* × 0.45; B, *c.* × 0.26.

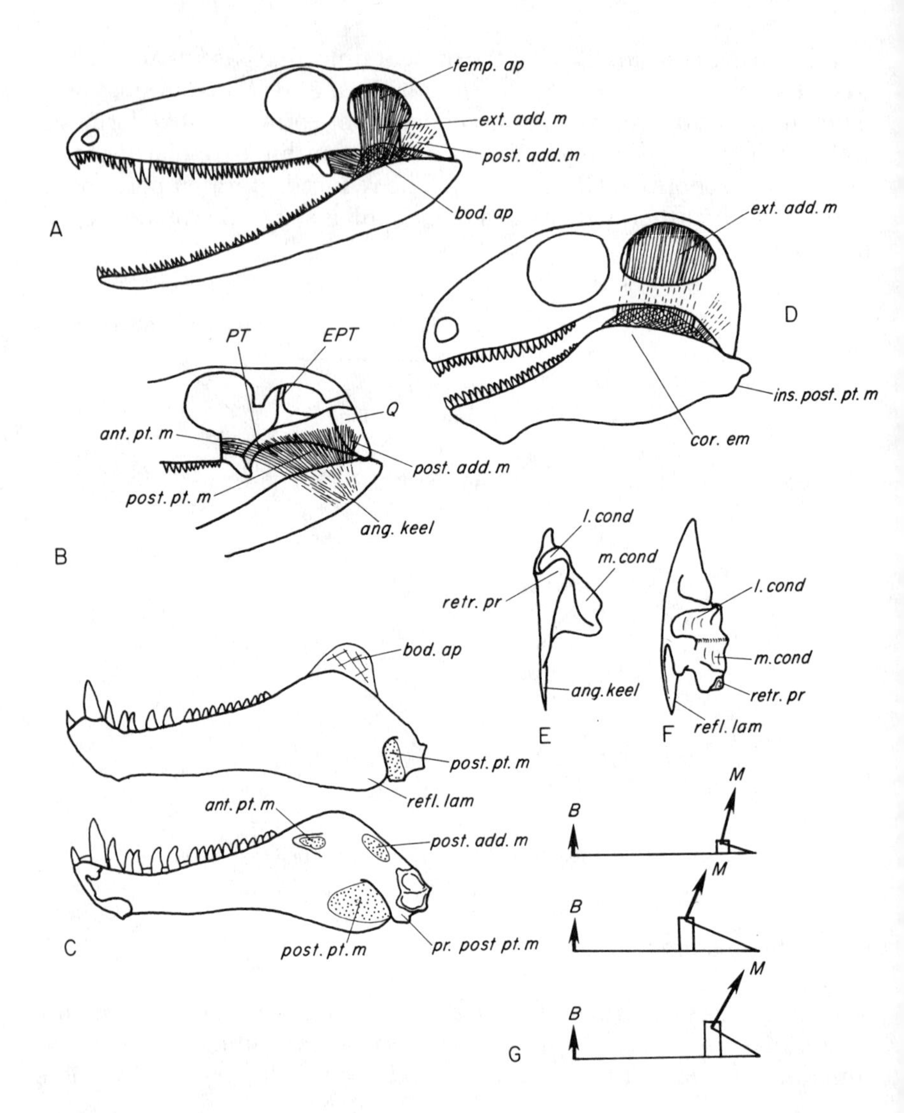


Fig. 18. Functioning of the pelycosaur lower jaw. A, superficial view of the adductor musculature of a primitive pelycosaur, lower temporal bar omitted. B, lateral view of the internal adductor musculature of a primitive pelycosaur. C, lateral and medial views of the lower jaw of *Dimetrodon* for muscle insertions. D, external adductor muscle of *Edaphosaurus*. E, posterior view of the lower jaw of a primitive pelycosaur *Ophiacodon*. F, posterior view of the lower jaw of *Dimetrodon*. *G*, jaw mechanics of a primitive pelycosaur (top), an edaphosaurid (middle) and a sphenacodontid (bottom). (Outlines redrawn from Romer and Price, 1940.)

*ang. keel*, angular keel; *ant.pt.m*, anterior pterygoideus muscle; *B*, bite force; *bod.ap*, bodenaponeurosis; *cor. em*, coronoid eminence; *EPT*, epipterygoid; *ext.add.m*, external compared with the great changes that were to come in the later mammal-like reptiles. A fairly generalised view of the functional anatomy of the group as a whole may therefore be taken as representing the primitive mammal-like reptile condition, from which the later developments took place. At the same time, the more obvious variations of the specialised groups may be noted.

## *Feeding mechanisms*

In living tetrapods, the main jaw-closing muscle is the adductor mandibuli, innervated by the fifth cranial nerve, the trigeminal. A somewhat arbitrary division of the muscle into component parts is recognised with reference to the course of the trigeminal nerve, there being respectively an external adductor, a posterior adductor and an internal adductor.

The structure of the posterior part of the pelycosaur skull, and the nature of the surfaces of the bones show that a broadly similar arrangement of the adductor musculature was present (Barghusen, 1973), although because the course of the nerve is unknown the divisions of the muscle cannot be delineated exactly. The largest part of the muscle was the external adductor (Fig. 18A), which was associated with two, important connective tissue aponeurotic sheets as in living reptiles. One, the temporal aponeurosis, attached to the upper and posterior edges of the temporal fenestra, and muscle fibres originated from both its internal and its external faces. Because the muscle was attached in this way it was strongly anchored to the skull as the aponeurotic sheet was continuous with the periosteal layer of connective tissue covering both the external and the internal surfaces of the bones of the skull roof (Frazzetta, 1969); as has been seen (p. 22), this was probably the very reason for the development of the temporal fenestra in the synapsids. The second connective tissue sheet attached to the dorsal edge of the hind part of the lower jaw and is called the bodenaponeurosis in modern reptiles. It acted as the main area in which the external adductor muscle fibres inserted and again gave a very strong attachment to the muscle as a whole. Thus, muscle fibres descending from the temporal aponeurosis, and also the more medial parts of the undersurface of the skull roof, inserted into both the external and the internal surfaces of the

adductor muscle; *ins.post.pt.m*, insertion of the posterior pterygoideus muscle; *l.cond*, lateral condyle; *M*, muscle force; *m.cond*, medial condyle; *post.add.m*, posterior adductor muscle; *pr.post.pt.m*, process for the posterior pterygoideus muscle; *PT*, pterygoid; *Q*, quadrate; *refl.lam*, reflected lamina of the angular; *retr.pr*, retroarticular process; *temp.ap*, temporal aponeurosis.

bodenaponeurosis. The whole of this muscle complex corresponds to the external adductor of modern reptiles. The posterior adductor of pelycosaurs originated from the back of the temporal fenestra, that is from the front face of the squamosal and quadrate bones (Fig. 18B). It inserted into the posterior part of the bodenaponeurosis, and also directly onto the inner surface of the lower jaw below and behind the bodenaponeurosis.

Finally, the internal adductor of pelycosaurs (Fig. 18B) appears to have had an origin from the dorsal surface of the palate, beneath the orbit, and was the equivalent of the anterior pterygoideus muscle of modern reptiles. It inserted towards the front of the adductor fossa, the large concavity occupying the inner surface of the lower jaw below the attachment of the bodenaponeurosis. A muscle equivalent to the posterior pterygoideus also occurred, originating from the back of the massive lateral pterygoid flanges that mark the posterior edge of the palate, and from the powerful bony struts of the pterygoid that lay alongside the base of the brain case. The area of insertion of the posterior pterygoideus muscle was mainly the internal surface of the deep ventral keel of the angular bone, this being, in fact, the reason why such a prominent keel was present.

The structure of the primitive pelycosaurian jaw hinge (Fig. 18E) reflects the action of the jaw muscles. The bulk of the muscle fibres pulled in a transverse plane, imparting no tendency to force the lower jaw either forwards or backwards. However, a greater proportion of the muscle lay internally to the jaw and only a few fibres lay in a plane external to the plane of the jaw. There was, therefore, a tendency for the jaw musculature as a whole to pull the back of the jaw inwards off the quadrate articulation. The articular surface of the articular bone consists of two concavities facing dorsally, which correspond to a pair of bulbous articulating facets on the quadrate. The ventrally directed reaction generated at the jaw articulation when the muscles contracted is therefore resisted. At the same time, the lateral facet of the articular bone is at a higher level than the medial facet, so that the tendency for the articular end of the jaw to move inwards is also resisted. The movement of the jaw about its hinge is purely orthal, that is, it rotates up and down about a simple hinge-like joint.

The dentition of the primitive pelycosaurs consists of a large number of sharp-pointed teeth, slightly enlarged in the canine region, particularly of the upper jaw. Such teeth clearly reflect the carnivorous diet, for they were designed for the capture and holding of small active prey, which was then swallowed whole. Enlargement of the teeth in the region shows a functional compromise. On the one hand, the enlarged teeth needed to be as far forwards as possible for the convenience of prey capture, but, on the other hand, they needed to be as far back as possible in order that the size of the bite force was as big as possible between them, bearing in mind that the jaw muscles producing the bite force were at the back of the jaw. One can only suppose that the actual position at which the enlarged teeth developed was the best compromise between these conflicting requirements.

From such a primitive arrangement of the feeding mechanism as described, and as found in ophiacodonts and varanopids for example, there arose two main kinds of modifications. One, towards a specialised herbivorous mode of life, occurred in *Edaphosaurus* and, probably independently, in the caseids. The other was adaptation to a much more highly predaceous life, as found in the sphenacodontids. In neither case, however, is there evidence of a radical change from the primitive condition. Rather, there was a series of changes in the proportions and shapes of various parts of the skull, reflecting relatively minor alterations in the pattern of jaw mechanism.

In the case of the herbivore groups (Fig. 18D) the facial region of the skull and the lower jaws were very much shortened, which increased the fraction of the muscle force actually applied as a bite force at the front (Fig. 18G). The effectiveness of the external adductor muscle was increased by a relative lowering of the position of the jaw hinge by a ventral extension of the squamosal and quadrate bones. At the same time, a broad coronoid eminence developed on the upper part of the jaw. Taken together, these changes had the effect of causing the external adductor muscle to incline backwards and upwards from the insertion on the jaw and thus increase the moment arm about the jaw hinge. The torque produced by the muscle on the jaw (the product of the muscle force generated and the moment arm) was consequently greater. Furthermore, the actual muscle force produced was increased by enlarging the size of the muscle. The occipital surface, which sloped antero-dorsally in the primitive pelycosaurs, became vertical, to increase the volume available for muscle fibres within the temporal region of the skull. Finally the bar of bone that defined the lower border of the temporal fenestra (eventually to become the zygomatic arch of advanced mammal-like reptiles and mammals) became bowed dorsally and allowed a greater volume of muscle to insert on the external surface of the bodenaponeurosis. In conjunction with these improvements of the jaw musculature, the skull became more strongly built in the temporal and occipital regions. Both the paroccipital processes and the supraoccipital developed strong attachments to the cheek region of the skull, especially the former in caseids and the latter in *Edaphosaurus* (Fig. 16).

An important new development in the herbivores was the ability of the lower jaw to move to and fro to some extent, as well as orthally. The articular facets of the articular bone of the jaw elongated, but the quadrate convexities remained short, so that antero-posterior sliding of the articular upon the quadrate was possible. Clearly this was of great advantage to a herbivore because it increased the grinding action of the teeth. The more an animal can break up the cellulose cell walls of its plant food before swallowing it, the more readily available is the cell content. The posterior inclination of the external adductor muscle was necessary, to cause the backward movement of the jaw. The forward movement was a result of the action of the internal adductor musculature, the anterior and posterior pterygoideus muscles. This muscle complex had increased in size, as indicated by the nature of the areas of insertion on the lower jaw. The angular keel is large and in the case of *Edaphosaurus*, it extends back alongside and below the jaw articulation. A new ventral process developed on the articular bone, immediately below the jaw hinge, for the posterior pterygoideus muscle.

Complex, occluding teeth were not possible in the herbivorous pelycosaurs because the jaws were incapable of precise or powerful enough movements to activate them. Instead, the teeth of the margins of the jaws became blunt and stout and, in the case of the caseids, coarsely crenulated (Fig. 16A). They were used for tearing off suitable pieces of vegetation. What pulping of the food did occur was performed by the elaborated palatal teeth. The caseids probably used a tough tongue in association with their palatal teeth, suggesting a relatively soft diet. *Edaphosaurus* (Fig. 16B), however, possessed massive upper and lower tooth plates, covered by small teeth and probably capable of masticating relatively hard foodstuffs.

The advanced sphenacodonts, notably *Dimetrodon* and *Sphenacodon*, showed an equivalent degree of specialisation of the primitive feeding apparatus, this time for their predaceous mode of life. Unlike the herbivores, the long snout and jaws of the primitive forms were retained although the facial region became much deeper and heavier to accommodate the very large teeth (Fig. 11). The force of the bite at the anterior end of the jaw was still only a small fraction of the size of the jaw muscle force because the jaws were long (Fig. 18G). However, an important component of the food capture mechanism of sphenacodonts was the kinetic energy imparted to the closing jaws by the contracting muscle, which was dissipated when they met. The kinetic energy of the jaw depends solely on the amount of muscular work done to accelerate the jaw, and not at all on the length of the jaw; therefore, as long as this kinetic method was used to disable the prey, long jaws did not matter.

On the other hand, long jaws are advantageous because of their prehensile ability, compared to short jaws. Correlated with this particular means of prey capture was the considerable enlargement of canine teeth, both upper and lower, so that dissipation of the kinetic energy was concentrated at a few teeth.

The size and effectiveness of the external adductor musculature increased, much as in the herbivores. Quite independently of the latter, the jaw articulation was depressed and a large coronoid eminence developed on the lower jaw (Fig. 18C), thus increasing the moment arm of the muscle. The whole postorbital part of the skull is very wide, the occiput vertical, and the lower temporal bar bowed upwards, all features increasing the space available for the external adductor muscle.

The pterygoideus musculature was also larger than in primitive pelycosaurs, particularly that part of the posterior pterygoideus which originated from the back of the now massive lateral pterygoid processes of the palate. A large ventral keel of the angular bone was retained for the insertion of the pterygoideus muscle, but the insertion had also spread to the articular bone. As in the herbivores, a distinct ventral process had developed on the ventral side of the articular, below the jaw articulation, for this muscle. The recess created between the angular keel laterally and the downturned articular medially was also probably utilised to expand the insertion of the posterior pterygoideus muscle (Fig. 18C), which could now wrap around the ventral edge of the jaw in the articular region and attach to the lateral jaw surface in and behind the recess (Barghusen, 1973).

The jaw articulation of sphenacodonts has changed slightly: the articular facets of the articular have elongated. However, the quadrate condyles are also elongated so that very little, if any, antero-posterior shift of the lower jaw was possible. The greater surface area between the upper and lower hinge bones does, however, provide a stronger joint, less liable to disarticulation by struggling prey. The articular facets also face more posteriorly, in response to the posterior inclination of the external adductor muscle (Fig. 18F).

The degree of anisodonty, or tooth differentation, reaches a higher level in this group than any other pelycosaur. The anterior incisor teeth of the premaxilla and front of the dentary are large and round in cross-section. They are designed for holding onto the prey, possibly shaking it. The enlarged canine teeth, of which there is usually one in each jaw, upper and lower, are for rapidly disabling the prey. The postcanine teeth are smaller, but very sharp with recurved tips to help retain the prey in the mouth. Each one is flattened from side to side so that both the front and the back edge are sharp. There is no indication

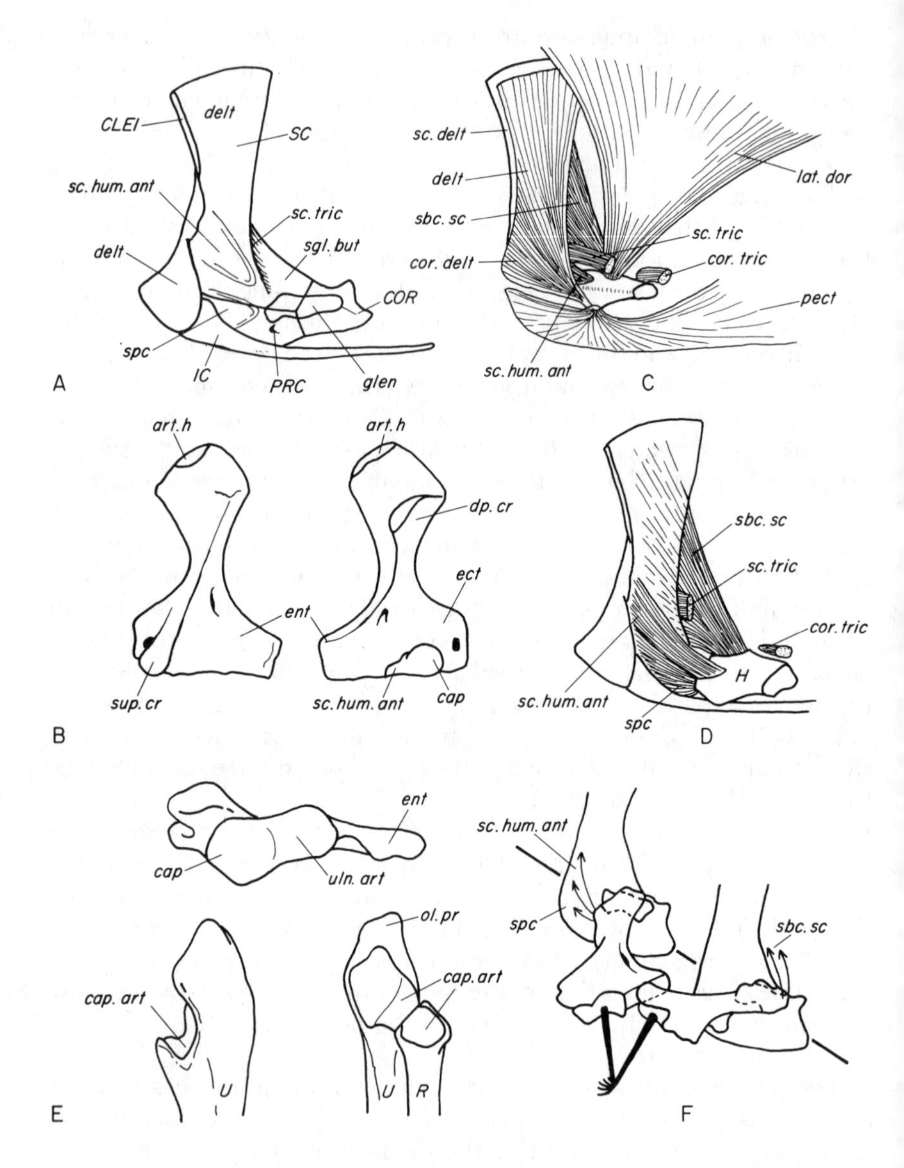


FIG. 19. Shoulder girdle and forelimb of pelycosaurs. A, lateral view of the shoulder girdle. B, dorsal and ventral views of the left humerus. C, superficial shoulder musculature. D, deep shoulder musculature. E, distal view of left humerus, anterior view of the left ulna, and medial view of the radius and ulna, to show the structure of the elbow joint. F, the forelimb stride viewed from the postero-lateral and dorsal aspect. (A–D redrawn from Romer, 1922; E, redrawn from Jenkins, 1973; F, redrawn from Jenkins, 1971a.)

that sphenacodonts were capable of any form of chewing of the food within the mouth. Palatal teeth are not particularly well-developed, except for a row along the posterior edge of each lateral pterygoid flange.

### *Locomotion*

The structure of the postcranial skeleton is much more constant than that of the skull amongst the various pelycosaurs; also it is very primitive because it resembles quite closely the postcranial skeleton of other early reptiles and amphibians.

The shoulder girdle (Fig. 19A) forms a massively constructed U-shaped arch around the thorax, which resisted the tendency of the humeri to compress the rib cage. The scapula blade follows the curvature of the ribs, to which it was attached by the serratus muscles, and the two coracoids form a ventral plate that curves inwards below to meet the interclavicle. Of the dermal shoulder girdle, there is a vestigeal, splint-like cleithrum still present but firmly attached to the scapula. The clavicle is a robust rod attached along the lower half of the front edge of the scapula, and expanding medially to form a broad plate attached to the underside of the interclavicle. As the various elements of the pectoral girdle are so firmly attached to one another, there can have been very little movement of the girdle relative to the vertebral column. The humerus (Fig. 19B) is also massive, with broadly expanded ends connected by a short shaft, and lying at a marked angle of typically around 50° to one another.

The shoulder joint is a complex structure which rigidly limited the movements of the humerus to one defined pattern. The glenoid cavity of the shoulder girdle is formed from all three of the primary bones, scapula above, procoracoid anteriorly and coracoid posteriorly. It is elongated antero-posteriorly and its articulatory surface is screw-shaped. The front part faces backwards and downwards, the middle part out-

*art.h*, articulating head; *bi*, biceps muscle; *cap*, capitulum; *cap.art*, surface for articulation with the capitulum; *CL*, clavicle; *CLEI*, cleithrum; *COR*, coracoid; *cor.br*, coracobrachialis muscle; *cor.tric*, coracoid part of triceps muscle; *delt*, deltoideus muscle; *dp.cr*, delto-pectoral crest; *ect*, ectepicondyle; *ent*, entepicondyle; *glen*, glenoid fossa; *IC*, interclavicle; *lat.dor*, latissimus dorsi muscle; *ol.pr*, olecranon process; *pect*, pectoralis muscle; *PRC*, procoracoid; *R*, radius; *sbc.sc*, subcoraco-scapularis muscle; *SC*, scapula; *sc.delt*, scapula part of deltoideus muscle; *sc.hum.ant*, scapulo-humeralis anterior muscle; *sc.tric*, scapula part of triceps muscle; *sgl.but*, supraglenoid buttress; *spc*, supracoracoideus muscle; *sup.cr*, supinator crest; *U*, ulna; *uln.art*, surface for articulation with the ulna.

wards, and the posterior part upwards, forwards and outwards. This shape corresponds to the articulatory surface of the humerus head, which is approximately a segment of a spiral. It commences as an antero-dorsally facing surface at the front, then winds postero-ventrally until it faces postero-ventrally at the back. Jenkins (1971a) has interpreted the humerus movements in a very finely preserved specimen of *Dimetrodon* (Fig. 19F). When the humerus was at its most protracted position, at the beginning of a stride, the antero-dorsal part of the head fitted against the front part of the glenoid, while the posterior part of the humerus head lay against the posterior part of the glenoid. To achieve this fit, the distal end of the humerus is slightly raised above the level of the proximal end, and the expanded distal end also faces antero-ventrally. The lower leg therefore projected forwards and downwards to the foot, the latter being in advance of the forearm. As the humerus was retracted the broad concave trough that formed the front part of the humerus articulating surface passed along the front part of the glenoid surface. Thus, the anterior part of the humerus head passed forwards beyond the confines of the glenoid. Because of the shapes of the respective articulatory surfaces, this caused retraction of the humerus, along with the lowering of the distal end to a level a little below that of the glenoid. At the same time, the posterior part of the humerus head was forced upwards, imparting to the humerus a small amount of rotation about its long axis. The distal end came to face directly ventrally, and the elbow joint now lay directly above the forefoot. The recovery stroke, whereby the humerus moved from its retracted position forwards in preparation for the next stride, was simply a reversal of these movements.

The necessary movements and the pattern of stresses occurring at the elbow and wrist joints were highly complex, as in sprawling-limbed tetrapods in general (Haines, 1946). The lower leg must flex and extend on the end of the humerus. Also, if the foot is to remain stationary during a stride while the humerus retracts, then there must be a relative rotation between the foot and the distal end of the humerus about a vertical axis, accommodated via the radius and ulna (Fig. 19F). Finally, the elbow joint must also resist torsion movements on the end of the humerus, so that when the humerus rotates about its own long axis the lower leg moves with it (Fig. 19F). No single joint could incorporate all these requirements; therefore use is made of several joints working in association, which is the basic reason why tetrapods have two bones in their lower limbs, in this case radius and ulna. It provides four joints between the humerus and the foot. The elbow joint (Fig. 19E) consists of a radius articulating with a hemispherical capitulum on the end of the humerus, which permits both flexion–extension and also rotation about its long axis by the radius, but offers no resistance to torsion. The ulna, however, has a complex articulation (Jenkins, 1973) which does resist torsion. The sigmoid notch of the ulna bears against the side of the capitulum as well as against a more dorsally placed facet of its own on the humerus. When the humerus rotates about its own long axis, the two contacts with the ulna force the ulna, and hence the lower leg, to move with it. Flexion–extension can also occur at the humero-ulna articulation, but rotation of the ulna about its own long axis is impossible. Instead, this function is left to the joint between the ulna and the foot alone. It will be seen that as far as the rotation of the humerus relative to the foot is concerned, the radius and the ulna behave independently of one another, the radius rotating on the humerus and the ulna rotating on the foot. This is probably a particularly strong arrangement.

The forefoot consists of eleven separate ossicles forming the manus and has a digital formula of 2.3.4.5.3. All the articulations between the foot bones are flat and hinge-like, indicating that the foot had a general flexibility as a whole, rather than specific, well-defined axes of bending.

The musculature of the shoulder girdle and forelimb (Fig. 19A–D) was reconstructed by Romer (1922), and this classic account still holds good to a considerbale extent. It is most convenient to consider the muscles from the functional point of view. The main movements of the humerus were protraction–retraction, elevation and depression, and rotation about its long axis. In addition the forearm was capable of extending and flexing at the elbow joint.

*Retraction* A complex of muscles originated from the inner face of the scapulo-coracoid, ran outwards behind the girdle, and inserted on the dorsal surface of the humerus, behind the articulating part of the head. It was the subcoraco-scapularis muscle and, by pulling the posterior part of the humerus head inwards and forwards, it imparted a retraction effect on the humerus. Another muscle with possibly some retraction function was the latissimus dorsi, a universal muscle amongst tetrapods which ran as a great fan from the connective tissue fascia of the back and flank to insert on a transverse ridge distal to the head of the humerus. The majority of its fibres must have had an elevating effect, but the more posterior ones would have tended to pull the humerus backwards.

*Protraction* The principal protractor of the humerus was the supracoracoideus muscle, originating on the lateral face of the procoracoid and lower part of the scapula. Its insertion was the anterior part of the proximal end of the humerus. By pulling this part of the humerus inwards, the bone as a whole would have moved forwards. A second muscle with some protraction function was the deltoideus, a fan of muscle fibres running from the clavicle and dorsal blade of the scapula to an insertion on the inner end of the prominent delto-pectoral crest. The more ventral fibres at least must have tended to pull the humerus forwards. A third protractor was the scapulo-humeralis anterior, lying just above the supracoracoideus.

*Elevation* Three of the muscles already mentioned also acted to elevate the humerus. These are the latissimus dorsi behind the bone, and both the deltoideus and the scapulo-humeralis in front of it.

*Depression* These muscles, as would be expected in a sprawling-limbed animal, were extremely well developed. The largest depressor, or adductor, of the humerus was the pectoralis. It originated from the interclavicle, clavicle and sternum, and inserted on the delto-pectoral crest. The coraco-brachialis occupied much of the lateral surface of the coracoid plate, and ran outwards to insert on the underside of the proximal end of the humerus, in the well-defined adductor fossa behind the delto-pectoral crest.

*Rotation* The function of rotating the humerus about its own long axis was not performed by muscles exclusively adapted for the purpose, but by the precise position of attachment of all the muscles, away from the axis of the bone. Thus, during the retraction phase, the subcoraco-scapularis tended to raise the hind margin of the humerus because it attached behind the axis of the bone and pulled partly upwards. Similarly, the pectoralis, attaching in front of the axis, tended to rotate the bone as well as depress it. In a similar way, the deltoideus and supracoracoideus muscles tended to rotate the humerus the opposite way during the recovery stroke.

*Forearm extension* The triceps muscle had three areas of origin, from a process on the coracoid behind the glenoid, from the scapula low down near to the posterior edge, and from much of the dorsal surface of the humerus shaft. All the muscle fibres converged upon the olecranon process of the ulna, creating a very powerful extensor of the forearm. A complex of extensor muscles also arose from the ectepicondylar region of the humerus, to insert on the radius, ulna and manus.

*Forearm flexion* The main flexor muscle of the forearm was the biceps, from the external surface of the coracoid to the radius and ulna bones. A brachialis muscle originated from the ventral surface of the humerus and inserted in close association with the biceps on the flexor surfaces of the radius and ulna.

The overall locomotory function of the forelimb and girdle may be summarised (Fig. 19F). At the commencement of the power stroke the humerus extended laterally from the glenoid while the radius and ulna extended antero-ventrally to the foot, which therefore lay in front of the level of the glenoid. Contraction of the pectoralis muscle and biceps-plus-brachialis muscles forced the distal end of the humerus down, and consequently raised the body off the ground. The subcoraco-scapularis muscle contracted, to draw the humerus backwards in a horizontal plane. As these muscles contracted, they also caused a rotation of the humerus about its own long axis, the left humerus, for example, rotating anticlockwise viewed from the side. The body was thus drawn forwards over the foot, adding to the effective length of the stride. Towards the end of the retraction phase, the triceps muscle contracted, extending the elbow joint and adding a further, probably small increment to the stride. The recovery phase was more or less a reversal of this sequence, the latissimus dorsi and deltoideus raising the humerus so that the foot left the ground, the biceps flexing the elbow, and the supracoracoideus drawing the limb forwards. Rotation about the long axis of the humerus in the opposite direction was a secondary result of the contraction of these muscles, forcing the foot forwards ready to be placed on the ground for the next power stroke. The whole of this cycle was pre-determined by the form of the shoulder joint, and elbow joint, with only very limited possibility of varying the stride pattern. This fact, along with the massiveness of the bones and the almost equal development of the recovery-phase and propulsive-phase musculature indicates that the forelimb functioned passively. It maintained the anterior part of the animal off the ground but contributed very little to the propulsive force. The requirements of such a limb were only that the stride length and frequency of striding were adequate.

The hindlimb contrasted sharply with the forelimb, for its movements were not nearly so predetermined, the bones are less massively built, and the power-stroke musculature was very much more developed than the recovery stroke musculature. Clearly the function of the hindlimb was the production of the propulsive forces. The pelvis (Fig. 20A) is supported by two, or, in the advanced forms, three sacral vertebrae. The ilium is a small plate extended posteriorly in primitive forms and sphenacodonts, and somewhat anteriorly in edaphosaurs. The pubis and ischium together form a very large plate that extends both backwards and forwards, and meets its fellow in a mid-ventral symphysis of some strength. The femur (Fig. 20B) is about 10% longer than the humerus, which compensates for the fact that the glenoid is situated further laterally than the acetabulum. Neither end of the femur is

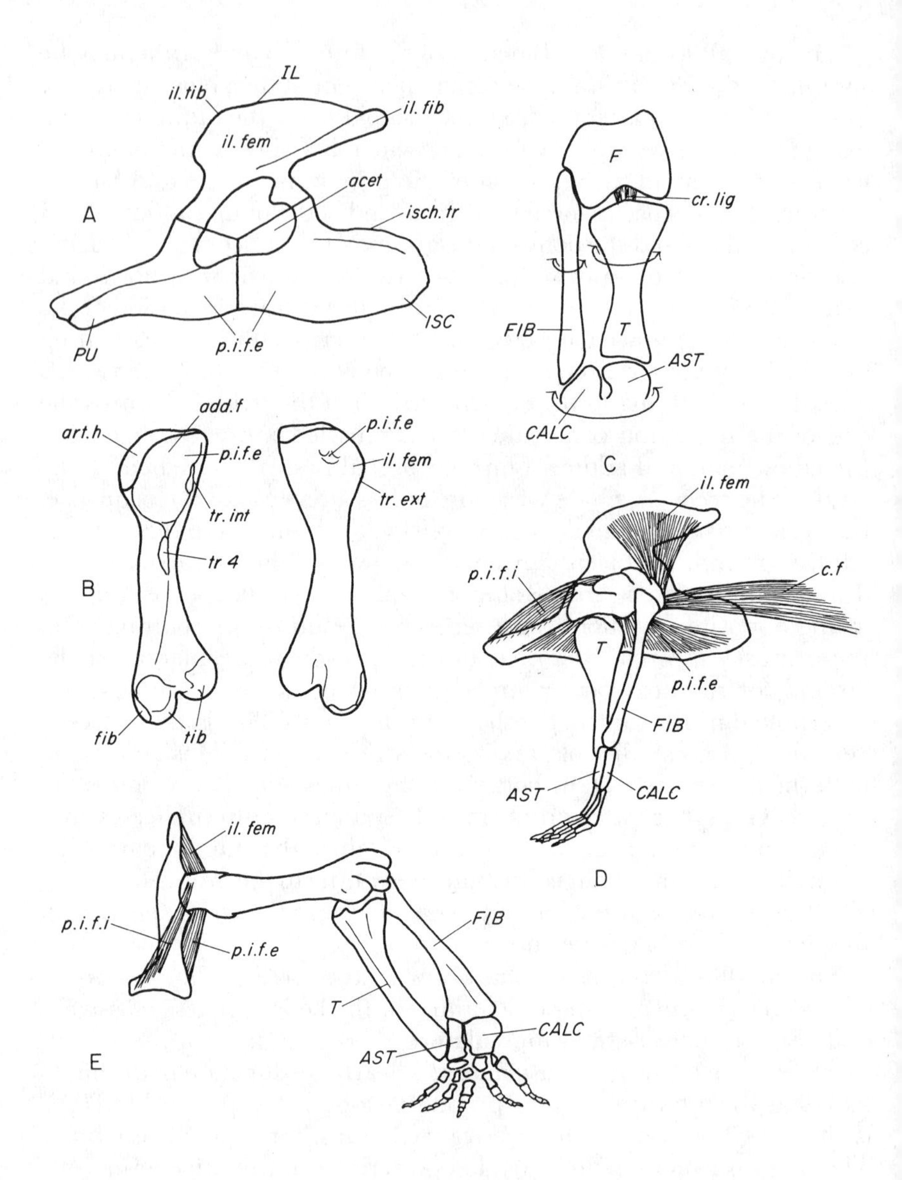


FIG. 20. Pelvis and hindlimb of pelycosaurs. A, lateral view of left pelvis. B, ventral and dorsal views of left femur. C, method of rotation of the foot relative to the femur, showing independent rotation of the tibia and the fibula about their respective long axes. D, lateral view of the main muscles of the hip. E, anterior view of the hindlimb and pelvis. (A and B redrawn from Romer, 1922; C redrawn from Haines, 1942; D and E redrawn from Jenkins, 1971a.)

*acet*, acetabulum; *amb*, ambiens muscle; *add.f*, adductor fossa; *art.h*, articulating head;

particularly expanded and therefore this bone is very difi
pearance from the humerus.

The hip joint is a simple, rather shallow ball-and-socket ty
with none of the elaborations for restricting movements se
shoulder joint. The articulating surface of the head of the fem , which
fits into the acetabulum, is entirely on the proximal end of the bone, and therefore the femur could only move in, or at least near to, the horizontal plane. It could, however, rotate about its long axis, and be raised or lowered to some extent.

The knee and ankle joints were subjected to the same complex of movements and stresses as the equivalent joints of the forelimb. The design of the hindlimb joints was, however, somewhat different. The tibia is larger than the fibula and articulates with the two distal condyles which span the whole breadth of the distal end of the femur (Fig. 20C). The fibula by contrast articulates only with a rather small facet on the posterior side of the femur. By comparison with living reptiles (Haines, 1942) the knee joint must have depended on a series of strong ligaments to maintain the bones together and control their movements, and particularly cruciate ligaments between the tibia and the intercondylar region of the femur. Both the tibia and the fibula could, necessarily, flex and extend to the femur. The relative rotation between the distal end of the femur and the foot that has to be accommodated by the tibia and fibula probably resulted from independent rotation of the tibia and fibula respectively about their long axes at the femoral end (Fig. 20C). The attachment of these bones to the foot appears to have been in the form of a simple hinge-joint (Haines, 1942). Like the hip joint, the pelycosaur knee must have relied on ligamentous and muscle attachments for control rather than osteological constraints of the kind seen in the forelimb. Presumably the advantage lies in a greater variety of possible movements, necessary for the thrust-producing hindlimb but not for the much more passive forelimb.

The pelycosaur hindfoot consists of a large calcaneum and astragalus, which were normally held almost vertically (Fig. 20D,E) with the proximal end high off the ground, two centrales, all five primitive distal

*AST*, astragalus; *CALC*, calcaneum; *c.f*, caudi femoralis muscle; *cr.lig*, cruciate ligaments; *F*, femur; *FIB*, fibula; *fib*, articulation surface for the fibula; *IL*, ilium; *il.fem*, ilio-femoralis muscle; *il.fib*, ilio-fibularis muscle; *il.tib*, ilio-tibialis muscle; *ISC*, ischium; *isc.tr*, ischio-trochantericus muscle; *p.i.f.e*, pubo-ischio-femoralis externus muscle; *p.i.f.i*, pubo-ischio-femoralis internus muscle; *PU*, pubis; *T*, tibia; *tib*, articulation surface for the tibia; *tr.4*, fourth trochanter; *tr.ext*, external trochanter; *tr.int*, internal trochanter.

tarsals, and a digital formula of 2.3.4.5.4. The ankle joint, between the crus and the calcaneum and astragalus was a simple hinge joint, the tibia and the astragalus on the one hand, and the fibula and a compound facet shared between the astragalus and calcaneum on the other. The whole pes formed an arch that gradually transmitted the animal's weight from the vertical crus to the horizontal toes, with none of the intervening joints specialised for particularly large movements (Schaeffer, 1941).

As with the forelimb, the classic acount of the limb musculature by Romer (1922) still holds good in general respects (Fig. 20).

*Retraction* The largest of the hip muscles was the caudi femoralis, as in modern reptiles. Its origin was the vertebrae and connective tissue fascia of the tail, and it inserted in the middle of the ventral surface of the femur, along the fourth trochanter. The more posterior part of the pubo-ischio-femoralis-externus, arising from the pubo-ischiadic plate and inserting in the adductor fossa of the underside of the proximal end of the femur would also have provided some retraction force.

*Protraction* The pubo-ischio-femoralis internus muscle originated from the internal surface of the pubis, which is in front of the femur. The fibres ran upwards, backwards and outwards to insert on the anterior and dorsal sides of the femoral head, providing virtually all the retraction force.

*Elevation* The ilio-femoralis muscle originated on the lateral surface of the ilium and inserted on the external trochanter on the posterior part of the femoral head.

*Depression* The pubo-ischio-femoralis externus, as well as its relatively small role as a retractor, was the main depressor, or adductor of the femur.

*Rotation* As in the case of the forelimb musculature, rotation of the femur about its long axis resulted from the exact manner in which these other muscles inserted, tangential to the actual axis of the bone.

*Extension of the crus* The triceps inserted on the outer side of the head of the tibia and consisted of three slips, the ilio-tibialis from the ilium, the ambiens from the pubis, and the femora-tibialis or vasti from the dorsal surface of the femur. Paralleling the triceps was the ilio-fibularis, running between the ilium and the outer side of the head of the fibula.

*Flexors of the crus* A complex series of muscles ran from the lower part of the pelvis out to the flexor surfaces of the tibia and the fibula.

The overall functioning of the pelycosaurian hindlimb was as follows. At the start of the power stroke the femur was horizontal and extended antero-laterally from the acetabulum. The crus sloped downwards and forwards from the knee to the foot. Contraction of the pubo-ischio-femoralis externus forced the femur downwards, thus raising the body off the ground, and contraction of the massive caudi femoralis retracted the femur, driving the body forwards. Because of the ventral position of the insertion of the latter muscle on the femur, a simultaneous rotation of the femur about its own long axis occurred, drawing the body over the foot and, as in the case of the forelimb, adding to the length of the stride. At the end of this phase, contraction of the triceps and ilio-fibularis muscles together actively extended the knee, giving a further propulsive thrust to the body. The recovery phase consisted of the ilio-femoralis muscle contracting to raise the leg off the ground, followed by the pubo-ischio-femoralis internus drawing it forwards.

The third and final part of the postcranial skeleton to consider is the vertebral column, and its role in locomotion. In most primitive tetrapods, and reptiles generally, the articulating surfaces of the zygapophyses are horizontal, which permits quite extensive lateral movement to occur between adjacent vertebrae. Among the pelycosaurs, this is true of the ophiacodont *Varanosaurus* alone. All the rest have developed zygapophyses whose articulating surfaces are inclined to some extent, the prezygapophyses partly facing each other and the postzygapophyses partly facing laterally. In typical ophiacodonts this tilting is around 30° from the horizontal, while in the more advanced groups it is more like 45°. The change reflects a marked tendency to reduce the extent of lateral undulation of the trunk during locomotion. There are several reasons why, unlike even modern lizards for example, lateral undulation might have been reduced. It is a very inefficient means of increasing the stride because much of the muscular effort producing the undulations is devoted to shifting parts of the animal's body sideways, rather than forwards. The larger the animal, the greater will this inefficiency be. It is also a potentially clumsy, poorly manoeuvrable type of terrestrial locomotion. Therefore, the pelycosaurs, as they evolved larger size, appear to have lost undulation as a significant component of the overall locomotory pattern, developing instead longer limbs and a greater stride. The tendency of pelycosaurs to reduce their intercentra, the crescentic bones that lie in between adjacent centra, may also relate to decreased lateral undulation.

### *Middle ear*

Much controversy surrounds the pelycosaur middle ear, centred upon the question of whether a tympanic membrane was present or not. The view that a tympanic membrane was present is based upon the fact that a stapes bone occurs, whose inner end fits into the fenestra ovalis, opening from the otic region of the braincase (Figs 8D and 16B). In typical members of most of the modern reptile groups, a similar basic arrangement is acompanied by a tympanic membrane, to which the outer end of the stapes attaches via a cartilaginous extrastapes. Airborne sound causes vibration of the tympanic membrane, and the stapes conducts the vibrations to the fenestra ovalis and hence to the fluid in the endolymphatic canal of the cochlea. It is argued that, because a tympanic membrane, presumed homologous, is present in the modern amphibians (the Anura), the modern reptile groups, the birds and the mammals, it must have been present in the common ancestor of all these. Its biological significance is such that its loss, except in occasional specialised cases, would be unlikely. Therefore all the primitive reptiles including pelycosaurs had one.

There are, however, certain difficulties inherent in this theory. The pelycosaur stapes is a relatively massive bone compared to the light, delicate rod of modern reptiles, and it is strongly attached to the underside of the paroccipital process by a massive dorsal process, which suggests that, at the very least, it functioned in a rather different manner to the stapes of modern reptiles. There is no indication in the form of the bones of the posterior region of the skull of where a tympanum attached. This again contrasts with the condition of the modern reptiles, where some part of the tympanum is invariably supported by a concave bony margin, the otic notch. There is also the problem, ultimately, of how and why the ear changed by incorporation of reduced postdentary jaw bones into the middle ear, between the stapes and presumed tympanic membrane in the line leading eventually to the mammals (p. 280). If the pelycosaur ear functioned as in modern reptiles, why did it not improve it, as they evidently did, by simply lightening the stapes? The functioning of the modern reptilian ear is not inferior to that of mammals, at least at moderate frequencies (Manley, 1972). Finally, there are certain anatomical difficulties in the way of accepting the homology of the tympanic membrane in all modern tetrapods (Lombard and Bolt, 1978).

Notwithstanding the normal occurrence of a tympanic membrane in modern reptiles, the frequency with which this structure is lost indicates perhaps that a tympanic-membrane operated stapes is less of an overwhelming biological advantage than is normally supposed. *Sphenodon*, all the snakes, various lizards, urodeles and apodans all lack a tympanum, and yet the ears of at least several of these are known to be sensitive to low-frequency airborne sound, as well as ground-borne sound. Only the higher frequency soundwave detection is seriously impaired. One possible correlation with the occurrence of non-tympanic ears is the absence of sound communication between individuals (Manley, 1972). There is, therefore, no particularly compelling reason to believe that pelycosaurs had tympanic membranes, and it seems on balance more probable that the stapes functioned only in the detection of low frequency sound, either air or ground borne. The physics of non-tympanic hearing is not well understood, but it is perfectly feasible.

Further support for the view that pelycosaur ears lacked a tympanic membrane is the small size of the cochlea recess, lying in the floor of the otic region of the braincase. In modern reptiles the sensitivity to high-frequency sound waves is limited by the sensitivity of the cochlea rather than the ability of the tympanic membrane to respond. The sensitivity of the pelycosaur cochlea cannot of course be known, but there was certainly inadequate space for a long cochlea, and therefore it may be presumed to have been relatively poor in its ability to discriminate high frequencies. Thus it would be incongruous to expect pelycosaurs to have had the highly sensitive tympanic middle ear.

## *General biology*

The anatomy of the pelycosaur skeleton indicates that they were thoroughly reptilian animals. They lacked such advances as the ability to masticate their food before swallowing it (with the partial and special exception of the herbivores perhaps), and their locomotion was of a primitive, sprawling pattern. One of the key features of modern reptiles is their temperature physiology, which is ectothermic. They can maintain a reasonably constant body temperature, but they do it by differential absorbtion and radiation of heat via the environment, rather than by the high metabolic heat production found in mammals and birds. Pelycosaurs appear to have a typically ectothermic method of temperature control. The most direct evidence is the huge dorsal sail supported by the enlarged neural spines of the vertebral column, which evolved independently in *Edaphosaurus* and in *Dimetrodon*. Whether or not it had some such function as a species recognition signal as well, it is certain that this great increase in the surface area of the animal must have had a profound effect on the temperature regulation. Holding the sail face on to the sun would cause a rapid increase in body temperature, while holding it edge on would cause radiation away of heat, even to the extent of causing a fall in body temperature when the ambient temperature was higher than the body temperature (Bramwell and Fellgett,

1973). Only an ectothermic animal would have so benefited from such a device, and the fact that these two pelycosaurs were therefore ectotherms implies that the group as a whole probably were. Another possible correlate of ectothermy is the histological structure of the bone, as investigated at length by Ricqles (e.g. 1974). Pelycosaurs generally have a lamellar-zonal pattern of primary bone tissue, with "growth rings" and little secondary Haversian replacement. All these are characteristic of modern ectothermic reptiles.

A second great physiological characteristic of modern reptiles is their high level of independence from water, based on the waterproof skin, water-reabsorbing cloaca, uric acid excretion, and shelled cleidoic egg. Scales, being formed of the protein keratin, rarely preserve in the fossil record, and there is no direct evidence of the extent to which pelycosaur skin was keratinised and therefore potentially waterproof. However, the sail of *Edaphosaurus* and *Dimetrodon* again is suggestive, for it is hardly conceivable that such an increase in surface area could be tolerated if the skin was freely permeable to evaporative water. As will be seen below, the distribution of the pelycosaurs was largely in moist areas, where at least some water-permeability of the skin would not have mattered. There is, however, evidence of some pelycosaurs living in dry, upland areas, suggesting that they at least possessed impermeable skin. Nothing is known of the physiology of excretion, or the structure of the cloaca, and one can only surmise that they were probably typically reptilian. A fossil egg has actually been found in the lowest part of the Wichita of Texas. It is about 59 mm in length, shelled and therefore presumably cleidoic, and statistically it is probable that it was laid by a pelycosaur, since these animals are far the commonest members of the fauna (Romer and Price, 1940).

The brain of pelycosaurs, as assessed from the size of the brain case, was about the same size as in modern reptiles.

## Palaeo-ecology

The great majority of pelycosaur specimens come from the north-central area of Texas, where the redbeds extend continuously through time from the very base of the Permian (Wolfcampian) through to the early part of the Late Permian (Guadalupian), a sequence of some 4500 feet of deposits. Other areas with a comparable fauna are known elsewhere in North America, but none have anything approaching the extent of the Texas series. E. C. Olson and his collaborators (Olson and Vaughn, 1970; Olson, 1971; 1975) have done a great deal of work on interpreting the palaeo-ecology of this and related faunas, and it is to them that we owe the concept of the Permo–Carboniferous chronofauna. A chronofauna is defined as an ecological system which is persistent and uninterrupted through time. Relatively minor changes in the fauna may occur, by the evolution or extinction of existing members, and by invasion from elsewhere of new members. Similarly there may be differences in the detailed structure of the fauna in different geographical areas, and the boundaries between adjacent contemporaneous chronofaunas are not sharp.

During this period, a mid-continental sea occupied what is now the Gulf of Mexico and the southernmost part of North America. The sediments laid down in northern Texas were formed close to the sea shore in a large deltaic floodplain where rivers flowing southwards met the sea. This area lay just to the south of the Permian equator, and at the beginning of Wichita times the climate was warm and humid. Permanent lakes, streams and swamps abounded and the rainfall was fairly evenly distributed throughout the year. The ecology was very much based on aquatic and semi-aquatic organisms. The freshwater shark *Xenacanthus* and lungfish *Sagenodus* as well as several crossopterygian and palaeoniscid fish were abundant, and there was a series of semi-aquatic amphibians like *Archeria* and *Trimerorhachis*. The more common terrestrial animals were the captorhinomorph reptile *Captorhinus*, a small insectivore, and the three well-known pelycosaurs *Ophiacodon*, *Edaphosaurus* and *Dimetrodon*. *Ophiacodon* was a fish-eater and probably spent much of its time in the water. *Edaphosaurus* was herbivorous, and *Dimetrodon* occupied the large carnivore niche. The base of the food web undoubtedly lay among the aquatic organisms, with aquatic and semiaquatic plants providing food for the fish either directly or indirectly via invertebrates. These in turn were eaten by *Ophiacodon*. The relative abundance of *Dimetrodon* indicates that it too must have eaten fish, as well as the other tetrapods, for there do not appear to have been sufficient of the latter to supply all its food.

During the succeeding period, when the deposits of the Clear Fork were laid down, changes in the sediments indicate a gradual alteration towards a rather drier climate with the rainfall becoming seasonal. Swamps were reduced, and complete drying out of the smaller lakes and streams occurred, presumably annually. Changes in the fauna followed which at least in part reflect the changing conditions. Most strikingly, the lungfish *Gnathorhiza* became common; this form was capable of aestivating in a mud burrow, much as the modern African lungfish *Protopterus* today. Occasional fossilised burrows have been found. Among the amphibians, small lepospondyls *Diplocaulus* and

*Lysorophus* became common, forms that were presumably more resistant to dry conditions. Among the pelycosaurs, only *Dimetrodon* persisted, indicating its position as the least dependent on water of the earlier pelycosaurs present. *Ophiacodon* disappeared, no doubt because during the dry season its food was unavailable. The lungfish aestivated, while *Xenacanthus* and the other fish probably migrated to deeper, permanent waters. The extinction of *Edaphosaurus* was perhaps because its food was the plants of the swampy areas, which would have been greatly reduced in the hot season. Conditions continued to deteriorate, and by the end of the Clear Fork evaporite deposits indicate very arid conditions. The fauna completely disappears for a time, and when it reappears in the Pease River deposits it has changed dramatically. In the Flower Pot and San Angelo Formations *Dimetrodon* is still present—further evidence of its physiological independence of water. However, a series of caseid pelycosaurs have appeared, including the giant *Cotylorhynchus* and *Angelosaurus* as the dominant herbivores, and a single form of varanopid sphenacodont *Varanodon* also represents a completely new element of the Permo-Carboniferous chronofauna. Most striking of all, however, is the appearance of several different kinds of therapsid mammal-like reptiles, the group which before long was to replace completely the pelycosaurs as the dominant terrestrial forms. These are discussed in the next section.

Several other localities in North America have yielded faunas contemporaneous with various sections of the Texas sequence, which indicate an essentially similar faunal composition. They include Oklahoma, Colorado, New Mexico and Utah close to Texas, and also, more distantly, the Dunkard Formation of Ohio, and Prince Edward Island off Nova Scotia. Hints of the same chronofauna are also known in parts of Europe. Despite minor faunal differences between these various localities, they suggest that the characteristic Permo-Carboniferous chronofauna was widespread within the tropics of the time. Its origin certainly lay in the Carboniferous, although detailed knowledge of the faunas of this period are lacking. At the Garnett locality of Kansas, for example, which is Stephanian (late Carboniferous) in age, *Edaphosaurus* occurs along with the primitive ophiacodontid *Clepsydrops* and a primitive sphenacodont *Haptodus* (Peabody, 1957; Currie, 1977). Even the earlier, Westphalian fauna of Nova Scotia, containing the earliest known pelycosaurs, probably had an essentially similar palaeo-ecology to the better known Permo-Carboniferous faunas (Reisz, 1972).

Olson (e.g. 1974, 1975) has adduced evidence for a quite separate chronofauna contemporaneous with the Permo-Carboniferous chronofauna, which he has termed the caseid-chronofauna. At a number of places in the Texas sequence, an odd occurrence of forms otherwise unknown has appeared. These, termed "erratics", are interpreted as the freak preservation in an area of animals normally existing elsewhere. In one particular spot in the Clear Fork there comes a single nodule containing two specimens of *Casea*, and at another a small pocket containing only *Casea*, *Varanops* and an otherwise unknown amphibian *Cacops*. These and a few similar cases suggest that there was a quite different fauna existing in geographic isolation from the Permo-Carboniferous chronofauna. There are two possible direct finds that may represent it. In the Cutler and Abo Formations of northern New Mexico, equivalent in time to the lower part of the Wichita of Texas, a number of otherwise unknown forms of both amphibians and reptiles are found. The pelycosaurs include the very *Varanops*-like *Aerosaurus*, and a possible caseid relative *Oedaleops*. The sphenacodontid *Sphenacodon* occurs in place of *Dimetrodon*, although a single specimen of the latter has been found (Berman, 1977). On the other hand, both *Ophiacodon* and *Edaphosaurus* species are present, and therefore the extent to which these New Mexico deposits contain a different chronofauna to that of Texas is not clear.

The second possible direct sampling of the caseid chronofauna is much less well known, but potentially far more interesting. At Fort Sill, in Oklahoma, there are infilled fissures in the Arbuckle Limestone, which were part of the uplands of Clear Fork times. Fragmentary remains of tetrapods occur within the material filling the fissures and the fauna is unique. It includes very terrestrially adapted amphibians (*Doleserpeton*, a possible relative of the modern amphibians—Bolt, 1969) and captohinomorphs. There is also a series of very small, presumably insectivorous pelycosaurs. One is a somewhat caseid-like form (Vaughn, 1958) and there is also a nitosaurid (primitive caseid?) and a sphenacodont (Fox, 1962). What the Fort Sill specimens indicate, despite their very incomplete nature, is that there was indeed a very different chronofauna living in the upland regions, well away from the deltaic regions where most pelycosaurs lived. The palaeo-ecology was not water-based, but consisted of insects and insectivorous pelycosaurs and amphibians. Possible primitive herbivorous caseids were present.

There is no further evidence of the supposed caseid chronofauna until Guadalupian times at the base of the Late Permian. As mentioned, the San Angelo and Flower Pot Formations of Texas, the Chickasha Formation of Oklahoma, and also the lowermost vertebrate-bearing beds of the Russian Permian contained caseids, in association with the remnants of the Permo-Carboniferous chronofauna. Olson sees this as a result of the invasion of the Permo-Carboniferous chronofauna by elements of the hitherto separate caseid chronofauna. It seems likely that the arid, seasonal climate of the start of the Guadalupian in Texas created suitable conditions for animals that were adapted for dryland, upland conditions, for the first time permitting extensive invasion of the Texas region by caseids and the varanopid *Varanodon*.

# 5 The Origin of Therapsids

THE FINAL STAGE of the fossil-bearing Permian sequence in North America is the San Angelo and overlying Flower Pot Formation of Texas, along with the more or less contemporary Chickasha Formation of Oklahoma. Pelycosaurs still occur, as the most specialised of the caseids, as *Dimetrodon*, and at least in the Chickasha, as the varanopid *Varanodon*. In addition to these, however, there is the abrupt appearance of a series of totally different mammal-like reptiles, the first of the therapsids (Olson, 1962). The specimens themselves consist only of poorly preserved fragments but there can be no doubt that they represent the beginning of the next great phase in the evolution of mammal-like reptiles. The San Angelo, Flower Pot and Chickasha beds lie at the very base of the Late Permian, known in North America as the Guadalupian. Approximately contemporaneous with the Guadalupian is the part of the Permian exposed in Russia termed the Kazanian. In and near the Copper-bearing Sandstones of the cis-Uralian region of European Russia there have been found a series of therapsid skeletons which resemble in general the North American forms, but are more numerous and vastly better preserved. The earliest of the cis-Uralian fossils are from what is termed Zone I, and historically a number of localities from within the Copper-bearing Sandstones were discovered in the course of mining operations. However, the richest locality occurs just to the north-west of the copper, at Ocher (Ezhovo), and it is to this fauna that we owe most of our knowledge of these earliest of therapsids (Chudinov, 1965).

The North American record ceases completely at this point, but still in the cis-Uralian area slightly younger horizons are known and referred to as Zone II. Again there have been discoveries in the Copper Sandstones, but the richest collecting area is nearby at Isheevo (Efremov and Vjushkov, 1955). The Isheevo fauna is very similar to the Ocher, except that the individual forms are generally rather more advanced in structure. Besides its obvious significance, this Zone II fauna is of the greatest importance because it links temporally with yet another continent. There are certain marked similarities between the fauna of Zone II of Russia and the earliest fossil-bearing part of the great South African Karroo sequence, the *Tapinocephalus*-zone. The later Permian and much of the Triassic part of the story of mammal-like reptiles is revealed in the Karroo, and therefore the Russian faunas play the role of linking the North American pelycosaur phase with the great therapsid radiation to come.

All the more advanced, non-pelycosaur mammal-like reptiles of these three continents are members of the Therapsida. Even at their earliest appearance, they had diversified into several distinct types, but the similarities that they all share, in terms of characters evolved from the pelycosaur condition, indicate that the Therapsida form a monophyletic group, having descended from a single hypothetical pelycosaur ancestor. Perhaps the single most characteristic feature of all the therapsids is the structure of the angular bone of the lower jaw (Fig. 21B). A thin, extensive sheet of bone, the so-called reflected lamina, lies just lateral to the main body of the angular, connected anteriorly but with a free dorsal, posterior and ventral margin. All the therapsids possess a

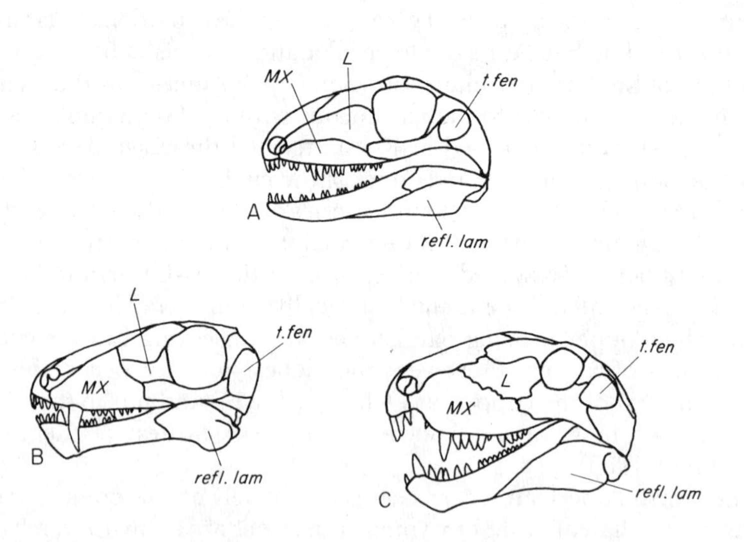


FIG. 21. Lateral views of the skulls of *Haptodus* (A), *Biarmosuchus* (primitive therapsid) (B), and *Dimetrodon* (C). (Redrawn from Currie, 1979.)

*L*, lachrymal; *MX*, maxilla; *refl.lam*, reflected lamina of the angular; *t.fen*, temporal fenestra.

temporal fenestra much larger than in pelycosaurs, indicating an increase in size of the jaw-closing musculature. Associated with this is the presence of a single, relatively very large canine in each jaw, sharply distinct from both the incisors and the postcanine teeth. The jaw hinge of therapsids is more anteriorly placed and the occiput consequently is closer to the vertical. The posterior part of the skull is even more robustly constructed than in advanced pelycosaurs, with massive processes of the supraoccipital and paraoccipital processes running laterally to brace the squamosal region of the skull. Other features possibly related to strengthening the skull are the complete absence of the supratemporal bone and the loss of the movable type of basipterygoid articulation. In pelycosaurs, the base of the braincase is attached to the upper jaw by a ball-and-socket joint between the basisphenoid and the pterygoid bones. In the therapsids this joint has become an extensive, immobile suture.

The therapsid postcranial skeleton also shows certain new features in common, reflecting an improvement in locomotion. The blade of the scapula is narrowed, and the shoulder joint is no longer the complex screw-shape of the pelycosaurs. The glenoid is shorter and simpler, and the articulating head of the humerus is bulbous rather than strap-shaped. An ossified sternum has appeared behind the interclavicle. In the hindlimb, the ilium has expanded and the femur has a slight S-shaped curvature. A true trochanter major has developed behind the femoral head, and the internal trochanter of the pelycosaur femur has shifted to the middle of the ventral surface of the bone. The feet, both front and back, show a reduction in size of certain of the phalanges, so that the digits are more nearly the same length as one another. In the vertebral column, the intercentra have been lost from the trunk region, although still present in the neck and tail.

That the therapsids are closely related to the pelycosaurs was established by Broom's (1910) classic paper comparing the two, and the affinity has never been doubted since. In a general sense, the ancestry of the former lay within the latter, more primitive group. Among all the known pelycosaur groups, it is the advanced carnivorous sphenacodontids that bear the closest resemblance to therapsids. The two groups share a number of derived characters not occurring amongst other pelycosaurs. The most immediately striking similarity concerns the dentition, for sphenacodontids already have an enlarged canine coupled with differentiation of incisor teeth in front and postcanine teeth behind. The number of postcanine teeth is reduced to around twelve in both groups, compared to the larger number in more primitive pelycosaurs. The lower jaw of sphenacodontids and primitive therapsids is remarkably similar in shape and construction. Both have a high coronoid eminence from which the posterior part of the jaw descends steeply to a jaw hinge well below the level of the tooth row. Correspondingly, the suspensorium curves ventrally so that the quadrate lies at a ventral level. Only the sphenacodontids among all the pelycosaurs have developed an angular notch between the ventral keel of the angular and the downturned articular complex, giving rise to an incipient reflected lamina of the angular. The occiput of sphenacodontids approaches the construction of that of therapsids, with well-developed supraoccipital and paroccipital processes. The postcranial skeletons of the two groups also show certain similarities, although the comparison is less dramatic than in the case of the skull. The intercentra of sphenacodontids tend to be reduced, although never lost, and the nature of the articulation of the ribs to the vertebrae is very similar to therapsids. The scapula blade is the narrowest of all pelycosaurs.

The precise origin of the therapsids from amongst the sphenacodontids is debatable, and the issue has been somewhat clouded by attempts to recognise the actual ancestral subgroup rather than to determine phylogenetic interrelationships. *Haptodus* (Fig. 21A), the most primitive of the sphenacodontids, is widely regarded as closest to the therapsids because it lacks any specialisations not expected to have been present in the therapsid ancestor (Romer and Price, 1940; Currie, 1979). The more advanced sphenacodontines, which include *Dimetrodon* (Fig. 21C), do have specialisations such as the reduction in the number of incisor teeth and the elongation of the neural spines of the vertebrae. However, these latter forms have developed larger canines, a reduced lachrymal bone and a more prominent reflected lamina of the angular, all more therapsid-like features and contrasting with *Haptodus*. Currie (1979) suggested that these particular characters evolved independently in sphenacodontines and in therapsids from a *Haptodus*-like common ancestor. He quoted a difference in shape of the lachrymal bone, and the exposure of the septomaxilla in therapsids (Fig. 21) as evidence for an independent retreat of the lachrymal in the two groups, which in turn is correlated with independent enlargement of the canines.

The possibility that dicynodonts, the main herbivorous therapsid group, arose independently from caseid pelycosaurs has been raised (Olson, 1962). However, the similarities between these two groups include such features as a very short skull, and depressed jaw articulation, which are adaptations common to several quite unrelated tetrapod herbivores. On the other hand, primitive dicynodonts possess the full suite of therapsid characters noted, including those common only to therapsids and sphenacodontids. It is probable therefore that convergent evolution alone was responsible for the resemblances between dicynodonts and caseids.

The actual therapsids found in the earliest therapsid-bearing rocks include members of at least three groups, and therefore the adaptive radiation of therapsids had already commenced by the beginning of Late Permian times, and had spread to areas at least as far apart as the North American and the Russian parts of what was then Pangaea. Indeed, there is no reason to doubt that their distribution was cosmopolitan at least within the tropical regions of the time. Whilst it must be stressed that the North American therapsids are exceedingly poorly preserved and their respective classification tentative, they do have an important bearing on speculation about the ecological conditions under which the therapsids arose.

The conditions under which the San Angelo, Flower Pot and Chickasha Formations containing the North American therapsids were laid down were those of a lowland, deltaic area bordering the sea (p. 65). The climate was semi-arid, with distinctly seasonal rainfall, and the flora was dominated by conifers. The cis-Uralian deposits of Russia were laid down in essentially similar circumstances, for a series of rivers flowed westwards from the higher land to drain into the sea (Olson, 1962).

Thus the earliest known therapsids were adapted to essentially lowland, rather arid conditions. If they had actually arisen in this kind of environment, it would be expected that transitional forms between pelycosaurs and therapsids would be found in slightly earlier beds laid down in similar conditions. There are no reptile-bearing beds underlying either the Russian Zone I of the Kazanian, or the South African *Tapinocephalus*-zone of the Karroo. In both cases earlier Permian terrestrial deposits are present and there is no obvious reason why they should not have contained land animals, unless it is presumed that the climate of the times was unsuitable for terrestrial life. However, the San Angelo and Chickasha Formations lie directly above the great Early Permian redbed sequence that occupied so much of the south-eastern part of North America, with its rich pelycosaur fauna. Climatic and geographical conditions were very similar between the Clear Fork, which represents the end of the Early Permian deposition, and the San Angelo Formations which occur at the start of the Late Permian. Indeed, the presence of *Dimetrodon* in both demonstrates at least a reasonable degree of environmental similarity. Despite many years of intensive collecting, in one of the largest and richest fossiliferous areas known, not a single trace of a therapsid or incipient therapsid has been found in the entire North American Early Permian. Although this is negative evidence, it nevertheless strongly indicates that the therapsids actually arose under different environmental conditions, and only subsequently invaded the lowlands, after they had evolved fully into therapsids and had already started to radiate.

Olson (1974, 1975) has suggested that the therapsids arose within the caseid chronofauna, that somewhat shadowy concept based upon the peculiar distribution of caseid pelycosaurs and certain associated forms in the Early Permian (p. 67). He cites as evidence the presence of caseids with both the North American and the Russian earliest therapsids, and of the varanopid *Varanodon* in the Chickasha Formation. Both these kinds of pelycosaurs are believed to be associated with the caseid chronofauna, a fauna which must be presumed to have occurred in areas other than lowlands. One may speculate that the area in question was the uplands of the continental interior, and that the earliest therapsids were relatively small insectivores or carnivores. The kind of fauna found in the Fort Sill fissure deposits, consisting of a variety of small, still poorly known pelycosaurs formed under upland conditions, may well prove to be the cradle of the therapsids.

On this interpretation of the origin of the therapsids, these newly formed animals were restricted to the upland regions during the Early Permian, and were prevented from invading the lowlands by the well-adapted pelycosaurs already existing there. The increasing aridity of the environment of the lowlands during Clear Fork times led to the extinction of these pelycosaurs. After a brief interval, the lowland conditions ameliorated sufficiently to permit a return of tetrapod life at the start of San Angelo times, whereupon the members of the caseid chronofauna, including the therapsids invaded the area. It was probably at this time that the therapsids evolved a dramatic increase in size, to buffer the animals from the excesses of temperature and dessication which they must have found.

# 6 *Dinocephalians*

A MAJOR PART of the earliest therapsid fauna of both North America and Russia were the dinocephalians, large forms which soon came to dominate the terrestrial fauna of the earlier part of the Late Permian, as represented by the Zone II Kazanian of cis-Uralian Russia and, more dramatically, the *Tapinocephalus*-zone of South Africa (Fig. 1). For all their early success, dinocephalians had become completely extinct by the close of *Tapinocephalus*-zone times, leaving no descendants.

Although retaining a basically primitive therapsid structure, dinocephalians were very large animals, and in course of their adaptive radiation both carnivores and herbivores were produced.

## Systematics

Many of the characteristics of dinocephalians are primitive for therapsids generally, such as the pattern of skull bones, the still relatively small temporal fenestra, and the structure of the lower jaw. There are, however, several derived characters possessed by them which show that they form a monophyletic group. These include the large size and thick skull bones, but most importantly the well-developed incisor teeth. The upper incisors interlocked with the lower incisors when the jaws closed and, at least in the more primitive forms, the anterior teeth were the most important part of the dentition. The exact way in which the temporal fenestra enlarged is also characteristic of the group. It expanded dorso-ventrally rather than posteriorly, and external adductor muscle fibres invaded the broad lateral faces of the intertemporal bar. Thus there was a tendency to develop a narrow intertemporal skull roof, although in certain of the advanced herbivorous dinocephalians, the roofing bones secondarily expanded, increasing the intertemporal width.

It was believed for a time (Watson and Romer, 1956) that the carnivorous dinocephalians were a separate group from the herbivores, but subsequent descriptions, particularly of the dentitions, show that the two types are actually closely related on one another (Boonstra, 1962).

The most primitive, pelycosaur-like dinocephalians are the brithopodids, which are included with the earliest known therapsids. *Archaeosyodon* occurs as a single, incomplete skull from the Russian Zone I locality at Ocher (Chudinov, 1960), and there are a series of fragments referred to as *Eosyodon* from the San Angelo Formation of Texas (Olson, 1962). The group is far better represented at the Russian Zone II locality of Esheevo by a series of beautifully preserved skeletons (Orlov, 1958), for example *Titanophoneus* (Fig. 22). Brithopodids were large, carnivorous animals with a skull length up to 600 mm. The very long, heavy snout of the skull (Fig. 23), the relatively short limbs and the long tail are reminiscent of the pelycosaurs but all the standard therapsid features had evolved. The canine teeth are much larger than the adjacent teeth, and the incisors are also very well developed. As in dinocephalians generally, the lower incisors intermeshed between the upper incisors, and each tooth had developed a slight internal shelf or heel at the base of the sharp crown. The postcanine teeth are reduced both in number and in the sizes of the individual teeth. They can have had no more than a holding function in the feeding behaviour of the animals. Compared to more evolved dinocephalians, the jaws of brithopodids are relatively long, although already the jaw articulation had shifted well forwards.

The features of the postcranial skeleton that distinguish therapsids generally from sphenacodontid pelycosaurs are fully expressed in brithopodids, such as the loss of the complex, screw-shaped shoulder joint, the narrower scapular blade, the enlarged ilium and the S-shaped femur. An early specialisation of brithopodids is the reduction in the number of phalanges in certain of the digits. The phalangeal formula of the forefoot is 2.3.3.4.3., and of the hindfoot 2.3.3.3.3.

Brithopodids are absent from the *Tapinocephalus*-zone fauna of the South African Karroo, where their ecological place is taken by the anteosaurids, a more advanced group of carnivorous dinocephalians (Boonstra, 1954). They are closely related to the brithopodids, and may be classified with them as the Brithopia (Boonstra, 1972). Anteosaurids are huge forms, with skulls 500–800 mm in length (Fig. 24). Like the brithopodids, the canines and incisors are enlarged and interlocking, while the postcanine teeth are even further reduced in both size and number. The postorbital region of the skull has deepened compared to

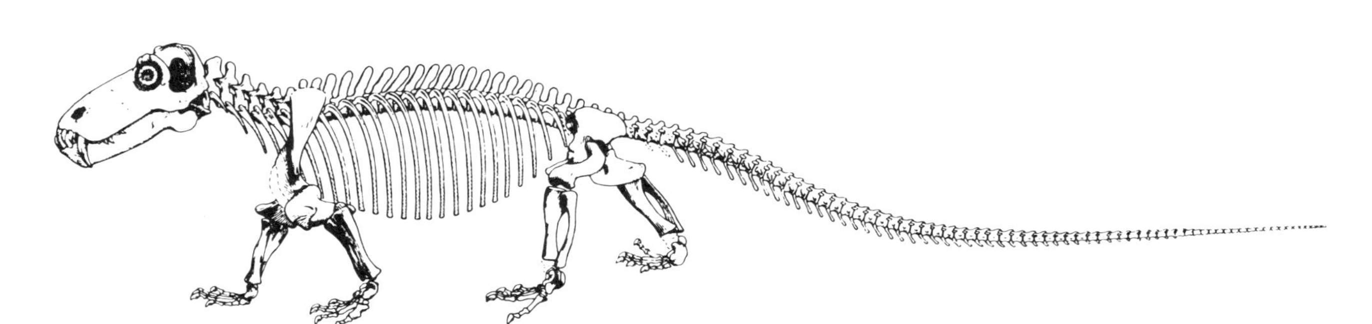

FIG. 22. The skeleton of the brithopodid *Titanophoneus*. (From Orlov, 1958.) Magnification *c.* × 0.05.

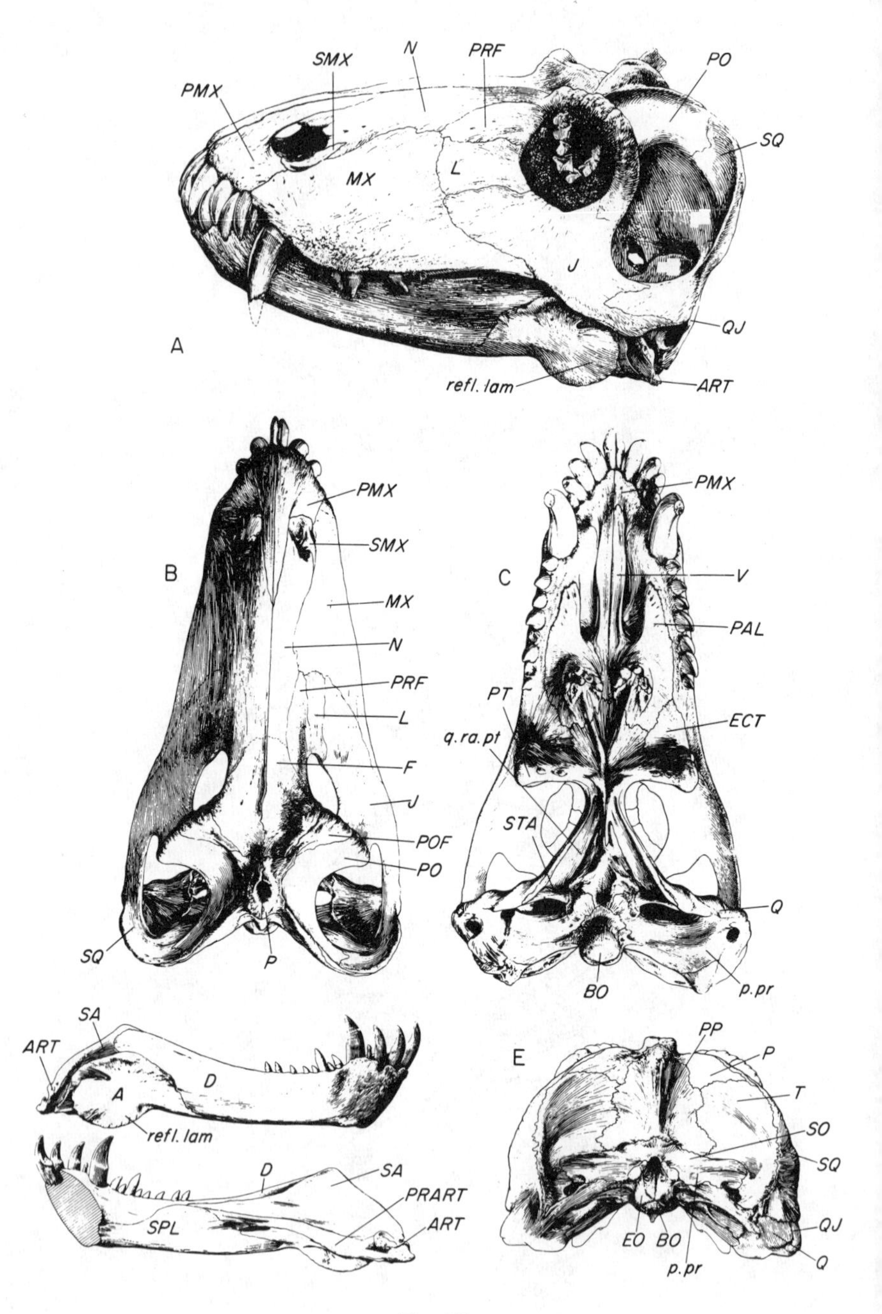

SMX
N
PRF
PO
PMX
SQ
MX
L
J
QJ
A
refl. lam
ART
PMX
SMX
B
MX
N
PRF
L
F
J
POF
PO
SQ
P
PMX
C
V
PAL
PT
ECT
q. ra. pt
STA
Q
BO
p. pr
SA
ART
D
A
refl. lam
D
SA
PRART
SPL
ART
E
PP
P
T
SO
SQ
QJ
EO
BO
p. pr
Q


FIG. 23.

brithopodids, so that the temporal fenestra is even more enlarged ventrally, and the jaw articulation has shifted further forwards to lie at a level anterior to the occipital condyle. The occipital surface consequently slopes slightly backwards and upwards. The bones of the skull roof, particularly the frontal and postfrontal are very much thickened, or pachyostosed, a feature incipiently present in the large brithopodid

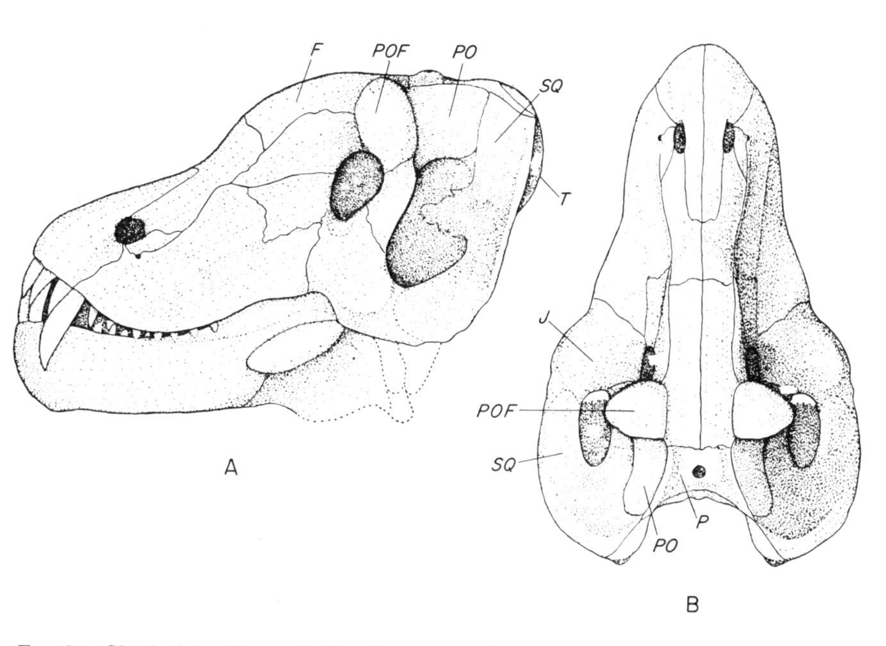


FIG. 24. Skull of the dinocephalian *Anteosaurus*. A, lateral view. B, dorsal view. (From Boonstra, 1954.)
*F*, frontal; *J*, jugal; *P*, parietal; *PO*, postorbital; *POF*, postfrontal; *SQ*, squamosal.
Magnification *c.* × 0.1.

FIG. 23. The skull of the brithopodid *Titanophoneus*. A, lateral view. B, dorsal view. C, ventral view. D, lateral and medial views of the lower jaw. E, posterior view. (From Orlov, 1958.)
*A*, angular; *ART*, articular; *BO*, basioccipital; *D*, dentary; *ECT*, ectopterygoid; *EO*, exoccipital; *F*, frontal; *J*, jugal; *L*, lachrymal; *MX*, maxilla; *N*, nasal; *P*, parietal; *PAL*, palatine; *PMX*, premaxilla; *PO*, postorbital; *POF*, postfrontal; *PP*, postparietal; *p.pr*, paroccipital process; *PRART*, prearticular; *PRF*, prefrontal; *PT*, pterygoid; *Q*, quadrate; *QJ*, quadratojugal; *q.ra.pt*, quadrate ramus of the pterygoid; *refl.lam*, reflected lamina of the angular; *SA*, surangular; *SMX*, septomaxilla, *SO*, supraoccipital; *SPL*, splenial; *SQ*, squamosal; *STA*, stapes; *T*, tabular; *V*, vomer.
Magnification *c.* × 0.68.

*Doliosaurus*. The postcranial skeleton of anteosaurids is not very well known (Boonstra, 1955). Compared to the other South African dinocephalians, it is relatively lightly built with small limbs. There are two South African genera, *Anteosaurus* which has a boss on the postfrontal bone, and *Paranteosaurus* which lacks it. The Russian Zone II form, *Deuterosaurus*, from the Copper Sandstones is very closely related to the South African anteosaurids (Boonstra, 1965a).

Members of the family Titanosuchidae (Boonstra, 1936) are much more common in the *Tapinocephalus*-zone, although there are no known Russian representatives. They show similarities to the Brithopia on the one hand, and to the herbivorous dinocephalians on the other, and are therefore important in demonstrating the monophyletic nature of the Dinocephalia. The canine is still distinct (Fig. 25) but much less prominent than in the anteosaurids. The incisors are also relatively smaller, but are specialised by the enlargement of the heel at the base of the inner surface of each crown. When the upper and lower incisors interlock, the opposing heels meet. The postcanine teeth, in marked contrast to the anteosaurids and brithopodids, are numerous, with around twenty in each jaw. Each of the teeth is leaf shaped, the crown being compressed from side to side and narrowing to a bluntly pointed tip, while both front and back edges are coarsely serrated. The titanosuchid snout is lower, broader and relatively longer than in anteosaurids, and the jaw articulation is even further forwards. While the postorbital region of the skull is still very large, the actual size of the temporal fenestra has been secondarily reduced by thickening of the squamosal below, and widening of the postorbital bar. However, the dorsal boundary of the fenestra is unaffected, so that adductor musculature still attached to the lateral and dorso-lateral areas of the fairly narrow intertemporal bar.

The postcranial skeleton of titanosuchids is much better known than that of anteosaurids, although mainly consisting of incomplete specimens (Boonstra, 1955). It is very similar to the herbivorous dinocephalians (Fig. 27), which together differ in several profound ways from the brithopodids and anteosaurids. Most obviously, the bones are very much more massive. The vertebrae are short but wide. The shoulder girdle is robustly built and a well-developed cleithrum persists, lying along the upper part of the front edge of the scapula blade. The humerus has enormously expanded, flattened ends, and the twist has been reduced to no more than 15–25° between the proximal and distal parts. The radius and ulna are similarly flattened and expanded at the ends. In the pelvis, the ilium is somewhat more expanded forwards, and the pubo-ischiadic plate is shortened, with the result that the pelvis as a whole appears very tall and short. It is supported by, probably, four sacral vertebrae. The femur is massive and flattened dorso-ventrally, although the S-shape and medially turned head of typical therapsids are still discernible. The internal trochanter is reduced to a rounded tubercle lying in the middle of the proximal end of the underside of the femur. The tibia and fibula are again heavy, flattened and expanded at the ends. The digial formula of the feet is reduced to the mammalian one of 2.3.3.3.3, there being no longer any vestigeal phalanges.

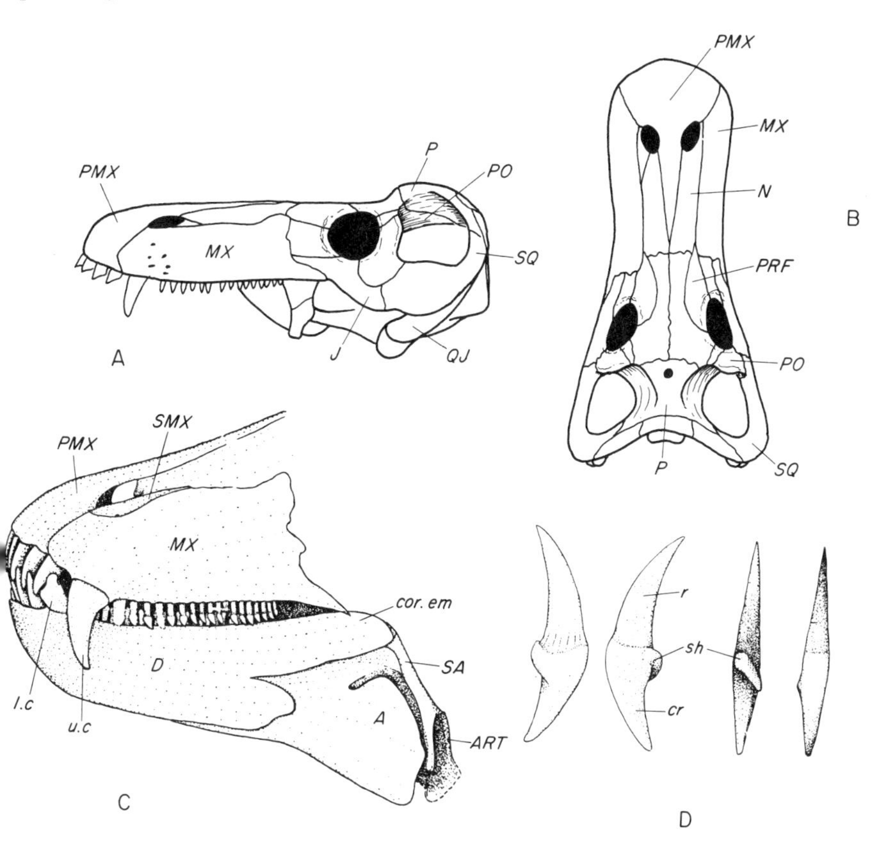


FIG. 25. Skull of the titanosuchid dinocephalian *Jonkeria*. A, lateral view. B, dorsal view. C, lateral view of the dentition. D, four views of a single incisor tooth. A and B redrawn after Boonstra, 1936; C and D from Boonstra, 1962.)

*A*, angular; *ART*, articular; *cor.em*, coronoid eminence; *cr*, crown of incisor; *D*, dentary; *J*, jugal; *l.c.*, lower canine; *MX*, maxilla; *N*, nasal; *P*, parietal; *PMX*, premaxilla; *PO*, postorbital; *PRF*, prefrontal; *QJ*, quadratojugal; *r*, root; *SA*, surangular; *sh*, shelf on incisor; *SMX*, septomaxilla; *SQ*, squamosal; *u.c*, upper canine.

Magnifications: A and B, *c.* × 0.08; C, *c.* × 0.12; D, *c.* × 0.25.

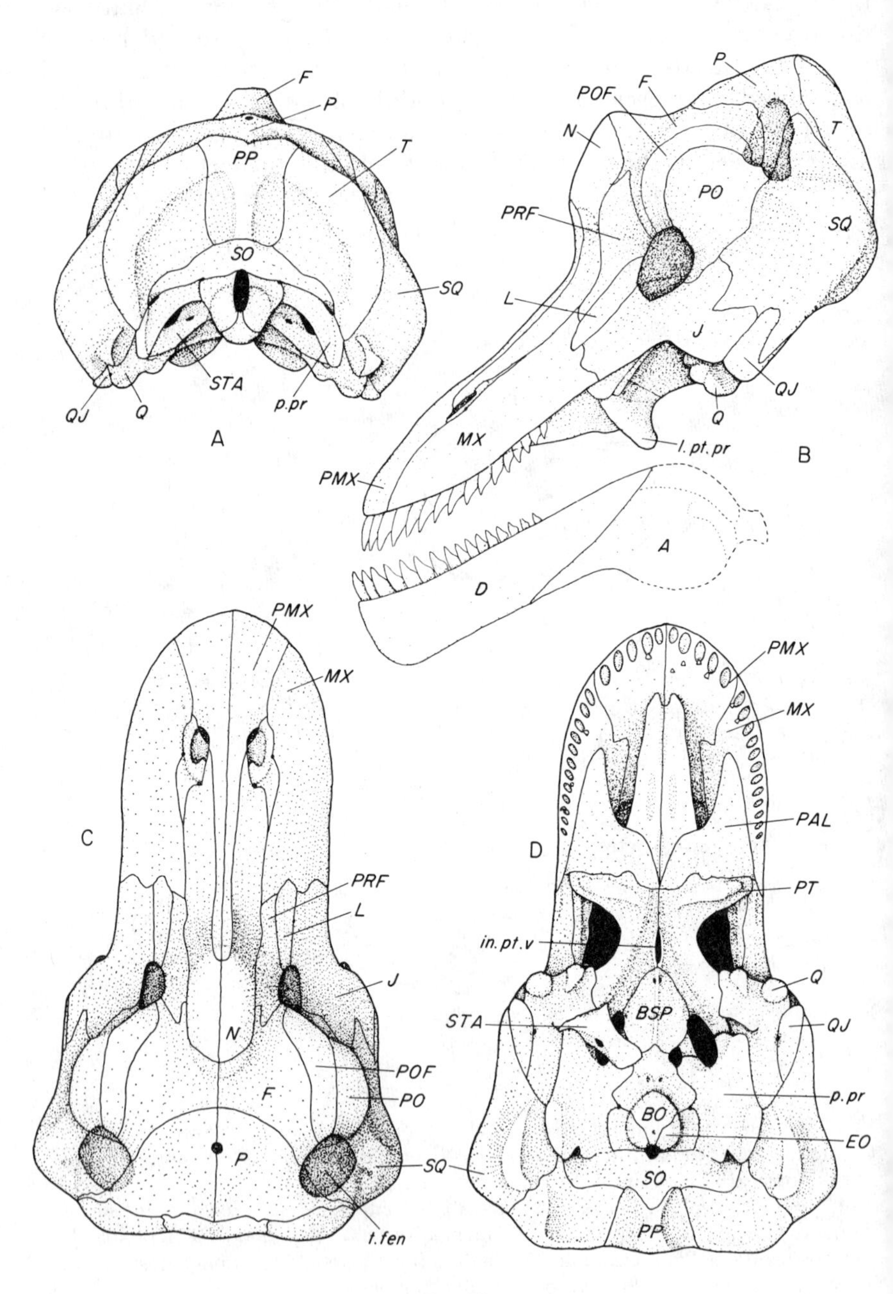
F
P
PP
T
SO
SQ
STA
p.pr
QJ
Q
A
P
POF
F
N
T
PO
PRF
SQ
L
J
QJ
Q
MX
l.pt.pr
B
PMX
A
D
PMX
MX
C
PRF
L
J
N
POF
F
PO
P
SQ
t.fen
PMX
MX
PAL
D
PT
in.pt.v
Q
STA
BSP
QJ
p.pr
BO
EO
SQ
SO
PP

FIG. 26.

Two well-established genera of titanosuchids occur, the long-limbed form *Titanosuchus* and the very similar but short-limbed *Jonkeria*. The Tapinocephalidae (Fig. 26) were the advanced herbivores. They are closely related to the titanosuchids, sharing with them the heeled incisor teeth, reduced temporal fenestra, anteriorly placed jaw hinge and the modified postcranial skeleton (Fig. 27). However the tapinocephalids carried even further the specialisations towards herbivory initiated in the titanosuchids. There is no distinct canine present, and the postcanines as well as the incisors have developed a heel. Interdigitation of the entire dentition occurs instead of just the anterior teeth. The size of the pointed crown of each tooth is reduced, but the heel is expanded inwards, creating a broad, horizontal shelf (Fig. 30B). The snout tends to become low and wide, in some cases looking almost like a duck-bill, and sharply set off from the still massive postorbital region. The tendency to reduce the size of the temporal fenestra seen in the titanosuchids is carried further in most of the tapinocephalids. The postorbital bar and ventral bar enlarge, and there is also overgrowth of the dorsal margin of the fenestra, as the intertemporal roof secondarily widens. In the most advanced forms, the temporal fenestra is all but obliterated (Fig. 28A). The anterior shift of the jaw articulation is also carried to an extreme, so that the occiput slopes strongly backwards from its ventral edge, and the paroccipital processes, greatly widened, are entirely ventral. Pachyostosis of the bones of the skull roof is carried to almost absurd lengths in some forms. The frontal and parietal bones of a 380 mm long skull of *Moschops* are no less than 11.5 mm thick, for example (Barghusen, 1975).

Taxonomic division (Boonstra, 1969) of the family is based on the different degrees of expression of the various specialisations of the skull (Fig. 28), such as snout shape, degree of reduction of the temporal fenestra, and degree of pachyostosis. *Avenantia*, for example (Fig. 28D), is relatively primitive, for it has retained a narrow intertemporal roof with external muscle attachment, and the degree of pachyostosis is slight. The snout is not sharply set off from the postorbital region.

FIG. 26. The skull of the tapinocephalid dinocephalian *Struthiocephalus*. A, posterior view. B, lateral view. C, dorsal view. D, ventral view. (From Boonstra, 1965b.)

*A*, angular; *BO*, basioccipital; *BSP*, basisphenoid; *D*, dentary; *EO*, exoccipital *F*, frontal; *in.pt.v*, interpterygoid vacuity; *J*, jugal; *L*, lachrymal; *l.pt.pr*, lateral pterygoid process; *MX*, maxilla; *N*, nasal; *P*, parietal; *PAL*, palatine; *PMX*, premaxilla; *PO*, postorbital; *POF*, postfrontal; PP, postparietal; *p.pr*, paroccipital proces; *PRF*, prefrontal; *PT*, pterygoid; *Q*, quadrate; *QJ*, quadratojugal; *SQ*, squamosal; *STA*, stapes; *T*, tabular; *t.fen*, temporal fenestra; *V*, vomer.

Magnification *c*. × 0.11.

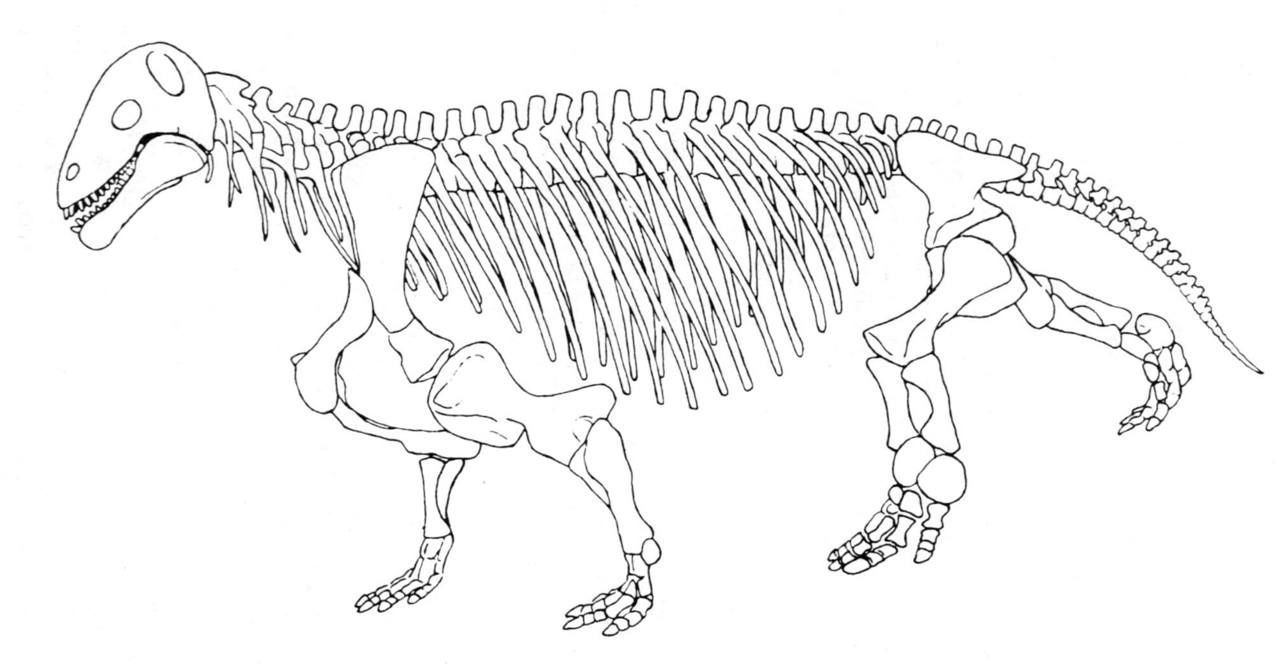

FIG. 27. The skeleton of the tapinocephalid dinocephalian *Moschops*. (From Olson, 1971, redrawn after Gregory, 1926.) Magnification *c.* × 0.07.

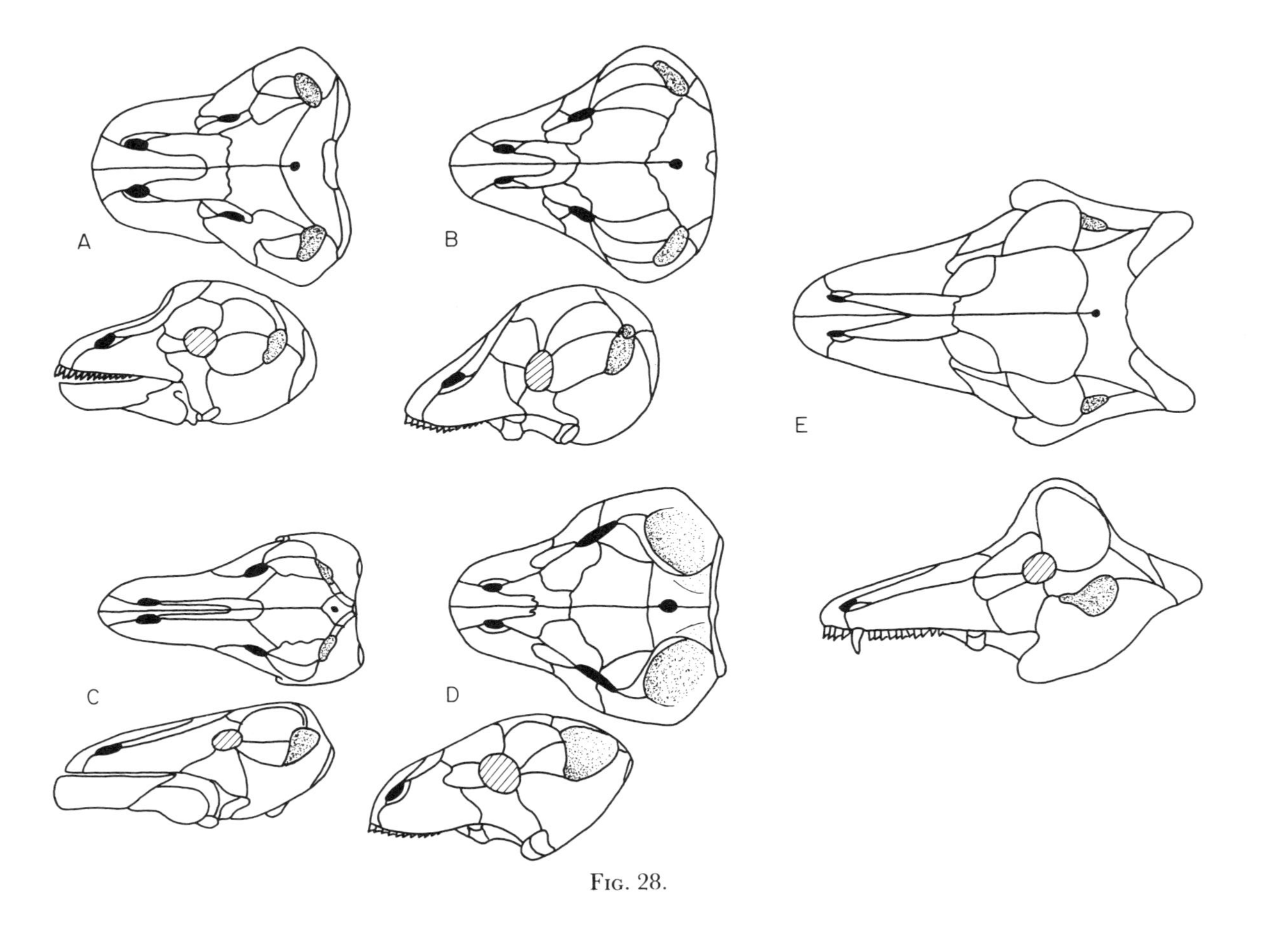


Fig. 28.

*Moschops* (Fig. 28B) and also the closely related Russian form *Ulemosaurus* (Efremov, 1940) have a similar rounded skull shape, but the temporal fenestra is very small, the intertemporal region wide, and pachyostosis heavy. *Riebeekosaurus* (Fig. 28C) combines a very long, low snout with a relatively narrow intertemporal region and only moderate pachyostosis. The most highly evolved form is *Tapinocephalus* itself (Fig. 28A), for it has a relatively low, short snout, greatly widened intertemporal region, and massive pachyostosis.

One form often put in a separate family of its own is *Styracocephalus*. It is a fairly small form, but possesses an advanced type of skull structure. However, a small but distinct canine is present, and the tabulars are produced posteriorly as a pair of horn-like processes (Fig. 28E).

The remaining dinocephalian family is the problematical Estemmenosuchidae (Fig. 29). These large carnivorous forms occur amongst the very earliest known therapsids, as *Estemmenosuchus* of the Zone I Kazanian deposits at Ocher (Chudinov, 1965). The canines are large, as are the interdigitating incisors, and the postcanines are reduced. The jaw hinge is sufficiently far forwards to give the occiput a backwards slope. Strange pairs of bony protuberances, perhaps bearing horns in life, are developed on the maxillae, frontals, parietals and jugals. To judge from the dentition this group evolved from a brithopodid-like form, but at a very early time in dinocephalian history.

## Functional anatomy

### *Feeding mechanism*

The particular way in which the feeding apparatus of the dinocephalians was modified compared to other therapsids provides the most impressive evidence that they were a monophyletic group. Briefly, they specialised the anterior teeth to provide the principal means of food collection by the intermeshing of the lower incisors with the uppers, and they enlarged the jaw-closing musculature in a characteristic fashion.

The specialisation of the anterior part of the dentition and reduction of the posterior part indicates that the carnivorous dinocephalians still relied on the kinetic action of the lower jaw, as in their pelycosaur

FIG. 28. Skulls of tapinocephalid dinocephalians in dorsal and lateral views. A, *Tapinocephalus*. B, *Moschops*. C, *Riebeekosaurus*. D, *Avenantia*. E, *Styracocephalus*. (Redrawn after Boonstra, 1969.)

ancestors (p. 50). The static force available between clenched teeth was less at the front of the jaws than further back. However, if the work done by the teeth on the food was derived from the kinetic energy of the

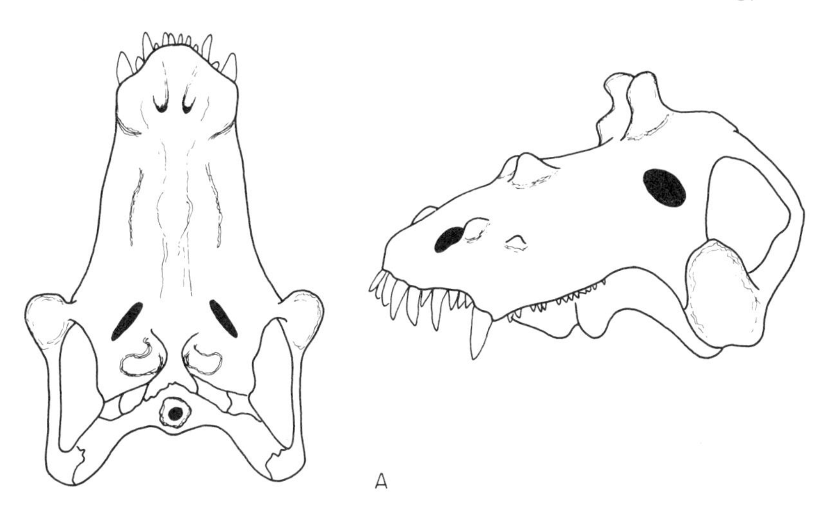


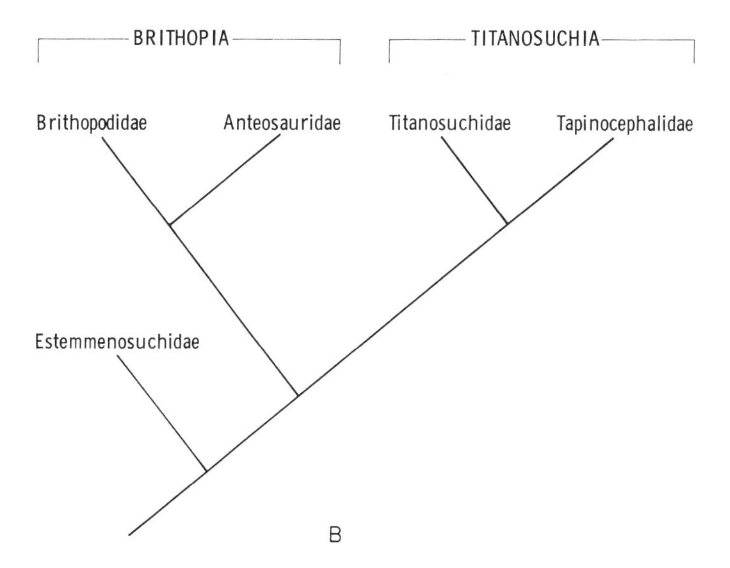


FIG. 29. A, skull of *Estemmenosuchus* in dorsal and lateral views. (Redrawn after Chudinov, 1960.)
Magnification *c.* × 0.09.
B, phylogeny of the main groups of dinocephalians.

moving lower jaw, then exactly the same amount of energy was available to any teeth, wherever they were situated. In these circumstances, the more anterior teeth, canines and incisors, were more conveniently placed for the capture and dismemberment of the prey. These animals therefore opened their jaws relatively widely, used the adductor musculature to accelerate the jaws towards the closed position, and sank the teeth into the food. The carnivorous brithopodids and anteosaurids used their enlarged canines in this kinetic fashion to disable their prey as rapidly as possible. The large, sharp and slightly procumbent incisors would similarly have been used kinetically, embedding themselves in the flesh. Acting as an anchor, the incisors then permitted the animal to tear off pieces of flesh of a swallowable size. The possession of five upper incisors and four lowers on each side allowed an accurate interdigitation of the lowers between the uppers, the front two uppers together passing medial to the front two lowers. Interdigitation of the incisors had the effect of producing a more nearly continuous slit in the meat, rather than a series of separate holes, which made it easier to remove mouthfuls.

The elaboration of the incisor teeth of the titanosuchids led to an arrangement whereby they actually did produce a continuous slit (Fig. 30A). The heel at the base of both upper (Fig. 25D) and lower teeth is little more than an oblique ledge on the lingual side. When the teeth occluded, the ends of the ridges of the uppers meet the ends of the ridges of the lowers (Boonstra, 1962), and the effect on the actual bite seems to be that the pairs of "nipping" ridges severed the food between the holes punched by the individual pointed crowns of the teeth. While the incisors of titanosuchids would still have performed adequately in flesh-tearing, this additional facility would now allow them to cut pieces of vegetation clean through as well. Leaves, stems, etc. appear to have been incorporated into their diet. The numerous spatulate postcanine teeth of these forms (Fig. 25C) do not occlude at all, and must have been used more crudely for gripping and tearing up vegetation, perhaps prior to cutting it up into swallowable sizes by the incisors. Titanosuchids were probably omnivorous animals.

The same principles of incisor action apply to the undoubtedly herbivorous tapinocephalids (Fig. 30B,C). The size of the crown was reduced and therefore their ability to tear up flesh was diminished. The heel, however, is elongated in a lingual direction, and now forms a relatively wide but short shelf. Efremov (1940) showed how the teeth of *Ulemosaurus* functioned (Fig. 30C). The side edges of the heels had a cutting action as in titanosuchids and the lingual elongation of the heel increased the length of the opposed cutting edges available, permitting the animal to cut up vegetation into finer pieces. In addition, the tapinocephalid incisors had a degree of crushing or grinding ability, both where the blunt points of the teeth met the sides of the opposing heels, and also possibly between the outer face of the lower incisors against the lingual faces of the upper incisors. The cutting function was nevertheless the most important, and these animals probably ate relatively soft vegetation. The modification of the canines and postcanine teeth to act in the same way as the incisors increased their cropping ability.

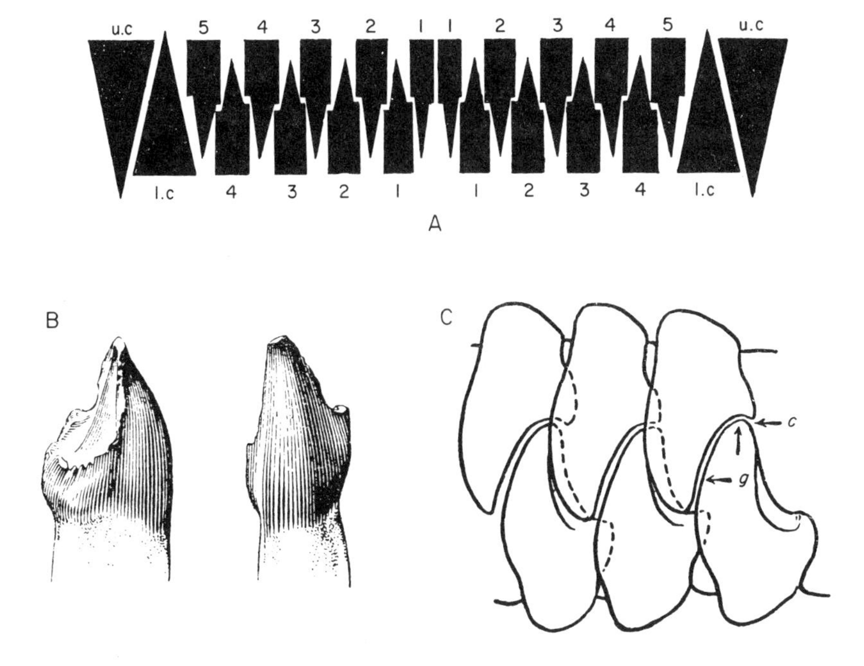


FIG. 30. Functioning of dinocephalian teeth. A, incisor and canine teeth of a titanosuchid, straightened out to show interlocking. B, lower incisor of the tapinocephalid *Ulemosaurus* in distal and buccal views. C, pattern of interlocking of tapinocephalid incisors showing points of contact between upper and lower teeth. (A from Boonstra, 1962; B and C from Efremov, 1940.)
*c*, cutting edge of upper tooth; *g*, regions of grinding action between upper and lower teeth; l.c, lower canine; u.c, upper canine.

The dinocephalian jaw musculature was enlarged compared to sphenacodontid pelycosaurs, although it was not radically changed in basic organisation. The expansion of the postorbital part of the skull and the accompanying enlargement of the temporal fenestra created two new areas of the skull for origin of external adductor muscle fibres (Barghusen, 1976). The first was the side of the skull roof behind the orbit, in other words the dorsal margin of the temporal fenestra, which was deepened and excavated on each side to form a broad, dorso-laterally facing surface (Fig. 31A). The second was the hind wall of the space within the fenestra, which is composed of the anterior surface of the squamosal and adjacent bones. A temporal aponeurosis probably covered the temporal fenestra, attaching to its dorsal and posterior margins as in pelycosaurs and giving origin to further muscle fibres. The area between the pterygoid bone and the ventral margin of the cheek, termed the subtemporal fenestra, was enlarged by shifting the pterygoid closer to the midline (Fig. 23C). This created enough space for the enlarged external adductor muscle to descend to the lower jaw. Its insertion on the jaw was very similar to pelycosaurs, the main area being a presumed bodenaponeurotic sheet attached to the prominent coronoid eminence. The adjacent dorsal and medial surfaces of the jaw were also utilised.

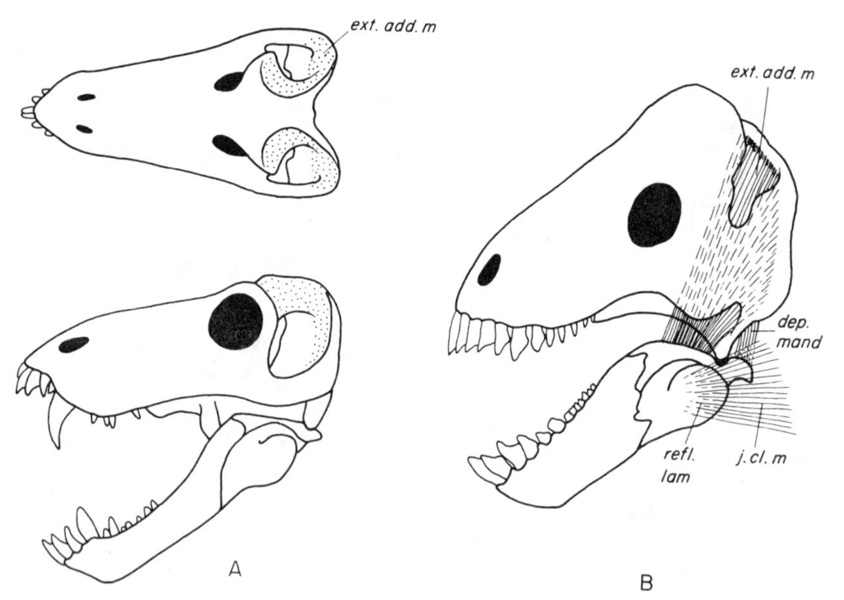


Fig. 31. Dinocephalian jaw musculature. A, dorsal and lateral views of the brithopodid skull showing areas of origin of the external adductor muscle (stippled). B, skull of the tapinocephalid *Ulemosaurus*. (Outlines based on Orlov, 1958; Efremov, 1940.)

*dep.mand*, depressor mandibuli muscle; *ext.add.m*, external adductor muscle; *j.cl.m*, jaw closing muscle; *refl.lam*, reflected lamina of the angular.

The internal adductor musculature was similar in arangement to pelycosaurs. The origin included the back of the large pterygoid processes of the back of the palate, and the pair of girders running back alongside the braincase (Fig. 23C), and it inserted onto the inner and ventral regions of the lower jaw.

The function of the reflected lamina of the angular of dinocephalians, and indeed of all therapsids presents a problem. The forerunner of the reflected lamina is believed to have been the sphenacodontid angular keel set off laterally from the articular region of the lower jaw, and associated with the insertion of pterygoideus musculature. However, the therapsid reflected lamina is very much larger, and the narrow recess between the lamina and the main body of the angular bone cannot be exlained in terms of the pterygoideus muscle alone. In view of the subsequent evolution of the angular bone, in which it becomes the mammalian tympanic bone supporting the tympanic membrane and surrounding the middle ear cavity, it has been suggested that even in the early therapsids an air-filled cavity filled the recess. Perhaps it was a secondary respiratory organ, or was possibly involved already in hearing (Westoll, 1945; Allin, 1975). However, the very narrow shape of the recess does not support this idea at all, and it seems far more likely that the angular complex was still related to the musculature. Parrington (1955) proposed that a muscle attached to the outer surface of the reflected lamina and ran antero-dorsally to an attachment close to the orbit. This would correspond to the muscle known as the masseter of mammals. The surface features of the lamina do not, however, offer any support to his interpretation (Barghusen, 1968). The reflected lamina of the dinocephalians has a series of fine ridges radiating downwards and backwards over the lateral surface. They indicate that a muscle was attached by a series of fine ligamentous sheets and ran in a postero-ventral direction (Fig. 31B). Since the head was carried with the nose to a greater or lesser extent pointing downwards, the muscle would project back towards the region of the animal's shoulder girdle, and it would be well placed to perform two important functions. First it would act to open the jaws. Modern reptiles open their jaws by means of a muscle, the depressor manibuli, which originates on the occiput and inserts on a process of the lower jaw, the retroarticular process, projecting backwards behind the jaw articulation. Most therapsids also have a retro-articular process and therefore are assumed to have possessed the depressor mandibuli muscle. However, jaw opening is also possible using ventral muscles of the throat and the mammals have elaborated this musculature to form their jaw closing muscle, the digastric. It is likely that therapasids possessed the forerunner of the digastric as well.

In the dinocephalians, the retroarticular process is very small, and yet with interlocking teeth jaw opening must have been difficult on occasion; the presence of a particularly well-developed jaw opening muscle from the reflected lamina is not therefore unexpected.

The second function of the muscle concerned the problem of prey struggling in the mouth. This would tend at times to pull the lower jaw forwards away from the quadrate. Yet the main jaw-closing musculature, the external adductor, pulled more upwards than backwards, and was not therefore well-placed to resist jaw disarticulation. A muscle pulling backwards from the reflected lamina would be better placed. Its enlargement in primitive therapsids correlates well with the enlargement of the canine teeth, the consequent ability to capture large prey, and therefore the subjection of the lower jaw to large disarticulatory forces.

The function of the angular recess, between the lamina and the main body of the angular remains unexplained. One possibility is that external adductor muscle fibres running down from the lower part of the squamosal gained an origin within the recess, which would account for the free dorsal edge of the lamina. It is also possible that the pterygoideus muscle wrapped around the ventral edge of the lower jaw from the medial side, and inserted some way within the recess. The point of having the lamina would then be to allow muscles running in very different directions to insert in the same region of the lateral side of the lower jaw.

This same general arrangement of the jaw muscles persisted throughout the dinocephalians, with one important modification in the more specialised, tapinocephalid herbivores. Here the jaw hinge has shifted anteriorly so that the postorbital region of the skull lies well behind the jaw hinge (Fig. 31B). The jaw itself is shortened, but has retained its height, so that it appears very robust. The dorsal edge of the surangular bone, behind the coronoid eminence, descends more steeply towards the jaw hinge. The majority of the external adductor muscle fibres therefore lie closer to horizontal, and they are aligned at right angles to the line between the jaw hinge and their points of insertion when the jaws are closed. The muscle now acts most effectively when the teeth are actually in occlusion rather than when partly open. The system has therefore changed from the more primitive kinetic arrangement of the carnivores to a static pressure one. The high posterior part of the jaw allows the muscle to possess a relatively long moment arm and coupled with the shortness of the jaws, a much larger bite force is available to the teeth. This is necessary for teeth which occlude with one another to produce the cutting, crushing and grinding actions. The reduction of the temporal fenestra of tapinocephalids is correlated with the reduced importance of the more anterior and dorsal part of the external adductor muscle, and the enhanced significance of the more posterior part of the muscle, in the newly evolved static pressure system.

The jaw articulation of dinocephalians is functionally fairly simple. The quadrate is held rigidly in place by a broad, direct suture to the front of the squamosal, and indirectly by the reduced quadratojugal which wedges between the quadrate medially and the squamosal laterally (Fig. 26D). The medial side of the quadrate possesses facets which receive respectively the broad distal end of the paroccipital process, the distal end of the relatively massive stapes, and the posterior end of the still substantial quadrate ramus of the pterygoid. Between them, these various contacts appear to have prevented any movement of the quadrate at all. The main body of the quadrate slopes somewhat posteriorly in brithopodids and the degree of posterior slope increases as the herbivorous adaptations of the skull become more marked in other dinocephalians (Fig. 28). In all cases the quadrate is in fact arranged so that it lies parallel to the principal direction of the pull of the external adductor muscle, thereby eliminating any tendency for the quadrate to rotate relative to skull when the jaw muscles contracted.

The actual articulating facets of the jaw hinge reflect the arrangement of the dentition and the jaw musculature. The teeth of dinocephalians require only a simple orthal opening and closing of the jaws, and the two condyles of each jaw ramus, lateral and medial, form simple roller-joints giving a hinge-like movement. However, the principal line of action of the external adductor muscle is directed backwards and inwards, as well as upwards. To resist the reaction force of the muscle most effectively, the articular facets face postero-dorsally, and the lateral condyle is above the level of the medial condyle. Together, therefore, the condyles face partly medially (Fig. 23D).

## *Pachyostosis*

The extraordinary thickening of the roofing bones of the skull of all but the brithopodids amongst dinocephalians has been explained by Barghusen (1975) as an adaptation to head-butting behaviour, presumably between members of the same species. In fact even the brithopodids, particularly the larger form *Doliosaurus*, show some degree of bone thickening, but pachyostosis reaches its most extreme expression in the tapinocephalids. The occipital condyle of these more specialised dinocephalians lies very far forwards, virtually on the ventral side of the skull. Therefore the skull was carried with the nose projecting ventrally, and the intertemporal region facing forwards. The pachyostosed region of the skull, centred on the frontal but extending to the surrounding postfrontal and parietal bones, would then be the region involved in head-to-head contact between two individuals (Fig. 32A). Furthermore, the increase in the width of the post-orbital bars and the massiveness of the paroccipital processes provided a mechanical arrangement whereby the force of the head butt was transmitted from the top of the head to the posterior part of the braincase and hence via the occipital condyle to the

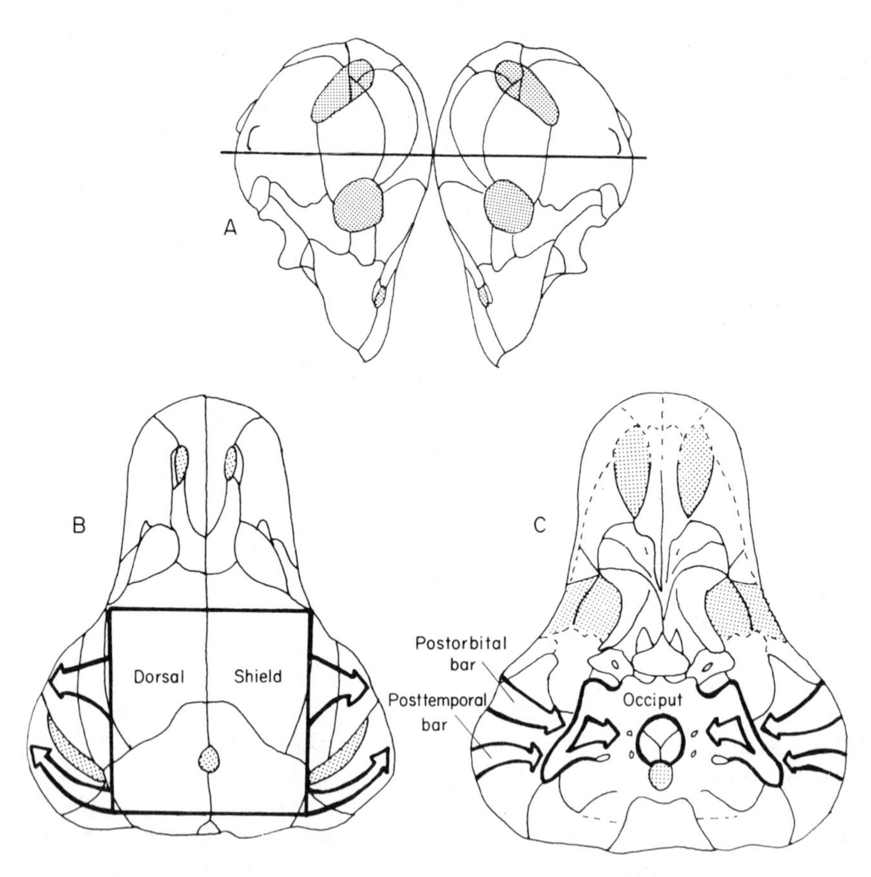


FIG. 32. Head-butting in *Moschops*. A, lateral view of the skulls of two butting individuals, showing that the respective occipital condyles and the point of contract are in line. B, dorsal and C, ventral views of the skull, showing how the thickened skull roof, postorbital bars, posttemporal bars and occiput form a powerful structure to resist the collision and transmit the forces to the vertebral column. (From Barghusen, 1975.)

vertebral column (Fig. 32B,C). The design of the skull for this kind of behaviour is clear, but the function of head-butting must of course remain speculative. Presumably it was a ritualised fighting for mates or territory.

## *Locomotion*

The primitive dinocephalian skeleton is represented by the brithopodids and probably the anteosaurids. The structure and presumably therefore the locomotion closely resembled other more or less unmodified therapsids and the detailed discussion of basic therapsid locomotion given in the context of the gorgonopsids (p. 115) probably applies equally well here. In fact, no analysis of brithopodid locomotion is yet available. Briefly, the main features of this form of locomotion may be summarised thus. Lateral undulation of the vertebral column, which had already been reduced in the sphenacodontid pelycosaurs, was virtually absent, and in compensation the length of the limbs and of their stride had increased. In the case of the forelimb some degree of mobility of the shoulder girdle on the ribcage probably added a new component to the overall forelimb stride, while the shoulder joint was simplified compared to pelycosaurs, permitting a greater excursion of the humerus. This bone still operated in a strictly horizontal plane and rotation about its long-axis was enhanced. A more radical change had affected the hindlimb, because, in addition to the primitive sprawling mode of action, it could also move in a plane closer to vertical as in modern mammals. The overall effect of these various modifications to the locomotion was a faster movement, and possibly a more efficient one. Equally important was the increased adaptability of the system to irregularity of the terrain and to varying requirements of speed and agility in the course of the animal's life.

The postcranial skeleton of the more modified dinocephalians, the titanosuchids and tapinocephalids (Fig. 27), can be derived from an ancestral condition like that of brithopodids, but several adaptations to increased size were superimposed. The relative size of most of the bones increased as would be expected. The forelimb retained its sprawling orientation (notwithstanding the classic reconstruction of a *Moschops* skeleton by Gregory (1926) which shows a more erect humerus; the orientation of the shoulder girdle required in order to achieve this is most improbable) but the hindlimb may have adopted a more or less permanently erect posture. The feet of these dinocephalians became almost plantigrade, with the carpus and the tarsus flat on the ground compared to their vertical orientation in pelycosaurs (Boonstra, 1966).

### *Middle ear*

No detailed study of the function of the stapes and hearing in dinocephalians is yet available, although Boonstra (1965b) has given a good description of the bones of this region. The stapes itself is massively constructed, with only a relatively small stapedial foramen (Fig. 26A,D). The distal end is expanded and makes a strong, rather complex attachment to the quadrate, a large dorsal process contacts the overlying paroccipital process, and the inner end, as well as covering the fenestra ovalis, also contacts both the prootic anteriorly and the paroccipital process posteriorly. The bone is therefore very firmly held in place and seems to have much more of a structural function than a delicate sound-conducting one. Boonstra in fact suggested that there was a tympanic membrane, supported in part by the quadrate, but its diameter could not have been much above the 30 mm he quotes, which is about the same dimension as the distal end of the stapes. Therefore, the mass and degree of attachment of the stapes makes it most unlikely that such a tympanum could have been used to transmit high-frequency vibrations to the stapes. Probably, as in pelycosaurs, ground borne vibrations and relatively low frequency air-borne sound were all that the animals could detect.

### *General biology*

There is evidence that the dinocephalians and indeed all therapsids had evolved a more advanced temperature physiology than their pelycosaur forerunners. Bakker (1975) regarded them as endotherms, defined as maintaining a constant high body temperature by means of a high rate of metabolic heat production, which is characteristic of mammalian temperature physiology. One of his lines of evidence concerns the histological structure of the bone (Ricqles, 1972). It differs markedly from that of pelycosaurs, for the cortex is well vascularised, Haversian canal systems extensively developed, and growth rings absent, and therefore closely resembles the bone of most modern mammals. In so far as it can be argued that there is a correlation between bone structure and temperature physiology, dinocephalians would appear to have indeed possessed a mammalian type of physiology. Further evidence for endothermy is that the Permian Karroo deposits of South Africa were formed at a time when that part of the world lay in high latitudes, around 60°S, which is well beyond the largely tropical range of modern, ectothermic reptiles of any size. Conditions must have been highly seasonal and the dinocephalians were presumably too large to have been capable of hibernating during the winters in suitably sheltered microhabitats. The absence of caseid pelycosaurs in South Africa is relevant. These presumably ectothermic remnants of the pelycosaur radiation are found with early therapsids, including dinocephalians in both the Russian and the North American Late Permian (p. 69). Both these areas, however, lay within the Permian Tropics, where large ectotherms would have been viable. On this argument, the South African deposits lay in an area not available to large ectotherms like caseids, and therefore the large dinocephalians which are found were not ectothermic. A third argument of Bakker is the relatively much greater number of herbivore dinocephalians compared to carnivores. Endothermic animals with their very high metabolic rates require relatively much more food than ectotherms, and therefore this low predator to prey ratio implies that dinocephalans were themselves endothermic.

An alternative view of dinocephalian temperature physiology (MacNab, 1978) is that they, like other large therapsids, were what is termed inertial homiotherms. This means that they had a low, reptile-like metabolic rate, but still maintained a constant body temperature by reducing the rate of heat loss to the environment. They achieved this by simply being large, compact animals, with a low surface area to volume ratio. They may even have evolved a fur covering, to reduce further the heat loss. MacNab's view is based on the fact that all the dinocephalians were of relatively very large size. He argues from first principals that the origin of endothermy was a two-stage process. The first stage was a reduction in the rate of heat loss, achieved by a simple increase in size. This is the inertial homiothermic stage, at which the dinocephalians and other large therapsids of the Permian are supposed to have reached. Only subsequently did an increase in metabolic heat production occur, and it was associated with a secondary decrease in bodv size, which caused the rate of heat loss to rise. In order to remain homeothermic at this smaller size, more internal heat had to be produced as compensation for the greater loss of heat. Dinocephalians lack the various adaptations of the skeleton which indicate increased metabolic rate. The dentition was not particularly elaborated to increase the rate of food processing, particularly that of the carnivores. There is no secondary palate to permit continuous breathing during feeding, and no sign that a diaphragm had evolved to increase the respiratory volume. MacNab argues that the bone histology of dinocephalians and other large primitive therapsids relates only to the homiothermic condition and not to the high metabolic rate associated with endotherms. He dismisses Bakker's predator–prey ratio argument as too unreliable, but does not comment upon the problems of the latitudinal distribution of dinocephalians.

These two interpretations, on the one hand that dinocephalians possessed a mammalian temperature physiology, and on the other that they were a large version of otherwise typical reptiles are the extreme views. The truth may well lie somewhere between. As an animal increases in size, the inertial homeothermic effect necessarily increases, and allows the animal to maintain a constant body temperature under a greater range of conditions. Equally, an increase in metabolic rate will permit homeothermy under wider conditions. There is no real reason to accept MacNab's (1978) assertion that the two are unlikely to occur simultaneously as adaptive responses to cooler conditions. A metabolic rate above that of typical modern reptiles, but still well below that of mammals where additional anatomical adaptations are needed to increase food and oxygen uptake, coupled with the characteristic body size of dinocephalians could well produce the ability to withstand high latitude circumstances. The accuracy with which the temperature needs to be maintained constant probably relates to the degree of complexity of the interrelated enzyme systems of the body and particularly of the central nervous system (p. 308). In the case of the dinocephalians this was probably not very great and therefore precise temperature regulation of the order seen in mammals was not necessary. Judging from endocranial casts (Efremov, 1954; Boonstra, 1968; Hopson, 1979), the dinocephalian brain was no larger than would be expected in a modern reptile of similar size, and differentiation of the various lobes was poor. The locomotory system does not appear to have been particularly highly coordinated. Therefore it is likely that a body temperature variation of several degrees would not have had a serious effect on the functioning of the animals, and evolution of a temperature regulatory mechanism as advanced as that of modern mammals is most improbable.

The climatic conditions under which the South African dinocephalians lived were not very severe. The extensive sediments that lie immediately below the *Tapinocephalus*-zone lack fossil reptiles, whose sudden appearance indicates that conditions had ameliorated sufficiently to permit reptile life. There is evidence from the nature of the sediments that a north–south running mountain range to the west of the region had eroded sufficiently to allow moist, warm westerly winds to penetrate (Hotton, 1967). At the same time the whole continental mass had been drifting northwards towards the equator, and there may also have been warm equatorial currents reaching the more southerly parts of Gondwanaland (Robinson, 1973). As a result of such effects, the huge lowland basin of what is now southern Africa became available for therapsids, and they migrated into the region from, possibly, the faunal complex already established in the Zone I times of Russia (Boonstra, 1969).

Fossil plants are very rarely found in association with animals for reasons not fully understood (Kitching, 1977), and therefore there is little direct evidence about the floral basis of the food web of which the dinocephalians were a major part. However, it is known from localities elsewhere that the whole of Gondwanaland was dominated during the Permian by a very characteristic flora, the *Glossopteris* flora, and no doubt the areas occupied by the dinocephalians were included. *Glossopteris* and related forms are relatively advanced plants with large, simple leaves and complex fructifications. They were associated with more or less temperate and cool conditions and occurred in great abundance along with a variety of club mosses, herbaceous horsetails, ferns, seed-ferns and conifers (Plumstead, 1973). The great success of the herbivorous dinocephalians in South Africa must surely be attributed to the presence of this rich flora, as well as their own ability to live under temperate conditions. Fish are present but rare in the *Tapinocephalus*-zone and therefore it is probable that a major shift in the basis of the food web compared to the pelycosaur faunas was away from fish and towards plants. This alone would satisfactorily account for the rise in the relative number of herbivorous forms compared to carnivorous forms noted by Bakker (1975).

# 7 | Primitive Carnivorous Therapsids

THE DOMINANT CARNIVORES of the upper Late Permian were the gorgonopsids, a well-defined group of rather conservative therapsids which nevertheless possessed several specialisations associated with feeding on large prey. There was also a range of more primitive therapsids, both contemporaneous with the gorgonopsids and also preceding them, which were closer in structure to the sphenacodontid pelycosaurs. The detailed interrelationships of these various forms are not yet clear, because they resemble one another in the possession of primitive therapsid features but clear-cut shared derived characters between the various groups have not been identified.

## Systematics

### *Eotitanosuchians (phthinosuchians)*

Historically the earliest of the primitive carnivorous therapsids to be described was a poorly preserved lower jaw of *Phthinosaurus* discovered in the Copper Sandstones of the Russian Urals. Eventually a partial skull was found and recognised as a similar form, named *Phthinosuchus* (Efremov, 1954). These two genera (Fig. 33I,J) were modest sized animals, with a skull length around 20 cm. Therapsid features included the enlargement of the canines and the temporal fenestra, a large reflected lamina of the angular, and an anterior position of the jaw articulation. However, much of the structure was still pelycosaur-like, particularly the curved dorsal margin of the skull and the wide intertemporal region. Unlike contemporary brithopodids there had been no invasion of the external surface of the skull roof by the jaw musculature.

For all the importance of *Phthinosaurus* and *Phthinosuchus* in linking the therapsids to the pelycosaurs, and their consequent prominent place in

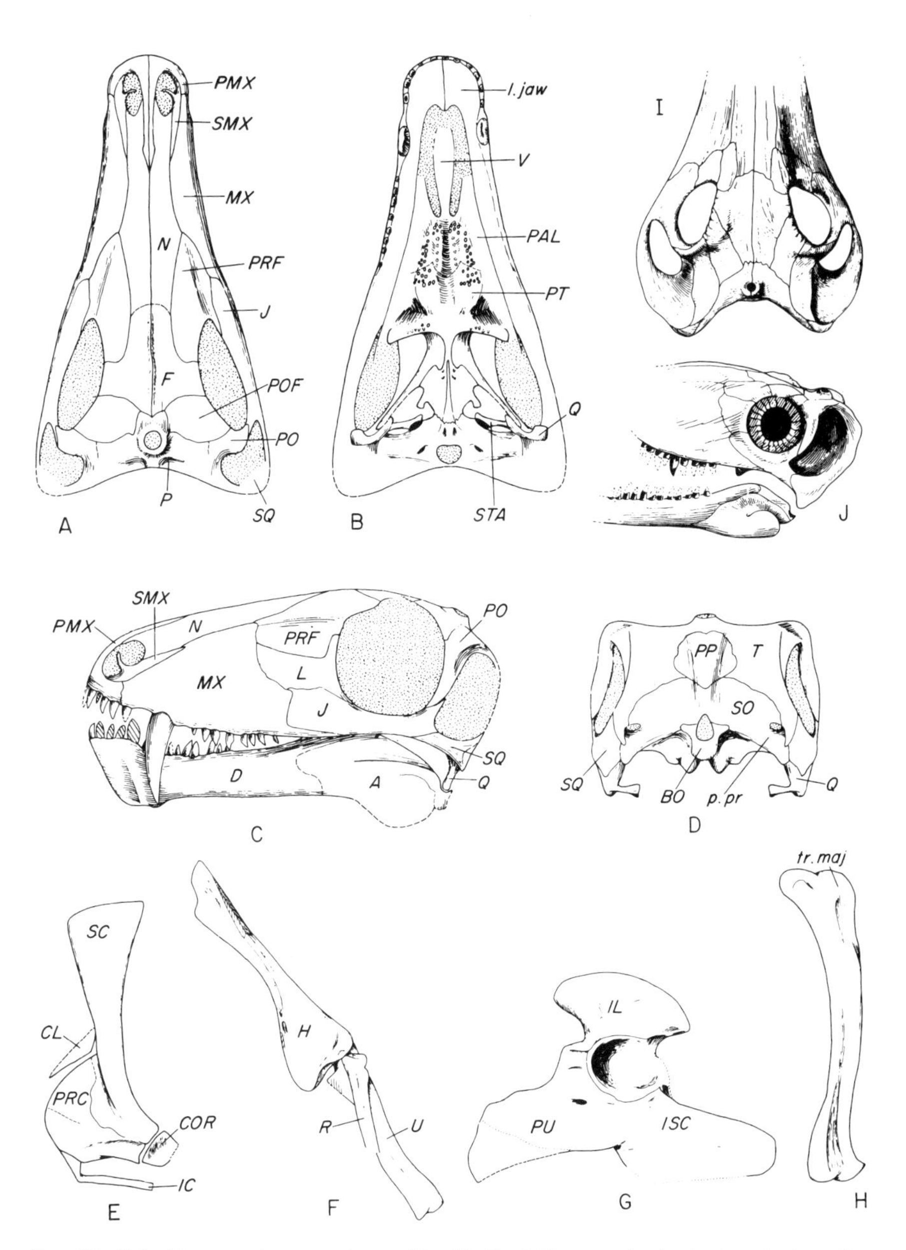


FIG. 33. Primitive carnivorous therapsids. Skull of *Biarmosuchus* in A, dorsal view; B, ventral view; C, lateral view; D, posterior view; E, shoulder girdle; F, forelimb; G, pelvis. H, femur of *Biarmosuchus*; I, dorsal view of skull of *Phthinosuchus*; J, lateral view of skull of *Phthinosuchus*, with lower jaw based on *Phthinosaurus*. (A–H from Sigogneau and Chudinov, 1972; I–J from Tatarinov, 1974.)

the literature of the subject, little is actually known of them, and no postcranial skeleton has been discovered. More recently, however, two forms of primitive carnivorous therapsids have been found in the Ocher deposits of the Russian cis-Uralian region, which date from Zone I of the Russian Late Permian (p. 69). Neither is referable to the earlier known genera, but both are at approximately the same level of evolution. The most primitive of the two, that is, the most sphenacodont-like, is *Biarmosuchus* (and the probably synonymous *Biarmosaurus*), which is known from several skulls and most of the postcranial skeleton (Chudinov, 1960, 1965; Sigogneau and Chudinov, 1972). It was an animal about the same size as *Phthinosuchus*, which it resembled closely, so far as is known (Fig. 33). The canine is relative very much larger than the other teeth, and the postcanines are reduced to about eight small ones. The lower incisors could probably interdigitate with the uppers when the jaws closed, as in some of the later carnivorous therapsids and in dinocephalians. The lateral profile of the skull roof is very sphenacodontid-like, with a gently convex snout and a steeply descending postorbital region. The temporal fenestra is particularly small, much smaller than the orbit for example, and the intertemporal region of the skull roof is wider than the interorbital width. The jaw musculature had not invaded the external surfaces of the intertemporal region. The occipital surface slopes forwards and downwards to the relatively anteriorly placed occipital condyle and jaw articulation. The lower jaw possesses a highly developed reflected lamina of the angular which reaches back almost to the jaw articulation, and bears a complex cross of strengthening ridges on its lateral face.

The postcranial skeleton has not yet been fully described. It is most remarkable in general for the relative slenderness of the limb bones, suggesting that *Biarmosuchus* was an agile animal. The shoulder girdle (Fig. 33E) has a very narrow scapula blade, and the glenoid, although not well known, was probably short as in all other therapsids. The humerus (Fig. 33F) is very slender and the radius is less robust than the ulna. The pelvis (Fig. 33G) is still rather pelycosaur-like, with a large pubis and only a slightly expanded ilium. The femur (Fig. 33H), again

*A*, angular; *BO*, basioccipital; *CL*, clavicle; *COR*, coracoid; *D*, dentary; *H*, humerus; *IC*, interclavicle; *IL*, ilium; *ISC*, ischium; *F*, frontal; *J*, jugal; *L*, lachrymal; *l.jaw*, lower jaw; *MX*, maxilla; *N*, nasal; *P*, parietal; *PAL*, palatine; *PMX*, premaxilla; *PO*, postorbital; *POF*, postfrontal; *PP*, postparietal; *p.pr*, paroccipital process; *PRC*, procoracoid; *PRF*, prefrontal; *PT*, pterygoid; *PU*, pubis; *Q*, quadrate; *R*, radius; *SC*, scapula; *SMX*, septomaxilla; *SQ*, squamosal; *STA*, stapes; *T*, tabular; *t.maj*, trochanter major; *U*, ulna; *V*, vomer.

Magnifications: A–D, *c*. × 0.35; E–H, *c*. × 0.25; I–J, *c*. × 0.23.

noticeably slender, nevertheless has the therapsid features of a trochanter major, small internal trochanter, and an inturning of the articulating head. The feet have retained the ancestral phalangeal formula of 2.3.4.5.?, but one of the phalanges of digit 2, and two of digit 3 are reduced to narrow discs, exactly as in the later gorgonopsians.

The second primitive carnivore from Ocher is *Eotitanosuchus* (Fig. 34) which is known from a single but almost complete skull, lacking the lower jaw and all the postcranial bones (Chudinov, 1960, 1965; Olson, 1962). It is larger than *Biarmosuchus*, with a skull length of 35 cm, and is more advanced in several respects. The dentition is similar in the two, except that there is a larger gap anterior to the upper canine in *Eotitanosuchus*, and no precanine maxillary tooth. The lower canine was consequently probably larger in this form. Whether the incisor teeth interdigitated is not known. The temporal fenestra is much larger and has expanded postero-dorsally, so that the postorbital part of the skull roof does not descend as in *Biarmosuchus*. However, the intertemporal skull roof is still very wide, and again there has been no invasion of its external surfaces by jaw musculature.

Several comparable types of primitive carnivorous therapsids occur in the San Angelo Formation of Texas. These fossils are extremely scarce and very scrappy, consisting for the most part of jaw fragments with teeth. Olson (1962) recognises three genera of generally phthinosuchid-like forms, *Knoxosaurus*, *Steppesaurus* and *Gorgodon*, although the possibility that at least some of them are aberrant sphenacodontids cannot be discounted yet. Another form is *Watongia* from the contemporary Chickasha Formation of Oklahoma (Olson, 1974), known from slightly better material, although its taxonomic relationships are still very doubtful. A fragment of the skull roof shows what might be a preparietal bone, a feature very characteristic of the gorgonopsids, and the limb bones also show possible indications of gorgonopsid features. However, until better skull material is found this possible relationship to the gorgonopsids is very tenuous.

## *Ictidorhinids (hipposaurids)*

A series of carnivores from the Beaufort Formation of the South African Karroo have traditionally been regarded as gorgonopsids, although as a distinctive subgroup. These are the ictidorhinids (Fig. 35) which are readily distinguishable from typical gorgonopsids by their possession of a series of primitive features (Sigogneau, 1968, 1970a, 1970b). Indeed they can be regarded as South African equivalents of the eotitanosuchians. The primitive features include the small temporal fenestra

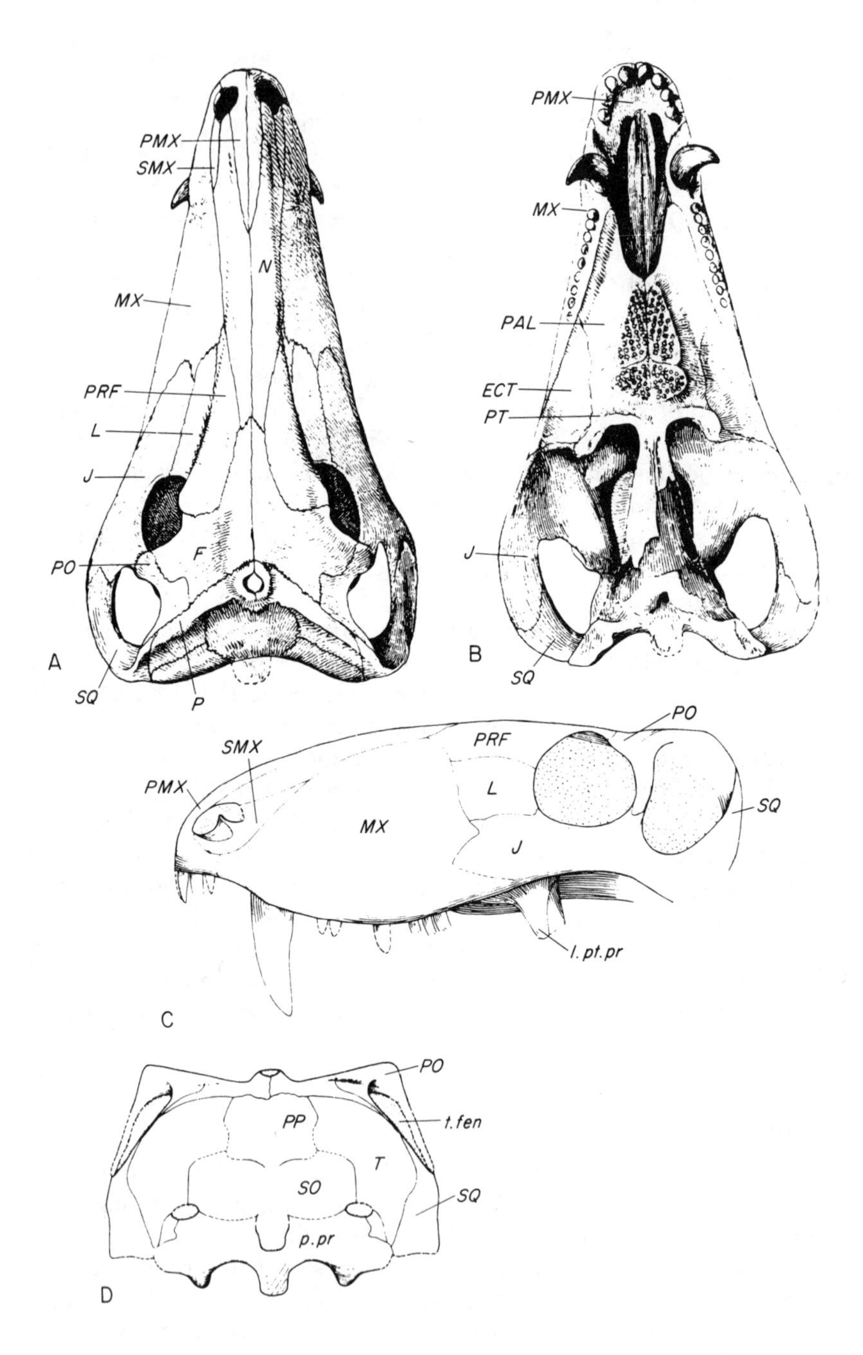

PMX
SMX
N
MX
PRF
L
J
F
PO
SQ
P
A
PMX
MX
PAL
ECT
PT
J
SQ
B
SMX
PMX
PRF
PO
L
MX
J
SQ
l. pt. pr
C
PO
PP
t. fen
T
SO
SQ
p. pr
D


FIG. 34.

and very wide intertemporal roof, the convex dorsal margin of the skull, the usually slightly greater number of postcanine teeth compared to the gorgonopsids, the absence of a coronoid process on the lower jaw, and the posterior extent of the reflected lamina of the angular. Certain details of the palate and base of the braincase are also eotitanosuchian-like, and the postcranial skeleton, where known, generally resembles *Biarmosuchus* (Boonstra, 1965c). The earliest of the ictidorhinids is *Hipposaurus*, which is from the *Tapinocephalus*-zone, the lower part of the Beaufort. All the other ictidosuchids are a little younger, from the *Cistecephalus*-zone, and are quite diverse in nature. However, none of them are particularly well known in detail yet.

A relationship between ictidosuchids and *Biarmosuchus* has been proposed (Sigogneau and Chudinov, 1972). The admittedly striking similarities between the two are, however, primitive therapsid features. Until the group is better known, it will remain uncertain whether ictidorhinids themselves constitute a monophyletic group, let alone whether they are related to any particular eotitanosuchian.

## *Gorgonopsids*

The gorgonopsids first appear in the *Tapinocephalus*-zone of South Africa as relatively rare and fairly small forms, but with the full suite of gorgonopsid characteristics present. By the later *Cistecephalus*-zone and *Daptocephalus*-zone, they had become the dominant carnivores, but the whole group went abruptly extinct at the close of the Permian, leaving no known Triassic descendants. A few forms occur in the Late Permian of the Russian sequence, in the horizon termed Zone IV, and they include the largest gorgonopsid, *Inostrancevia*, which had a skull length of about 45 cm. Although a large number of genera have been described, they remained very conservative and all of them can be contained in a single family Gorgonopsidae. Most of their characteristic features (Fig. 36) are related to the adoption of feeding on large prey, certainly prey of a similar size to themselves (Kemp, 1969a). The most immediate feature is a further exaggeration of the size of the canines, to

FIG. 34. Skull of *Eotitanosuchus*. A, dorsal view. B, ventral view. C, lateral view. D, posterior view. (A and B from Chudinov, 1960; C and D from Sigogneau and Chudinov, 1972.)

*ECT*, ectopterygoid; *J*, jugal; *L*, lachrymal; *l.pt.pr*, lateral process of the pterygoid; *MX*, maxilla; *N*, nasal; *P*, parietal; *PAL*, palatine; *PMX*, premaxilla; *PO*, postorbital; *PP*, postparietal; *p.pr*, paroccipital process; *PRF*, prefrontal; *PT*, pterygoid; *SMX*, septomaxilla; *SQ*, squamosal; *SO*, supraoccipital; *T*, tabular; *V*, vomer.

Magnification *c*. × 0.23.

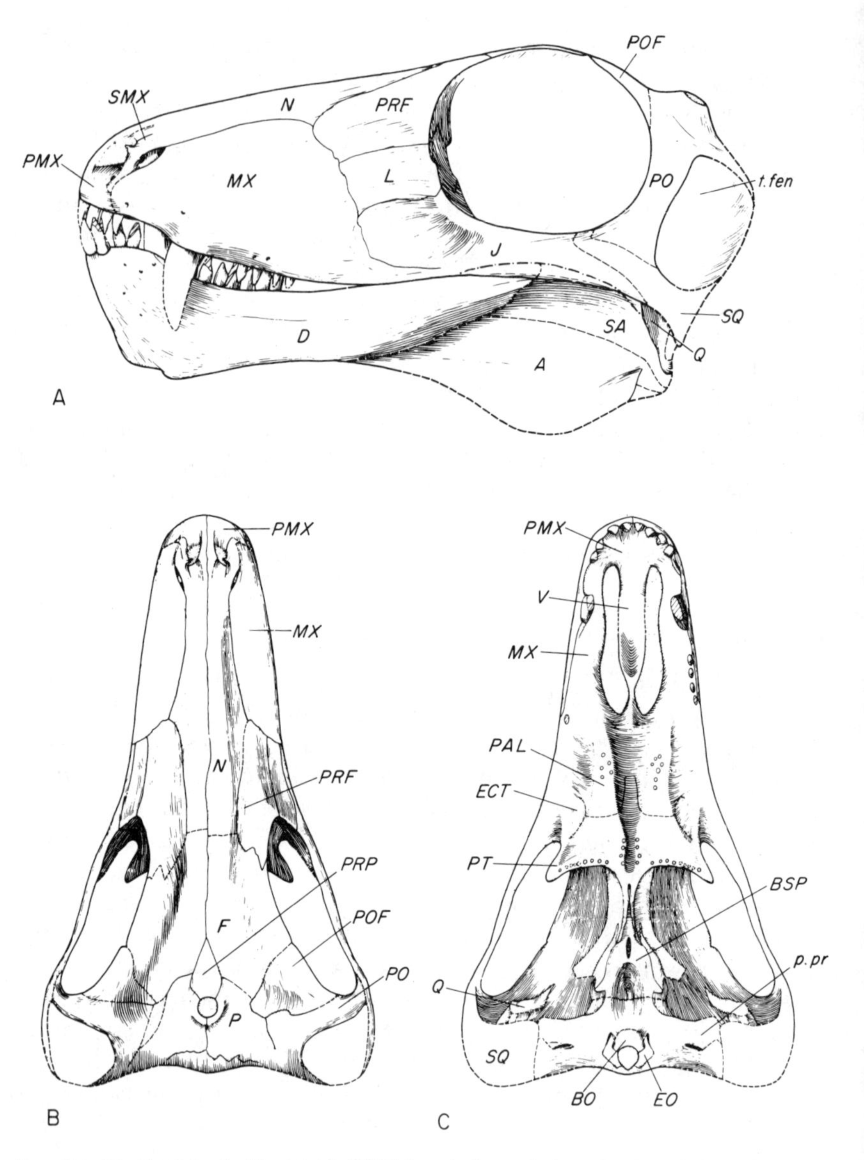


FIG. 35. Skull of the ictidorhinid *Rubidgina*. A, lateral view. B, dorsal view. C, ventral view. (From Sigogneau, 1970.)

*A*, angular; *BO*, basioccipital; *BSP*, basisphenoid; *D*, dentary; *ECT*, ectopterygoid; *EO*, exoccipital; *F*, frontal; *J*, jugal; *L*, lachrymal; *MX*, maxilla; *N*, nasal; *Q*, quadrate; *P*, parietal; *PAL*, palatine; *PMX*, premaxilla; *PO*, postorbital; *POF*, postfrontal; *p.pr*, paroccipital process; *PRF*, prefrontal; *PRP*, preparietal; *PT*, pterygoid; *SA*, surangular; *SMX*, septomaxilla; *SQ*, squamosal; *t.fen*, temporal fenestra; *V*, vomer.

Magnifications: A, *c.* × 0.8. B and C, *c.* × 0.65.

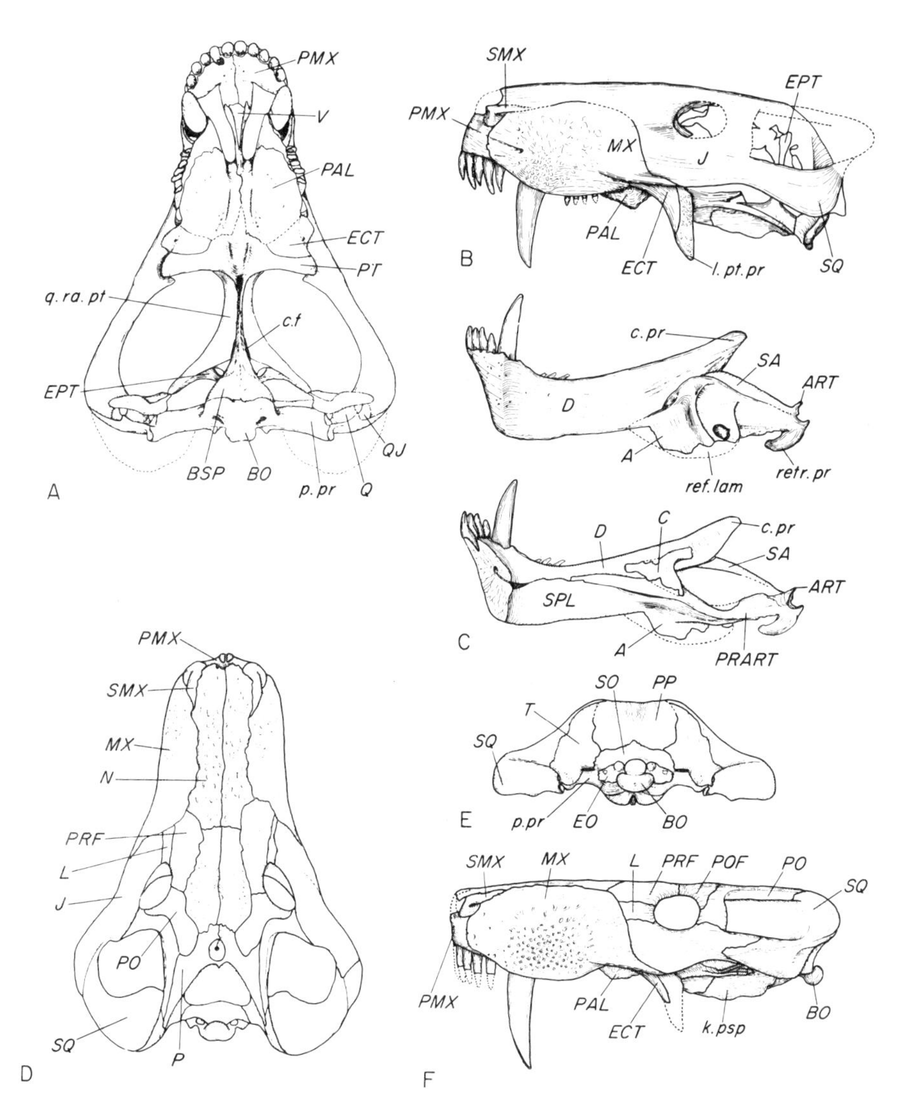


Fig. 36. The gorgonopsid skull. Skull of *Arctognathus* in A, ventral view; B, lateral view with lower jaw; C, medial view of lower jaw. Skull of *Leontocephalus* in D, dorsal view; E, posterior view; F, lateral view. (From Kemp, 1969a.)

*A*, angular; *ART*, articular; *BO*, basioccipital; *BSP*, basisphenoid; *C*, coronoid; *c.f*, carotid foramen; *c.pr*, coronoid process; *D*, dentary; *ECT*, ectopterygoid; *EO*, exoccipital; *EPT*, epipterygoid; *F*, frontal; *J*, jugal; *k.psp*, keel of the parasphenoid; *L*, lachyrmal; *l.pt.pr*, lateral process of the pterygoid; *MX*, maxilla; *N*, nasal; *P*, parietal; *PAL*, palatine; *PMX*, premaxilla; *PO*, postorbital; *POF*, postfrontal; *PP*, postparietal; *p.pr*, paroccipital the extent that the group has been referred to as the "sabre-toothed" reptiles. The five upper and four lower incisors are also well developed, and could be optionally interdigitated, depending on whether the lower jaw shifted anteriorly during the bite. The postcanines are reduced to a maximum of about five very small teeth. The snout is relatively short but very deep, in order to accommodate the huge upper canine roots, and to house the lower canines when the jaws were closed. In consequence, the convex dorsal margin of the skull seen in the more primitive carnivores has become straight. The temporal fenestra is considerably enlarged over the more primitive condition, particularly by a posterior and dorsal extension. The squamosal bones extend as a pair of great flared processes behind the level of the posterior edge of the skull roof. There was no medial extension of the fenestra, however, and therefore the intertemporal region of the skull is still wide and flat, and the jaw muscles had not invaded the external surfaces of the skull roof. The jaw articulation is at a relatively high level, and not particularly far forwards. The lower jaw is equally characteristic, for it has evolved a coronoid process of the dentary, rising freely above the post-dentary bones. The reflected lamina of the angular is far forwards, and it has a powerful, vertical ridge on the lateral face. Two particular features of the skull are both very characteristic of the group, and yet difficult to account for in functional terms. The first is the preparietal bone, present in most forms and lying in the midline just in front of the parietals. Only some ictidorhinids and the dicynodonts amongst other therapsids possess this bone. The second is that, at least in those forms where it is known, the palatine bones of the palate meet in the midline.

The gorgonopsid postcranial skeleton (Fig. 37A) is typically and fully therapsid, without any clear specialisations. The limbs tend to be relatively lightly built and long. Like *Biarmosuchus*, reduced phalanges are still present in both fore and hind feet (Fig. 41G).

Taxonomic variation amongst the gorgonopsids is small, generally differing in such apparently trivial features as size, snout shape, relative width of intertemporal and interorbital regions of the skull roof and number of postcanine teeth (Sigogneau, 1970a). One group, which may be separated as a sub-family, are the rubidgeines, which became larger, with broad, heavy skulls. Even in them, however, the maximum skull

process; *PRART*, prearticular; *PRF*, prefrontal; *PT*, pterygoid; *Q*, quadrate; *QJ*, quadratojugal; *q.ra.pt*, quadrate ramus of the pterygoid; *ref.lam*, reflected lamina of the angular; *retr.pr*, retroarticular process; *SA*, surangular; *SMX*, septomaxilla; *SO*, supraoccipital; *SPL*, splenial; *SQ*, squamosal; *T*, tabular; *V*, vomer.

length achieved was only around 40 cm, and gorgonopsids were generally much smaller than the dinocephalian carnivores which they eventually replaced.

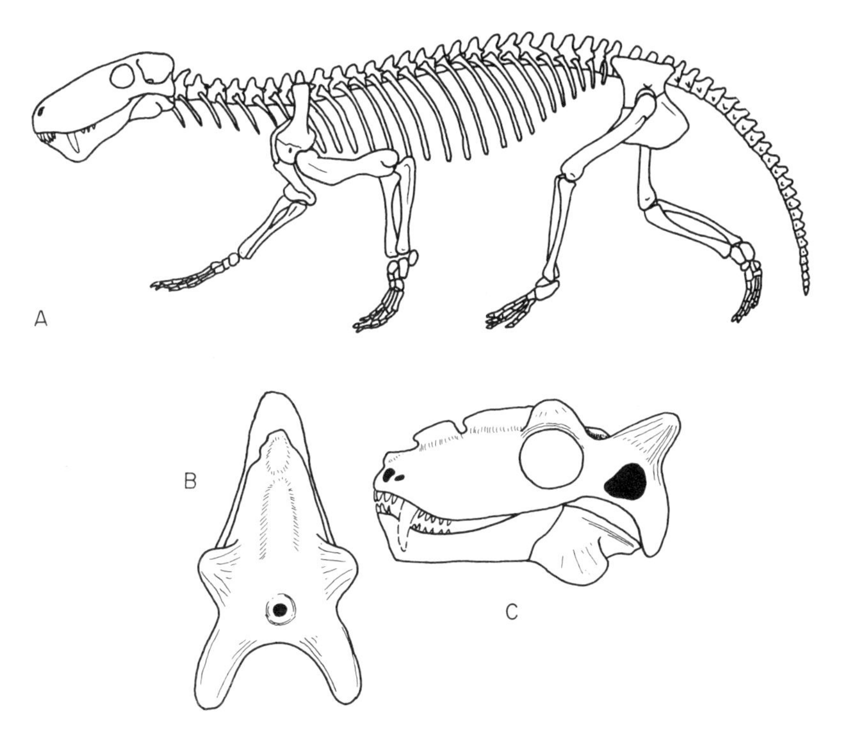


FIG. 37. A, skeleton of the gorgonopsid *Lycaenops*. B, dorsal view and C, lateral view of *Proburnetia*. (A redrawn after Colbert, 1948; B and C, redrawn after Tatarinov, 1974.) Magnifications: A, *c*. × 0.15; B and C, *c*. × 0.2.

The origin of the gorgonopsids from a primitive carnivorous therapsid of the general structure of the eotitanosuchians and ictidorhinids is clear, but which of the known primitive forms is most closely related to gorgonopsids is disputable (Fig. 38) (Sigogneau and Chudinov, 1972). *Eotitanosuchus* is almost certainly closer than *Biarmosuchus* because it has enlarged the temporal fenestra in a fashion that is reminiscent of the gorgonopsid fenestra, and its palatine bones meet in the midline of the palate. However, at least certain of the ictidorhinids may have a still closer relationship to gorgonopsids (Boonstra, 1963). *Hipposaurus* for example possesses a preparietal bone on the skull roof and the vomer is high in the skull, both gorgonopsid characteristics. Other ictidorhinids have an incipient coronoid process of the dentary. Until these various genera are better known, however, the exact interrelationships will remain obscure.

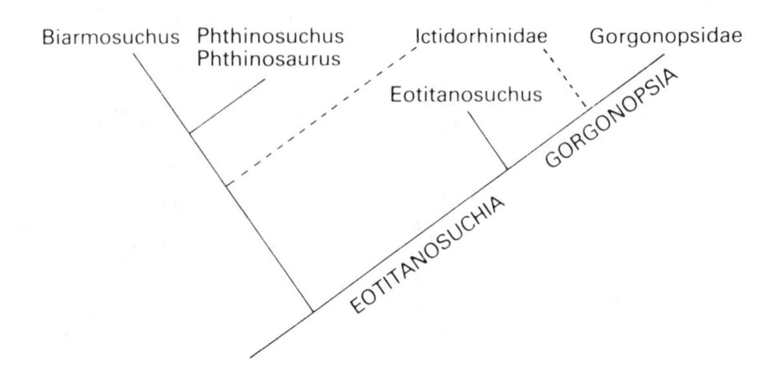


FIG. 38. Phylogeny of the primitive carnivorous therapsids.

### *Burnetiids*

The family Burnetiidae is known from the single skulls of two forms occurring late in the Permian. *Burnetia* (Sigogneau, 1970) is from the *Daptocephalus*-zone of South Africa, and the similar *Proburnetia* (Tatarinov, 1974) from the Zone IV Permian of Russia. They are usually regarded as aberrant relatives of the gorgonopsids although their phylogenetic relationships are not really clear. The bone of the skull roof (Fig. 37B,C) is heavily thickened and the sutures obliterated. Massive bony protuberances occur on the roof, and the temporal fenestra is very small. However, the structure of both the dentition and of the palate is similar to that of gorgonopsids.

## Functional anatomy

### *Feeding mechanisms*

Little is yet known about the functional anatomy of the skull of the most primitive carnivorous therapsids. Their dependance upon large canines indicates that they had a kinetic jaw system analogous to that of the more primitive dinocephalians, relying upon the kinetic energy of the motion of the lower jaw to deliver a lethal bite. The incisors probably interdigitated, lower between the uppers, in order to dismember the prey. The jaw muscles were enlarged but still arranged basically as in the sphenacodontid pelycosaurs.

The gorgonopsid skull, however, has been studied in considerable detail from this point of view. The whole of the feeding apparatus became specialised for dealing with relatively large prey, and the dentition was capable of two distinct modes of action (Kemp, 1969a). The extremely large canines, upper and lower, were used in a kinetic fashion to kill the prey, the jaw being opened widely then accelerated shut. The work done by the jaw muscles was thus transformed into kinetic energy of the lower jaw, and in turn dissipated as the canines sank into the food. During this stage, the lower incisors passed internal to the uppers without meeting them. Yet the incisors have a serrated edge on either side and in well-preserved specimens these edges are seen to have developed wear-facets. In fact, the lower jaw was capable of shifting anteriorly, so that as the lower incisors approached the upper incisors they could interdigitate (Fig. 39A–C). Each lower incisor then passed inwards, between two uppers, producing a cutting bite which functioned to bite off swallowable pieces of flesh from the prey once it had been disabled. The very reduced postcanine teeth and small palatal teeth had no more than a minor gripping role.

The jaw articulation was specialised to permit the very wide gape necessary to allow the lower canines to clear the upper canines by an adequate amount, a gape which must have approached 90°. At the same time the hinge needed sufficient stability to resist the tendency of struggling prey to cause disarticulation (Parrington, 1955). The lateral quadrate condyle is a narrow cylinder, which fits into the correspondingly deep, groove-like lateral condyle of the articular (Fig. 39E–G). The effective depth of the articular condyle is increased by a high process on its posterior edge. This provided a strong articulation, resistant to disarticulation, but not allowing a wide gape because the dorsal process meets against the back of the quadrate. However, the medial quadrate condyle is a short section of a screw-shape which fits against a correspondingly shaped medial articular condyle. As the jaws opened, the articular condyle screwed sideways on the quadrate condyle, causing a slight lateral shift of the whole of the back end of the lower jaw. This in turn caused the dorsal process of the lateral articular condyle to shift clear of the back of the quadrate and it rotated into a lateral notch above the quadrate condyle. Therefore a wide gape was possible even though the hinge was very stable with the jaws closed. A hinge of this description would not permit the lower jaw to slide antero-posteriorly upon the quadrate, yet this kind of movement was necessary to allow the interdigitation of the incisors. However, the

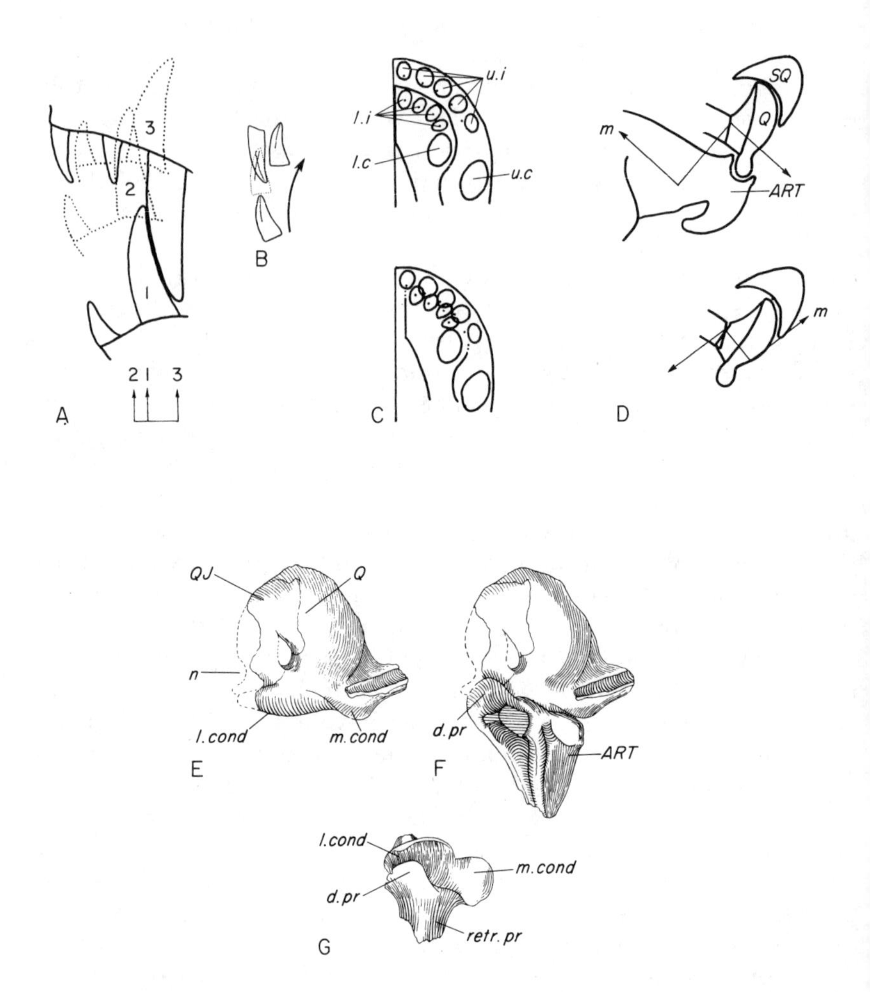


FIG. 39. Tooth and jaw-hinge function in gorgonopsids. A, movement of the lower incisor and canine teeth relative to the uppers during the interdigitating bite. The successive level of the tip of the lower canine is indicated by 1, 2 and 3. B, movement of a single lower incisor relative to an opposing upper incisor. C, plan of the relative positions of the teeth in a horizontal plane during a non-interdigitating bite (above) and an interdigitating bite (below). D, antero-posterior shift of the quadrate condyles as seen in lateral view. The appropriate direction of a muscle force to produce the movement is indicated by m. E, posterior view of the left quadrate complex. F, the same but with the articular in place and the jaws half-open. G, posterior view of the articular bone. (A–D from Kemp, 1969a; E–G from Parrington, 1955.)

*ART*, articular; *d.pr*, dorsal process; *l.c*, lower canine; *l.cond*, lateral condyle; *l.i*, lower incisor; *n*, notch in quadratojugal; *Q*, quadrate; *QJ*, quadratojugal; *retr.pr*, retro-articular process; *SQ*, squamosal; *u.c*, upper canine; *u.i*, upper incisor.

quadrate itself was capable of moving relative to the skull, carrying the jaw with it (Kemp, 1969a). The back of the quadrate is swollen and fits into a deep concavity in the squamosal (Fig. 39D). Together these bones form a ball-and-socket joint, whereby the quadrate could rotate about a transverse axis. This caused the lower end of the quadrate, the condyles, to move forwards and impart an anterior shift to the lower jaw, sufficient to cause the incisor teeth to meet.

To motivate this complex series of jaw movements, the jaw musculature was specialised, but in a quite different manner to that seen in the dinocephalians. The external adductor muscle increased in size considerably (Fig. 40). That part equivalent to the bulk of the muscle in pelycosaurs originated from the underside of the skull roof, and presumably from a temporal aponeurosis covering the temporal fenestra. It inserted on the dorsal and medial faces of the postdentary region of the jaw, and possibly into a bodenaponeurotic sheet rising from the surangular. In addition to this, two new parts of the external adductor muscle had evolved. The first originated from the anterior face of the squamosal, but this bone had expanded posteriorly behind the level of the back of the skull roof. At the same time, a coronoid process of the dentary had evolved, as a dorsal extension of that bone into the anterior

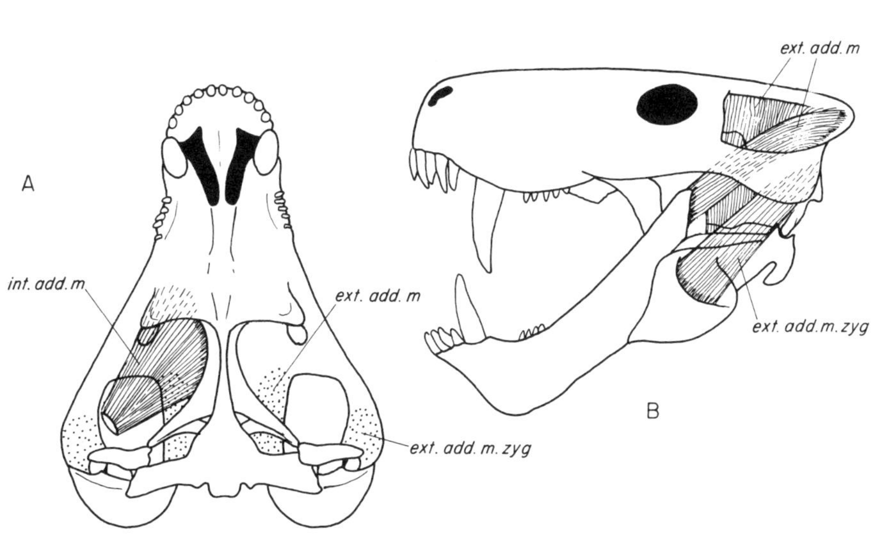


FIG. 40. Gorgonopsid jaw musculature. A, ventral view showing area of origin of external adductor muscle (stippled), and internal adductor muscle. B, lateral view of the external adductor muscle complex.

*ext.add.m*, external adductor muscle; *ext.add.m.zyg*, zygomatic part of the external adductor muscle; *int.add.m*, internal adductor muscle.

region of the bodenaponeurosis. The coronoid process provided the insertion for the squamosal part of the external adductor and this radically new development of the jaw musculature had certain important properties. It was oriented almost horizontally and therefore tended to pull the jaw backwards, yet it still provided a large torque for jaw-closing by virtue of its point of insertion on the jaw, high above the level of the hinge. The second new part of the external adductor similarly represented an important functional innovation. Its origin was the inner surface of the lower temporal bar, or zygomatic arch as it may be termed. Unlike the dinocephalians, the zygomatic arch remained high, well above the level of the lower jaw, and flared laterally to the jaw. Sufficient separation betwen arch and jaw permitted the zygomatic arch to become a new area for attachment of jaw musculature. The insertion of this part of the external adductor was the powerful, near-vertical ridge on the reflected lamina of the angular, which is found only in gorgonopsids. Like that part of the muscle attaching to the coronoid process, this new, zygomatic part tended to pull the jaw backwards as well as to close it but it differed in that it also tended to pull the lower jaw laterally instead of medially.

There was also a particularly large and powerful internal adductor, or ptergoideus musculature in gorgonopsids. The massive lateral pterygoid processes of the palate are relatively further forwards than in more primitive therapsids. Most of the pterygoideus musculature originated from the back of these processes, and from the dorsal surface of the palate in front of them. It ran backwards and downwards to insert on the lower region of the postdentary bones, probably also wrapping around the ventral edge of the jaw to gain insertion on the outer face both behind and probably within the reflected lamina. The geometrical arrangements are such that the pterygoideus muscle had two important properties. First, it was most effective as a jaw-closing muscle when the jaws were very widely open, and by shifting its origin from the skull forwards the length of the fibres was enough to prevent the muscle from limiting the gape. Second, the pterygoideus muscle tended to pull the jaw forwards, as well as closing it.

Taking the musculature as a whole, certain principals are apparent. The posteriorly directed component of much of the external adductor musculature is opposed by the anterior component of the pterygoideus musculature, and therefore antero-posterior movements of the lower jaw (propaliny) could occur. The posterior component is potentially much larger than the anterior, since a greater bulk of muscle was involved, and therefore a powerful bite involving a backwards shift of the jaw could occur, thus drawing the lower incisors backwards between the upper incisors. The pterygoideus muscle was required only to pull the jaw forwards ready for this power stroke. The lateral component of the zygomatic arch musculature would have tended to cancel the medial component of the rest of the musculature, reducing the tendency to force the lower jaw inwards during the bite. Finally, the muscles were so arranged that different ones became most effective at different gapes, the pterygoideus at very wide gapes, and the external adductor at smaller gapes. The tendency to balance the muscle forces acting on the jaw, and to increase the range and adaptability of the jaw action were taken no further in the gorgonopsids. As we shall see, however, similar principles were involved in the evolution of the cynodonts, and ultimately played an important part in the origin of the mammals themselves.

In addition to adequate jaw closing muscles, gorgonopsids also required more effective jaw opening muscles, in order to achieve the large gape often necessary. At least part of the opening musculature was a depressor mandibuli, running from the back of the occiput onto the large, downturned retroarticular process of the articular bone. In many modern reptiles such as crocodiles, a retroarticular process is present, but it simply extends posteriorly behind the jaw hinge. The ventrally reflected process in gorgonopsids permitted the muscle to produce a larger gape before the process moved too close to the occipital surface (Parrington, 1955). As was suggested in the case of dinocephalians, it is likely that throat musculature attached to the reflected lamina of the angular, and acting perhaps via the hyoid apparatus and shoulder girdle, also had a jaw-opening function.

## *Locomotion*

Gorgonopsids had a typical therapsid postcranial skeleton (e.g. Broili and Schröder, 1935; Colbert, 1948). It can be used as a model of the therapsid skeleton in general to illustrate the functional significance of the various therapsid features compared to the ancestral, pelycosaurian structure (p. 53).

The most striking change in the shoulder girdle and forelimb concerns the shoulder joint (Fig. 41). The glenoid is short and simple compared to the restrictive screw-shape of pelycosaurs, and the humerus was capable of a greater magnitude and variety of movements. Nevertheless, the humerus was still forced to work in a more or less horizontal plane and the forelimb gait was probably still as sprawling as in pelycosaurs. Looked at in detail, the shoulder joint is most peculiar because the articulating surface of the head of the humerus does not fit

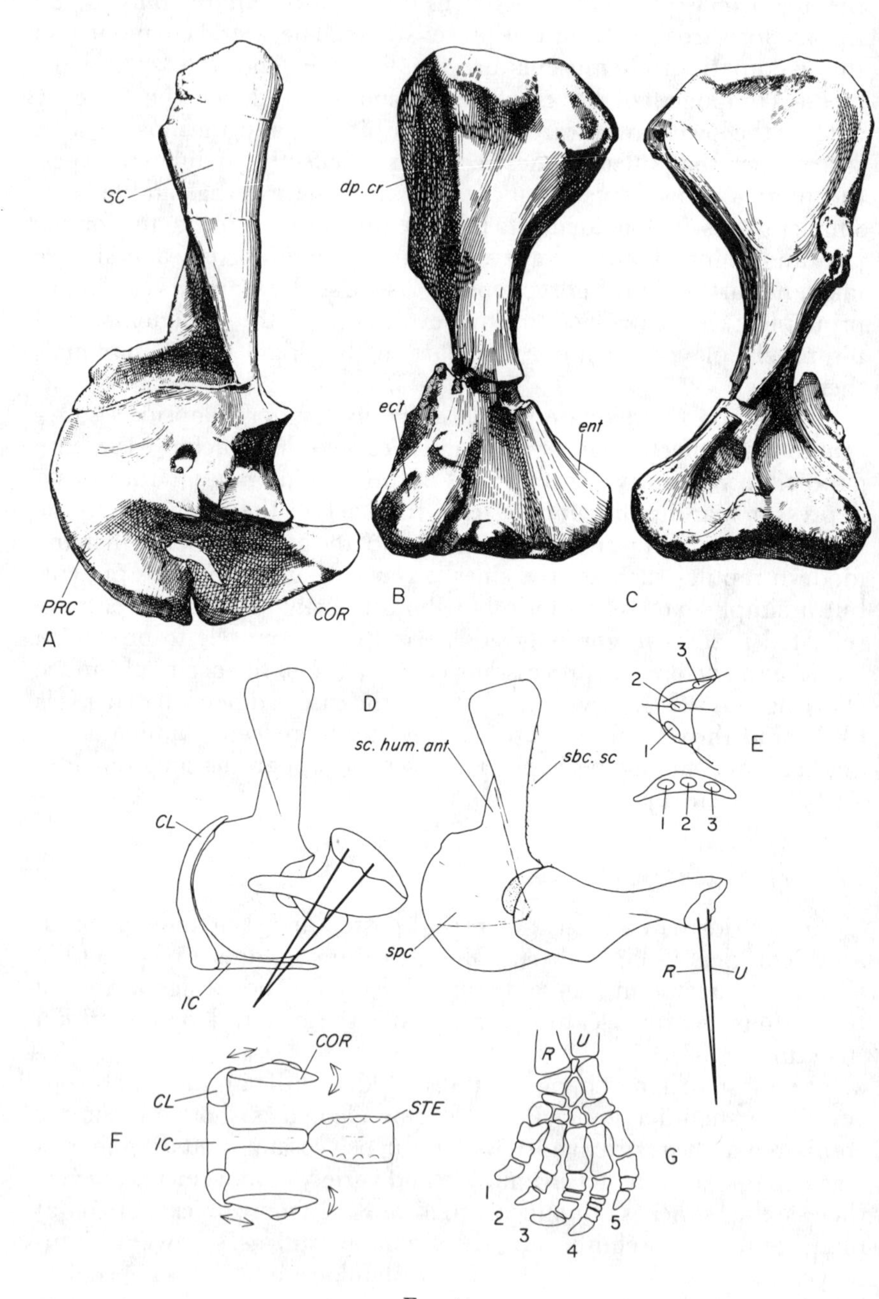
SC
PRC
COR
A
dp. cr
ect
ent
B
C
D
CL
IC
sc. hum. ant
sbc. sc
spc
3
2
1
E
1
2
3
R
U
COR
CL
STE
F
IC
R
U
G
1
2
3
4
5

FIG. 41.

at all well against the glenoid, unlike the pelycosaurs and also the mammals. The glenoid (Fig. 41A) is actually a deep notch, which is concave from top to bottom. However, it is convex from front to back, so that while the anterior part faces outwards, the posterior part faces backwards. In contrast, the articulating surface of the head of the humerus (Fig. 41B,C) resembles part of a long, narrow cylinder occupying the proximal end of the bone. This incongruity between the glenoid and humerus head is such that the maximum contact between them at any moment is at just two points. What appears to have happened (Kemp, 1980b) is that at the beginning of a stride the humerus extended laterally from the glenoid. The front part of the cylindrical humerus head contacted the front part of the glenoid (Fig. 41E). The humerus was then pulled backwards until it extended almost posteriorly alongside the body. As it did so, the straight articulating surface of the humerus rolled backwards and inwards across the convex glenoid surface until, at the end of the stride, the posterior part of the humerus head contacted the posterior-facing part of the glenoid. Simultaneously with the retraction, the humerus also rotated about its long axis as in pelycosaurs. At the start of the stride, the lower part of the forelimb ran downwards and forwards to the foot which lay anterior to the end of the humerus (Fig. 41D). By the end of the stride long axis rotation of the humerus had brought the end of the humerus over the foot. The distal end of the humerus could also be raised, and therefore the recovery phase of the stride must have consisted simply of a reversal of the power stroke, but with the humerus elevated to lift the foot off the ground.

This curious arrangement whereby the opposing joint surfaces do not slide relative to one another but roll, analogous to a wheel rolling on the

FIG. 41. The shoulder girdle and forelimb of gorgonopsids. Unidentified gorgonopsid specimen (Cambridge University Museum of Zoology No. T.883): A, lateral view of the scapulo-coracoid. B, dorsal view of the left humerus. C, ventral view of the left humerus. D, lateral view of the forelimb protracted (left) and retracted (right), showing both the retraction of the humerus in a horizontal plane, and its rotation about the long axis. E, the movement of the head of the humerus in the glenoid. Successive areas of contact between the head and the glenoid are indicated by *1–1*, *2–2* and *3–3*, while the lines indicate the orientation of the antero-posterior line of the head in the glenoid at each phase. F, ventral view of the complete shoulder girdle, indicating possible modes of movement of the scapulo-coracoid. G, dorsal surface of the left forefoot. (A–C, original drawings by David Nicholls; G redrawn after Boonstra, 1934.)

*CL*, clavicle; *COR*, coracoid; *dp.cr*, delto-pectoral crest; *ect*, ectepicondyle; *ent*, entepicondyle; *IC*, interclavicle; *PRC*, procoracoid; *R*, radius; *sbc.sc*, subcoracoscapularis muscle; *SC*, scapula; *sc.hum.ant*, scapulo-humeralis anterior muscle; *spc*, supracoracoideus muscle; *STE*, sternum; *U*, ulna.

Magnification A–C, *c.* × 0.37.

ground, has the apparent disadvantage that very high pressures must have occurred at the limited points of contact between them. On the other hand, compared to the pelycosaur system, it allowed a much greater swing of the humerus, thus increasing the length of the stride. It also permitted the humerus to vary both the extent of rotation about the long axis and the degree of elevation, and was therefore important in making the animal more manoeuvrable, over rough ground for example.

Another possibly very important consequence was that the main muscles causing movements of the humerus could attach almost exclusively to the scapulo-coracoid, and did not require an origin from regions of the neck or trunk of the animal. The retractor muscles, which insert at the posterior end of the head of the humerus, must pull in a medial direction to retract the bone. Similarly the protractor muscles, inserting on the anterior part of the head, must also pull almost medially in order to cause the humerus to move forwards. Both sets of muscles therefore project towards the scapulo-coracoid. The reason why it should be necessary to arrange the main locomotory muscles in this way probably relates to another innovation of the therapsids, a mobile shoulder girdle. Reduction of the width of the scapula blade, and of the attachment of the coracoids to the interclavicle resulted in a more loosely attached scapulo-coracoid. It could probably shift backwards and forwards, and also rotate about its point of attachment to the clavicle and therefore produced an increase in the overall length of the forelimb stride (Fig. 41F). A relatively loosely attached shoulder girdle would not have been possible if large locomotory muscles ran between the humerus and other parts of the body, for when they contracted they would have tended to cause disarticulation of the shoulder girdle.

The musculature of the shoulder was not radically different from that of the pelycosaurs, except for the main retractor muscle, the subcoracoscapularis (Fig. 41D). In the pelycosaurs retraction of the humerus was caused by an almost forwards pull of this muscle on the posterior end of the humerus head (p. 55). In gorgonopsids, the alteration to the shoulder joint makes a more medial pull necessary. To achieve the required change in muscle orientation, the glenoid migrated anteriorly, almost to the level of the scapula blade and the characteristic supraglenoid buttress of the pelycosaurs was lost. The principal complex of protractor muscles included the supracoracoideus and scapulohumeralis anterior which were still aranged as in the pelycosaurs, although perhaps somewhat enlarged. The scapula blade had not yet been invaded by protractor muscles, as occurred in the cynodonts and mammals, and still gave origin to a large deltoideus muscle.

The gorgonopsid elbow joint was not modified from the primitive, pelycosaurian type (p. 54), designed to convert long axis rotation of the humerus into a change in the angle between the radius–ulna and the ground. The radius and ulna were themselves shorter than the humerus, and both these anatomical points indicate that the forelimb continued to operate in a sprawling fashion. The digits of the forefoot were, however, modified (Fig. 41G). One of the phalanges of the third digit and two of the fourth digit were reduced to short discs (Parrington, 1939) and the digits were therefore much nearer to equal in length. This probably indicates that the rate of stepping had increased, so that the feet had to be picked off the ground at the end of each stride more rapidly.

The gorgonopsid hindlimb, in common with other typical therapsids (Kemp, 1978) had evolved a radically new ability. Instead of being restricted to a sprawling mode of action, it could also be used in a much more erect, mammalian fashion as well. This dual-gait is found today in crocodiles (Schaeffer, 1941; Cott, 1961) and large lizards and probably represents the initial stage in improvement of the locomotion. The evidence for the dual-gaited condition of the gorgonopsid hindlimb is derived from the structure of the bones and joints, from the disposition of the main locomotory muscles, and from a number of striking points of similarity to crocodiles.

The gorgonopsid hip joint accommodated both gaits with equal facility. The head of the femur is set off at an angle to the main shaft, and so the femur can extend almost directly forwards while still keeping in articulation with the acetabulum socket of the pelvis. From this position, the femur can either swing backwards in a horizontal plane until it lies transversely (Fig. 42D–F), or else swing back in a near vertical plane until it projects downwards from the acetabulum (Fig. 42G,H). Indeed, it appears to have been able to move in any plane between these two as well. The hip musculature was adapted to move the femur in either mode. The ilio-femoralis muscle was enlarged compared to the pelycosaurs, as indicated by the expanded ilium (Fig. 42A) and by the development of a prominent trochanter major on the back of the femoral head for its insertion (Fig. 42B,C). In the pelycosaurs, this muscle functioned to rotate the femur about its long axis during the power stroke, as well as to elevate it during the recovery phase, and it was well placed to perform the same role during the sprawling gait of the gorgonopsids (Fig. 42D). During the more erect gait, however, the trochanter major lay behind and below the acetabulum, and the ilio-femoralis muscle would have caused retraction of the femur when it contracted (Fig. 42G). This muscle, which was destined eventually to

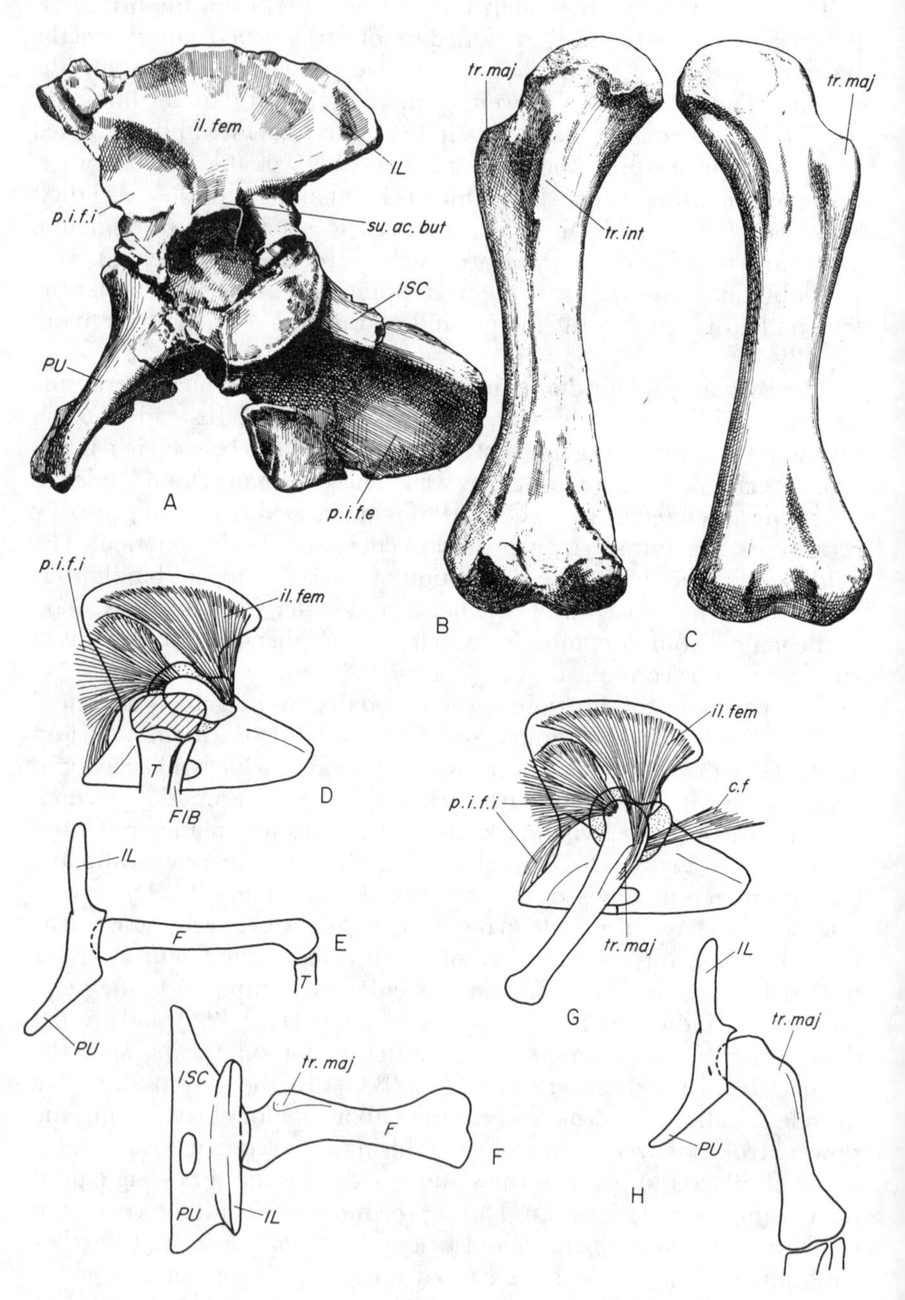
il. fem
IL
p.i.f.i
su. ac. but
ISC
PU
p.i.f.e
A
tr. maj
tr. int
B
tr. maj
C
p.i.f.i
il. fem
T
FIB
D
IL
F
E
T
PU
il. fem
p.i.f.i
c.f
tr. maj
G
IL
tr. maj
PU
H
ISC
tr. maj
F
F
PU
IL

FIG. 42.

become the main retractor muscle of mammalian locomotion, had just adopted this function in gorgonopsids and other therapsids, but still in conjunction with its ancestral function. The protractor muscle of the pelycosaurs was the pubo-ischio-femoralis internus, which originated from the inner side of the pubis and ran over the anterior edge of that bone, then backwards to the femur. The gorgonopsid pubis is reduced in relative size, indicating that the origin of some of the fibres of the pubo-ischio-femoralis internus muscle must have migrated elsewhere. Some presumably originated from the body fascia in the region of the lumbar vertebrae and ribs, and others may have achieved an origin from the ilium itself. This muscle retained its protraction function in gorgonopsids and worked equally well for both the gaits (Fig. 42D,G).

The knee and ankle joints of the gorgonopsids have not been analysed in detail, but what is known of them indicates that they functioned in the same manner as in at least one other therapsid group, the therocephalians (Kemp, 1978). The most remarkable specialisation concerns the ankle (Fig. 43), which has evolved a totally new mobile joint not present in the pelycosaurs, but actually analogous to a similar joint in modern crocodiles (Charig, 1972). More than any other single feature, the ankle supports the dual-gait hypothesis of therapsid locomotion. The two proximal bones of the hindfoot are the astragalus, which articulates with the end of the tibia and the calcaneum which articulates with the fibula (Fig. 43A. cf. Fig. 68). A special pair of facets between these two foot bones allowed them to rotate relative to one another, about a transverse axis. As in the pelycosaurs, the sprawling gait required a relative rotation between the distal end of the femur and the foot, which had to be accommodated by the tibia and fibula bones (Fig. 43C). The lower end of the large tibia rotated on the astragalus, while the upper end of the fibula probably rotated on the femur. It was the manner in which the foot flexed and extended on the lower leg that differed from the pelycosaurs (Fig. 43B). Instead of a simple hinge joint

Fig. 42. Unidentified gorgonopsid specimen (Cambridge University Museum of Zoology No. T.883): A, lateral view of left pelvis. B, ventral view of left femur. C, dorsal view of left femur. D, lateral view of the pelvis and femur during the sprawling gait. E, the same in anterior view. F, the same in dorsal view. G, lateral view of the pelvis and femur during the more erect gait. H, the same in anterior view. (A–C, original drawings by David Nicholls.)

*c.f*, caudi femoralis muscle; *F*, femur; *FIB*, fibula; *IL*, ilium; *il.fem*, ilio-femoralis muscle; *ISC*, ischium; *p.i.f.e*, pubo-ischio-femoralis externus muscle; *p.i.f.i*, pubo-ischio-femoralis internus muscle; *PU*, pubis; *su.ac.but*, supra-acetabular buttress; *T*, tibia; *tr.int*, trochanter internus; *tr.maj*, trochanter major.

Magnification: A–C, *c.* × 0.37.

connecting the astragalus and calcaneum on the one hand with the tibia and fibula on the other, the astragalus remained immoveably attached to the tibia. These latter two behaved as a single unit, and the flexion–extension of the foot occurred by the hinge joint between calcaneum and fibula, plus the rotation joint between calcaneum and astragalus. Thus there was a separation of the various necessary movements of the foot, each movement becoming the specialised function of one particular joint, and no single joint having more than one function. It seems to be a general rule that the more different kinds of movement, or degrees of freedom a single joint posseses, the more the amplitude of

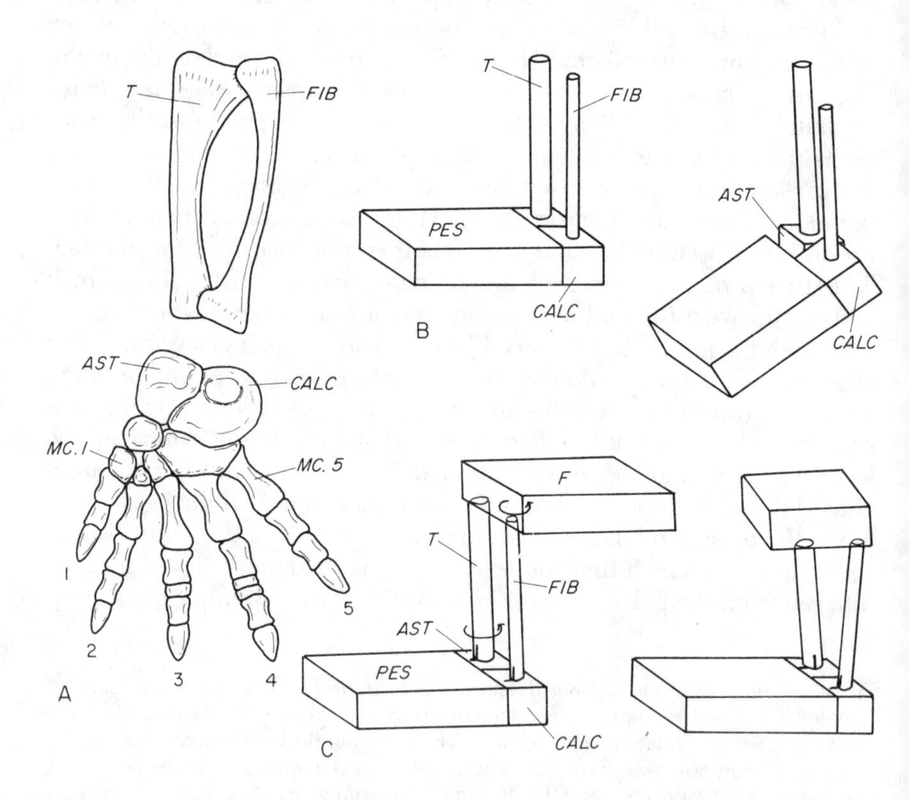


Fig. 43. Gorgonopsid ankle function. A, anterior view of the left tibia and fibula, and dorsal view of the hindfoot. B, flexion and extension of the hindfoot, showing the rotation of the calcaneum on the astragalus. C, rotation of the hindfoot relative to the femur during the sprawling gait, showing that the tibia rotates on the calcaneum while the fibula rotates on the distal end of the femur. (A redrawn after Colbert, 1948.)

*AST*, astragalus; *CALC*, calcaneum; *F*, femur; *FIB*, fibula; *MC*, metacarpal; *T*, tibia.

each of the individual movements is restricted. Conversely, by increasing the total number of moveable joints in a structure, the amplitude of any particular movement can be greater. In the case of the therapsid ankle, the new astragalo–calcaneum joint must have increased the extent and also the power with which the foot could extend on the lower leg. This movement is of greatest significance in the more erect gait. When the femur pointed more or less ventrally at the end of its stride, powerful extension of the foot would have added a significant increment to the total length of the stride and to the total thrust produced per stride. Two other features of the foot are important in this context. First, the foot was now plantigrade, the whole of the sole being in contact with the ground except at the end of the stride. Second, a large lever arm, the tuber calcis, extends posteriorly from the calcaneum, behind the lower end of the fibula. The extensor musculature of the foot inserted into it. For all its apparently highly specialised nature, this type of therapsid ankle joint was a necessary stage in the ultimate evolution of the mammalian ankle.

The axial skeleton of gorgonopsids was less obviously modified than the limbs compared to pelycosaurs. The first two vertebrae, the atlas and the axis (Fig. 44A–D) respectively are more complex, in association with more extensive movements of the head compared to pelycosaurs (Kemp, 1969b). The occipital condyle has widened to a kidney shape, and there was an equivalent widening of the intercentrum of the atlas to accommodate it. A greater rotation of the head about a longitudinal axis was therefore possible. Also dorso-ventral, or nodding, movements were greater because the pair of articulations between the paired neural arches of the atlas and the lateral parts of the occipital condyle respectively were higher up, nearer the level of the centre of the foramen magnum. The tendency to stretch the nerve cord as the head bent downwards was reduced. Movements of the head from side to side were greater, not by virtue of the structure of the atlas-axis complex, but by the development of horizontal zygapophyses between the neck vertebrae, which allowed extensive lateral movements.

Intercentra are still present in the region of the neck, assisting in lateral movements between the vertebrae by behaving as double-sided sockets into which fit the ball-like lower parts of the adjacent vertebral centra. In contrast, no intercentra remain in the rest of the trunk, indicating a reduction, indeed probably virtually a loss of lateral movements. The articulating faces of the trunk zygapophyses are almost vertically aligned, which would prevent all but the smallest of lateral movements between adjacent vertebrae. It is apparent that the facility of lateral undulation of the body, reduced in the sphenacodonts, was virtually absent in the gorgonopsids, and played no significant part in the locomotory mechanism.

There are three sacral ribs supporting the sacrum; the tail, although not well known, appears to have been considerably reduced compared to pelycosaurs.

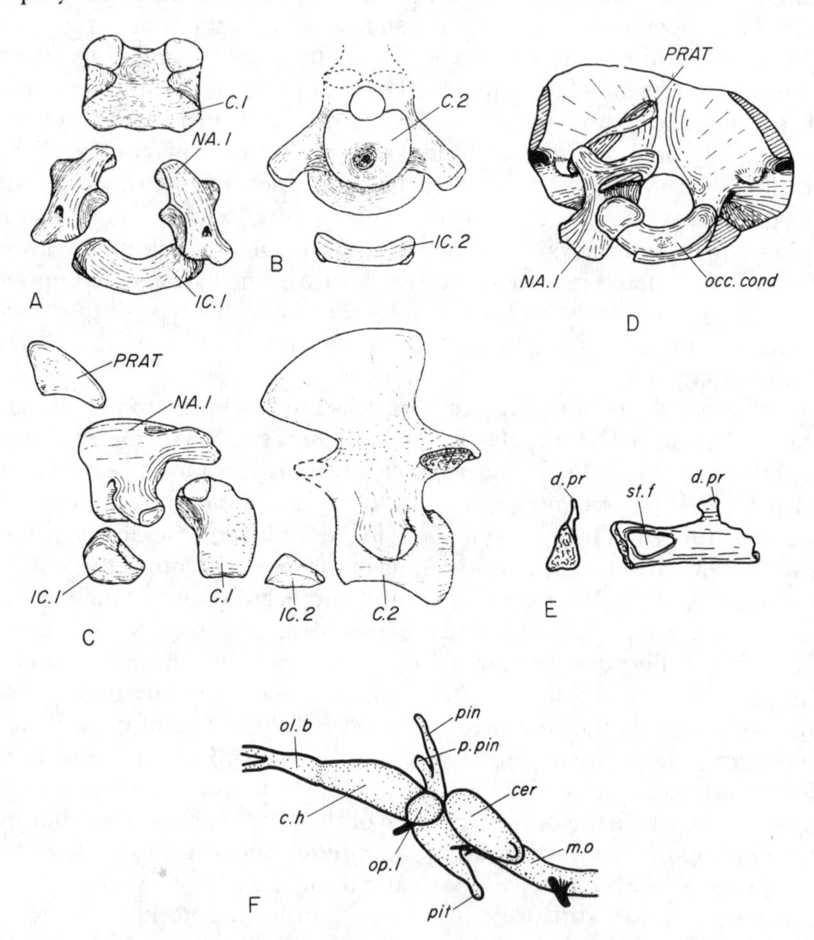


FIG. 44. A, gorgonopsid atlas elements in anterior view. B, gorgonopsid axis elements in anterior view. C, exploded lateral view of the atlas–axis complex. D, part of gorgonopsid occiput with left proatlas and atlas neural arch in posterior view. E, gorgonopsid stapes in proximal and anterior view. F, reconstruction of gorgonopsid brain in lateral view. (A–C from Kemp, 1969b; E and F from Kemp, 1969a.)

*C.1*, atlas centrum; *C.2*, axis centrum; *cer*, cerebellum; *c.h*, cerebral hemisphere; *d.pr*, dorsal process; *IC.1*, atlas intercentrum; *IC.2*, axis intercentrum; *m.o*, medulla oblongata; *NA*, atlas neural arch; *occ.c*, occipital condyle; *ol.b*, olfactorybulb; *op.l*, optic lobe; *pin*, pineal; *pit*, pituitary; *p.pin*, parapineal; *PRAT*, proatlas; *st.f*, stapedial foramen.

To summarise basic therapsid locomotion, as manifested by the gorgonopsids, the inefficiencies associated with lateral undulation of the vertebral column had been abandoned, and the increment of the stride that this had provided was replaced by longer striding limbs. The forelimb worked in a sprawling fashion, with the humerus moving in a horizontal plane. The length of the stride was increased by a less restrictive shoulder joint, mobility of the scapulo-coracoid part of the shoulder girdle, and increase in limb length. The forelimb did not produce a significant part of the locomotory thrust, however. The hindlimb could also work in a primitive, sprawling fashion with the femur horizontal. However, it could also adopt a much more erect gait, with the femur moving in a plane approaching the vertical (parasagittal). By analogy with modern dual-gaited animals, it is probable that the more erect hindlimb gait generated a greater thrust and therefore a higher speed. It was probably necessary for the animal to rest with its body on, or near, the ground in a sprawling stance. The sprawling gait therefore continued to be the most convenient for slow, leisurely walking, and was hence retained.

### *Middle ear*

The gorgonopsid stapes (Fig. 44E) is a relatively lightly built bone, an impression increased by the presence of a large stapedial foramen. Its inner end fits into the large fenestra ovalis, and it extends laterally to lie within a recess in the quadrate bone, although it does not seem to have been firmly attached to the latter. The dorsal process which supports the stapes from the overlying paroccipital process is also delicate (Kemp, 1969a). As with the pelycosaurs, there is an active debate about whether the gorgonopsids had a tympanic membrane. Certainly it is much more reasonable in this case, for the lighter build of the stapes suggests that it might be capable of being activated by a sufficiently large tympanum. On the other hand, there is no clearly indicated site for such a structure. The matter remains unresolved.

### *General biology*

Many of the comments already made about the dinocephalians (p. 96) apply equally to the various primitive carnivorous therapsids discussed here. The bone histology is of an identical, advanced kind and the early members lived under the same environmental conditions in both the cis-Uralian Russian deposits and in the *Tapinocephalus*-zone of the South African Karroo, including therefore temperate regions of the Late

Permian world. The only consistent difference in the context of temperature physiology is the relatively smaller size of these forms compared to the dinocephalians. This could imply that the metabolic rate was higher, to counter the greater rate of heat loss resulting from the greater surface area to volume ratio, although the presence of greater skin insulation is also a possibility. No bony secondary palate was developed in any forms, although the gorgonopsids possessed a deep median channel on the palate, leading from the internal nostrils to the presumed position of the epiglottis (Fig. 36A). If, as seems likely from the anatomy, the channel was floored by a sheet of soft tissue in life, then the air passage would have been isolated from the buccal cavity and breathing could have continued during feeding. On this evidence, it may be assumed that the metabolic rate of gorgonopsids had indeed increased compared to more primitive forms.

As has been seen, the sense of hearing in gorgonopsids may have become more acute, and a tympanic membrane was possibly present. What is more certain is that the sense of olfaction played an important part in the life of these animals. A series of fine ridges in the roof of the very large nasal cavity were probably supports for sheets of cartilage, which hung down within the cavity and increased the surface area of olfactory epithelium (Kemp, 1969a).

The gorgonopsid brain (Fig. 44F) was still long and tubular, and had not increased significantly in relative size over pelycosaurs. The main lobes, including the cerebral hemispheres may have been rather better differentiated however (Kemp, 1969a; Hopson, 1979).

# 8 Anomodonts

THE ANOMODONTS were exclusively herbivorous therapsids, and in terms of numbers were the most successful of all the mammal-like reptiles. The most primitive forms are amongst the earliest known therapsids, occurring in the Russian Zone I deposits. By the final stage of the Late Permian, they dominated terrestrial faunas, accounting for well over 90% of specimens, and represented by many genera. After a brief decline in diversity, although not of relative abundance, a second but lesser adaptive radiation occurred in the Triassic, persisting almost until the close of that period.

## Systematics

### *Venjukovoids*

The relationships of the anomodonts as a whole to the other therapsid groups are not at present clear. They are the group most modified from the primitive therapsid structure, and satisfactory derived characters shared with other therapsid taxa are absent. An extreme hypothesis is that anomodonts are most closely related to the caseid pelycosaurs, and that their therapsid features, such as the reflected lamina of the angular, enlarged temporal fenestra and various postcranial features evolved independently (Olson, 1962; but see Olson, 1971). The similarities between caseids and anomodonts are, however, very superficial; they include, for example, the shortened skull and ventral position of the jaw articulation. They are probably no more than common adaptations of the skull to herbivory in the respective groups. Certainly this view has not found favour. The usual hypothesis is that the anomodonts should also include the dinocephalians (e.g. Romer, 1966). Again the similarities are rather trivial, such as the depressed jaw hinge and a tendency to narrow the intertemporal region of the skull. All that can really be said is that the dicynodonts diverged very early on from the ancestral therapsid stock, and that relationships to other groups cannot be resolved yet (Boonstra, 1971, 1972).

Two similar primitive anomodonts occur in the Russian Late Permian. *Otsheria* is known from a single, fairly complete skull (Fig. 45A–C)

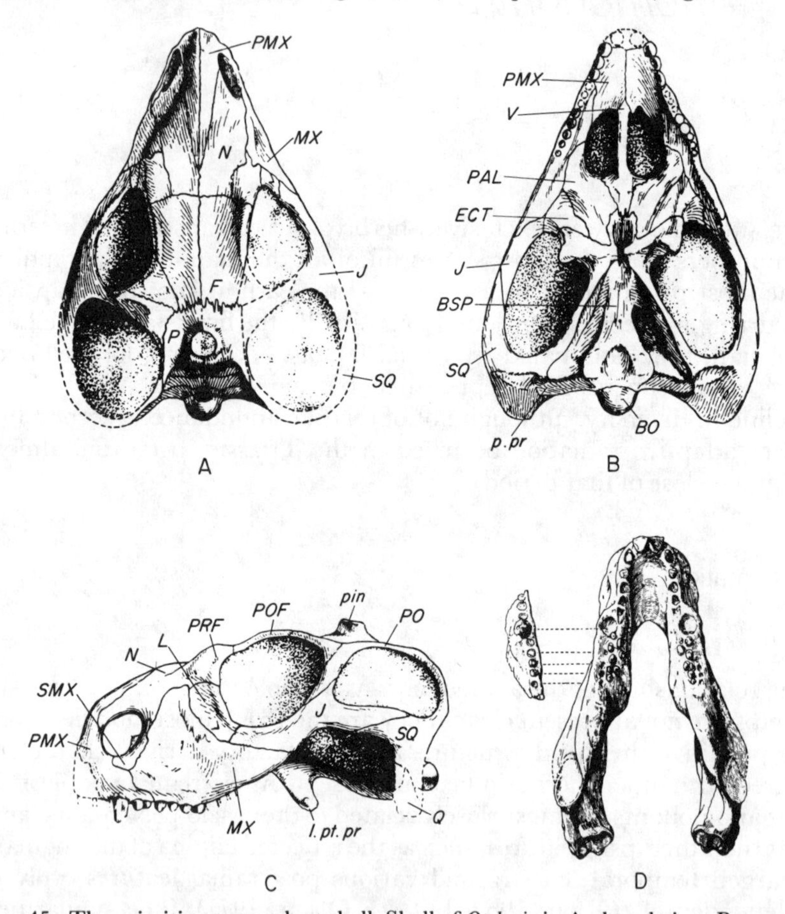


FIG. 45. The primitive anomodont skull. Skull of *Otsheria* in A, dorsal view. B, ventral view. C, lateral view. D, dorsal view of the lower jaws of *Venjukovia*, with a corresponding fragment of the upper jaw. (A–C from Chudinov, 1960; D from Watson, 1948.)

*BO*, basioccipital; *BSP*, basisphenoid; *ECT*, ectopterygoid; *F*, frontal; *J*, jugal; *L*, lachrymal; *l.pt.pr*, lateral proces of the pterygoid; *MX*, maxilla; *N*, nasal; *P*, parietal; *PAL*, palatine; *pin*, pineal foramen; *PMX*, premaxilla; *PO*, postorbital; *POF*, postfrontal; *p.pr*, paroccipital process; *PRF*, prefrontal; *Q*, quadrate; *SMX*, septomaxilla; *SQ*, squamosal; *V*, vomer.

Magnifications: A–C, *c.* × 1.1; D, *c.* × 0.33.

which lacks both the lower jaws and the postcranial skeleton, from the Ocher locality of Zone I (Chudinov, 1960, 1965).

It was a very small therapsid, with a skull length of about 10 cm. Several of the diagnostic therapsid characters are present, including an enlarged temporal fenestra, an anteriorly positioned jaw articulation, and the structure of the basicranial axis. Nevertheless, *Otsheria* was the most modified of all the early therapsids known, for it possessed a number of adaptations for a herbivorous mode of life. The enlargement of the temporal fenestra had occurred by posterior and lateral expansion, producing a shape distinct from that of other therapsids. The intertemporal region of the skull is only slightly narrowed, and there was no invasion of the external surfaces of this region by the jaw musculature. The squamosal extends posteriorly behind the level of the occipital condyle, in a manner reminiscent of the gorgonopsids, and the lower temporal bar, the zygomatic arch, is very high and arch-like. The suspensorium, carrying the quadrate, extends ventrally well below the level of the zygomatic arch. The snout of *Otsheria* is extremely short, and bluntly pointed. The dentition is unfortunately poorly preserved. There were probably four incisors and nine maxillary teeth, all of which were short and stout, and with laterally compressed points. No very distinct canine is present. The palate is also very distinctive, for the palatal processes of the premaxillae extend a considerable distance posteriorly, where they meet a narrow but deep vomer (Fig. 45B). The palatine bones together form a median vaulting of the palate. *Otsheria* is therefore approaching the condition of a secondary palate, with the internal nostrils carried backwards, leaving a broad premaxillary shelf at the front.

The second primitive Russian anomodont is *Venjukovia* (Fig. 45D) which is known from the Copper Sandstones and also the Zone II locality at Isheevo. A skull which is well preserved except for the posterior part as well as several lower jaws have been found (Efremov, 1940; Watson, 1948; Barghusen, 1976). It was a larger animal than *Otsheria*, for the skull is around 20 cm long, but has a similar shape, with the high, posteriorly extended temporal fenestra, very short snout, and triangular shape as seen from above. The lower jaw possesses a characteristically therapsid reflected lamina of the angular, and there is a prominent coronoid eminence but no discrete coronoid process of the dentary. As a whole, the lower jaw is very deep and short. *Venjukovia* differs from *Otsheria* primarily in its highly specialised dentition. There are four or five short, stout incisor teeth in the premaxilla, which are large anteriorly but decrease in size backwards. In the maxilla there are three small teeth followed by a short but very wide canine, and then a row of about five small teeth. The massive dentaries, fused at the symphysis, bear a row of six incisiform teeth, larger at the front and decreasing in size posteriorly but there is no lower canine. The more posterior teeth are small, peg-like, and form an irregular double row. All the teeth are heavily worn. The anterior part of the dentition occluded, lower teeth directly with upper teeth, to form a simple grinding surface. However, the more posterior teeth, including the upper canine, could not have met, since the lower teeth lay too far medial to the uppers. Instead, special horny tooth pads had developed. The dentary forms a broad shelf lateral to the lower teeth on which there are small pits corresponding to the positions of the upper teeth. The latter, including the canine, bit into their respective pits, which in life were no doubt covered by a layer of horn. Similarly the lower post-canines must have bitten against a horny pad covering the palatine bone on the palatal surface.

The close relationship between these two Russian genera is clear, although the differences in the dentition have led to their separation into two families.

### *Dromasaurs*

Four very small specimens from the Beaufort Beds of the South African Karroo constitute the dromasaurs (Fig. 46). None of the material has been restudied since the early descriptions by Broom (Broom, 1932), and the relationships of the group are not at all clear, not even whether they are, in fact, therapsids. At least one of them, *Galeops*, is from the *Tapinocephalus*-zone, the lowest of the fossil bearing part of the Karroo, but the horizons from which the others, *Galechirus* and *Galepus*, come is uncertain. Therapsid characters described by Broom are the relatively anterior position of the jaw articulation, the possible presence of a reflected lamina of the angular and the possible absence of a large quadratojugal bone. The scapula blade of the shoulder girdle is narrow, and the structure of the glenoid appears to be of the therapsid rather than the pelycosaurian type. The limbs are very slender and relatively long. On the other hand, there is no sign of an enlarged canine, for the dentition consists of a simple row of equal-sized little teeth (*Galeops* has no teeth, but this may be an artifact of preservation). The temporal fenestra is much smaller than the orbit, and the ilium only very slightly expanded. The postcranial skeleton also possesses several unexpectedly primitive features, especially notochordal vertebrae and ventral ribs or gastralia.

On balance, it seems likely that dromasaurs will prove to be very primitive and possibly juvenile therapsids. Of the known therapsid groups, they have certain similarities to anomodonts. The jaw articulation is carried ventrally below the tooth row in a dicynodont fashion and there is no coronoid process of the dentary. A preparietal bone is possibly present in at least *Galepus*. Boonstra (1972) included the dromasaurs as a family of the Venjukovoidea, although almost equally reasonable is Watson and Romer's (1956) erection of a separate group for them within the Anomodontia. Haughton and Brink (1954) actually classified them as edaphosaur pelycosaurs, which seems extremely doubtful.

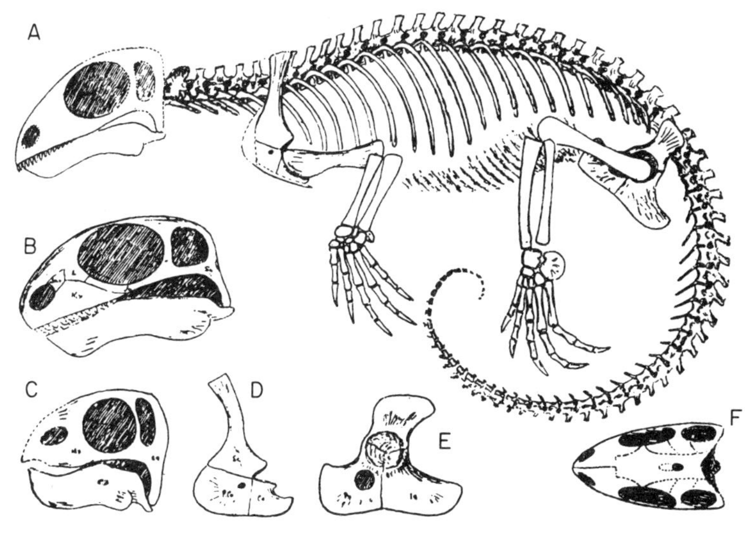


Fig. 46. Dromasaurs. A, lateral view of the skeleton of *Galechirus*. B, lateral view of the skull of *Galepus*. C, lateral view of the skull of *Galeops*. D, scapulo-coracoid of *Galeops*. E, pelvis of *Galepus*. F, dorsal view of the skull of *Galepus*. (From Broom, 1932.)

## *Dicynodonts*

The dicynodonts form a very coherent group with a number of unique specialisations not found in any other therapsids. They are presumed to have descended from an *Otsheria*-like ancestor. The earliest known dicynodont is *Eodicynodon* (Fig. 47), which is claimed to be from rocks underlying the *Tapinocephalus*-zone of the South African Karroo (Barry, 1974). This horizon is termed the Ecca, and is otherwise devoid of fossil reptiles. However, the geology is complex and the boundary between the Ecca and the *Tapinocephalus*-zone is not precisely delineated, so that there is some suggestion that *Eodicynodon* is actually from the younger horizon (Kitching, 1977). Even if true, *Eodicynodon* is from the very base of the *Tapinocephalus*-zone and remains the earliest South African therapsid. At any event, dicynodonts were to prove the longest living of all the main mammal-like reptile groups, with a span of some 55 million years.

The characteristic features of the dicynodonts include extreme modifications to the skull for the herbivorous diet. All the incisors and the lower canine teeth are completely absent, but the upper canines are retained in most forms as a pair of enlarged tusks, round in section and possibly used as much for interspecific display as for grubbing for food. The postcanine dentition is reduced to, at most, a few small teeth lying medial to the tusks, and along the inner edge of the dentary. In most forms, even these few postcanines are absent. Instead, the jaws must have been covered with a hard, horny layer rather as in turtles. The premaxillae form a sharp ridge around the anterior edge of the snout. The dentaries which are fused together at the front are normally beak-like, providing a structure which bit within the margin of the premaxillae.

The temporal fenestra is very large and extends posteriorly well behind the level of the occipital condyle. The dicynodont squamosal bone is of a unique shape, for it extends very far downwards and forwards, providing a large new area over its front face for the origin of much of the external adductor muscle. It has also expanded posteriorly for the origin of further muscle fibres. The zygomatic arch remains high above the level of the lower jaw, to provide yet another muscle-bearing surface. The jaw articulation is placed far antero-ventrally, and the jaws themselves are very short.

Like most of the gorgonopsids, but unlike other therapsids, a preparietal bone is present, lying in the midline anterior to the pineal foramen. It is presumed to have evolved in parallel in the two groups.

The postcranial skeleton of dicynodonts is, like the skull, recognisably therapsid but somewhat distinct from that of any of the other therapsid groups known. The body was bulky, as would be expected of a herbivore, and the limbs relatively short and stout with the claws at the ends of the digits flattened. The anterior edge of the scapula is reflected outwards as a distinct acromion for the attachment of the clavicle, and the procoracoid is excluded from the glenoid fossa (Fig. 59A). In the hindlimb, the ilium tends to be extended further anteriorly than in other therapsid groups, and the femur is distinctive for it lacks any trace of the internal trochanter on its ventral side. Otherwise, however, the femur is therapsid-like with an S-shaped curvature, inturned head and well-developed trochanter major.

Until recently, the Late Permian dicynodonts have been divided into two families, the Endothiodontidae for the forms retaining post-canine teeth, and the Dicynodontidae for those which have lost these teeth. However, it has become clear that neither of these two families is monophyletic, and the loss of the already insignificant postcanine dentition occurred independently in several separate lineages (Cluver and Hotton, 1981; King and Cluver, in preparation). The earliest dicynodont, *Eodicynodon* (Fig. 47), is also the most primitive for it possesses several ancestral features, modified in all the other dicynodonts known. Neither the premaxillae nor the vomers are fused, the lateral pterygoid process of the palate is still well developed, the secondary palate formed from the premaxillae is short, and the stapes is still perforated by a stapedial foramen. The dentition is also primitive, for there are still about four small postcanine teeth behind the canine tusk. Two of these occur in the typical dicynodont position postero-medial to the tusk, but the other two still lie on the lateral margin of the

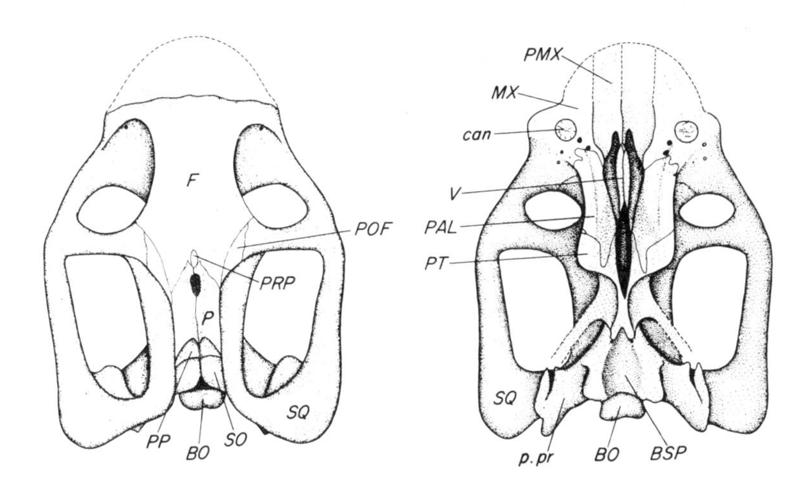


Fig. 47. The skull of *Eodicynodon*. A, dorsal view. B, ventral view. (From Barry, 1974.)
*BO*, basioccipital; *BSP*, basisphenoid; *can*, canine; *F*, frontal; *MX*, maxilla; *P*, parietal; *PAL*, palatine; *PMX*, premaxilla; *POF*, postfrontal; *PP*, postparietal; *p.pr*, paroccipital process; *PRP*, preparietal; *PT*, pterygoid; *SO*, supraoccipital; *SQ*, squamosal; *V*, vomer.
Magnification *c.* × 0.63.

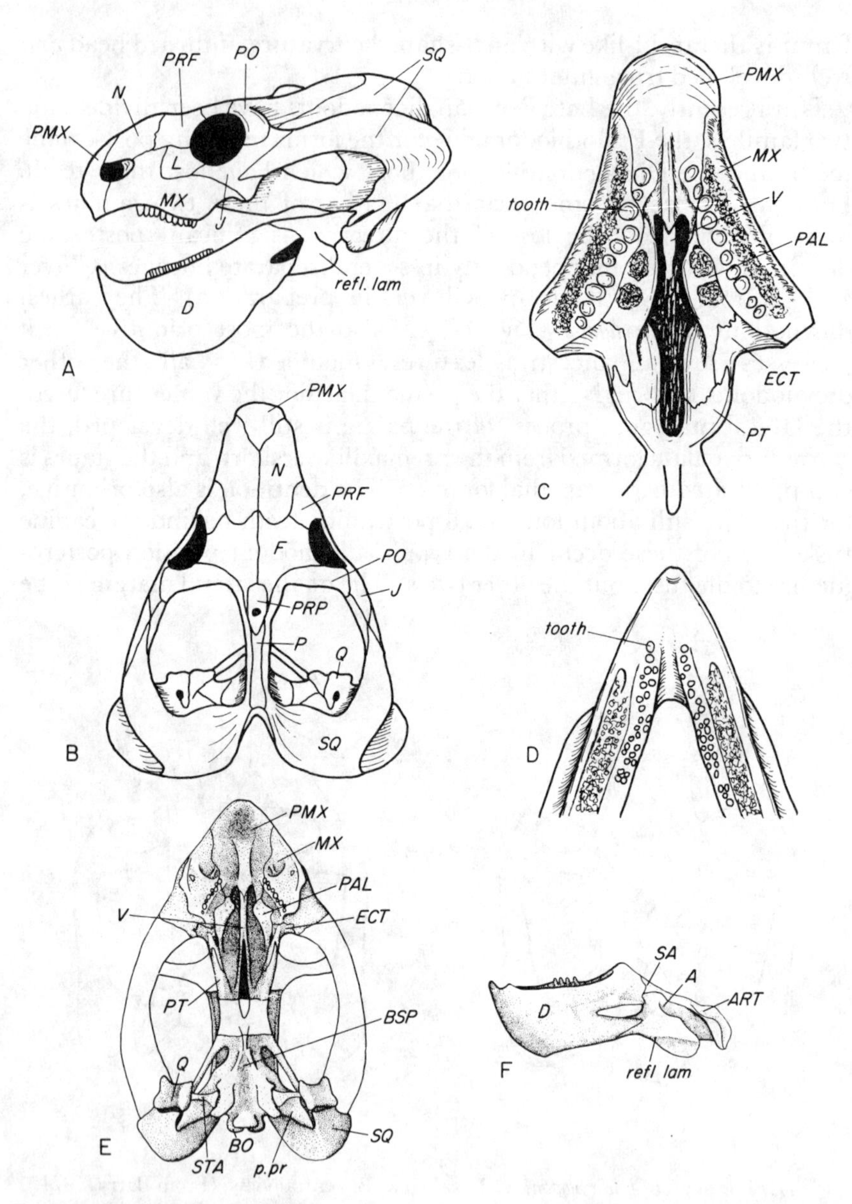


FIG. 48. Skull of *Endothiodon*. A, lateral view. B, dorsal view. C, ventral view of the palate and upper dentition. D, dorsal view of the front part of the lower jaw. Skull of *Chelydontops*: E, ventral view. F, lateral view of the lower jaw. (A–D redrawn after Cox, 1964; E–F from Cluver, 1975.)

*A*, angular; *ART*, articular; *BO*, basioccipital; *BSP*, basisphenoid; *D*, dentary; *ECT*, ectopterygoid; *F*, frontal; *J*, jugal; *L*, lachrymal; *MX*, maxilla; *N*, nasal. *P*, parietal; *PAL*, maxilla (King and Cluver, in preparation). In most other respects, *Eodicynodon* is a typical, small dicynodont, and it must have lain close to the origin of all the later forms.

*Endothiodon* (Fig. 48A–D) is a large form that combines typical dicynodont features with a highly atypical dentition quite unlike that of other dicynodonts. It is known from the *Cistecephalus*-zone of South Africa and from Brazil, while a possible relative occurs in deposits of the same age in India (Kutty, 1972). The skull reaches a length of over 50 cm in some individuals, and it has a high, narrow intertemporal region separating the large, characteristically dicynodont temporal fenestrae (Cox, 1964). The lower jaw is distinguished by the very deep anterior part attached to a relatively slender postdentary part. The upturned front tip of the fused dentaries worked against the broad, deeply concave plate formed from the fused premaxillae. Neither anterior, nor canine teeth are present, but the dentition is well developed posteriorly. There are about ten long, slender upper teeth, forming a single row which does not occlude with the lower teeth, but with a longitudinal groove on the dentary lateral to the lower teeth. The lower dentition consists of similar but somewhat smaller teeth forming two or three irregular rows. Only the more lateral ones were functional, biting against a horny pad presumably present in life on the large, heavily rugose palatine, internal to the upper tooth row. The more medial lower teeth are young replacement teeth just erupting and preparing to move laterally in the jaw as the functional teeth were discarded. There was evidently another area of vascularised bone in *Endothiodon*, on the maxilla lateral to the upper tooth row. Possibly it marks the attachment of a muscular, food-retaining cheek (Cluver, 1975).

*Chelydontops* (Fig. 48E–F) is an earlier and smaller form possibly related to *Endothiodon* (Cluver, 1975). It is known at present only from two rather poorly preserved skulls from the *Tapinocephalus*-zone. Like *Endothiodon*, the anterior part of the dentaries forms a high, sharp blade that worked against the vaulted premaxillae, while the cheek teeth are well developed compared to those of typical dicynodonts, with nine upper teeth and ten lower teeth in the type specimen. There are, however, important differences between the two genera, such as the broad intertemporal region of *Chelydontops*, and the presence of a small

palatine; *PMX*, premaxilla; *PO*, postorbital; *p.pr*, paroccipital process; *PRF*, prefrontal; *PRP*, preparietal; *PT*, pterygoid; *Q*, quadrate; *refl.lam*, reflected lamina of the angular; *SA*, surangular; *SQ*, squamosal; *STA*, stapes; *V*, vomer.

Magnifications: A–B, *c.* × 0.12; C–F, *c.* × 0.3.

coronoid process on the dentary, a feature unique amongst dicynodonts.

Endothiodontids have lost the ancestral features of *Eodicynodon*, in common with all the other dicynodonts, and therefore they presumably arose after the divergence of the latter form (King and Cluver, in preparation). There is, however, some similarity between the dentition of *Endothiodon* and that of venjukovoids, for both have a well-developed set of cheek teeth working against horny pads, the upper teeth lateral to the lower teeth. It seems likely, therefore, that this pattern of dentition is primitive for dicynodonts, and that *Eodicynodon* reduced its dentition independently of the reduction that occurred in the line leading to the rest of the dicynodonts (Fig. 57).

All the rest of the dicynodonts form a very well-defined group. Like *Endothiodon*, they have fused premaxillae and vomers, extensive secondary palate, reduced lateral pterygoid processes of the palate and (with one exception) imperforate stapes. Unlike *Endothiodon*, the dentition is further reduced, and even those forms which still have postcanine teeth have reduced them in both size and number. The upper teeth come to lie entirely postero-medial to the canine tusk, while the lower teeth form a short row well towards the front, on the medial side of the dentary. The confusion that has existed about the evolution and classification of the dicynodonts arose largely from the fact that often only skull roofs were described in any detail. Enormous numbers of forms were made into species of *Dicynodon*, and even those attributed to other genera tended to be diagnosed on very doubtful characters. In recent years it has become clear that the various dicynodonts differ most significantly from one another in details of their palates and lower jaws, features which reflect differences in feeding adaptations. As more and more specimens have been studied in terms of these features, a much more coherent outline of their interrelationships is becoming apparent (Fig. 57). Although the Late Permian radiation of dicynodonts is by far the best documented in the *Tapinocephalus*, *Cistecephalus* and *Daptocephalus*-zones of the Karroo of southern Africa, specimens from contemporaneous deposits in Russia, Scotland, China and India indicate that the phenomenon was world-wide.

Three distinct groups of dicynodonts (Fig. 57) can be recognised in the Late Permian (King and Cluver, in preparation). The first group is known from *Kingoria* (Fig. 49), which occurs in the *Cistecephalus*-zone (Cox, 1959) and is one of the forms that independently lost the postcanine teeth and has also lost the postfrontal bone. The horny biting surfaces of *Kingoria* differed considerably from those of the other dicynodonts, for the secondary palate formed from the premaxillae is smooth

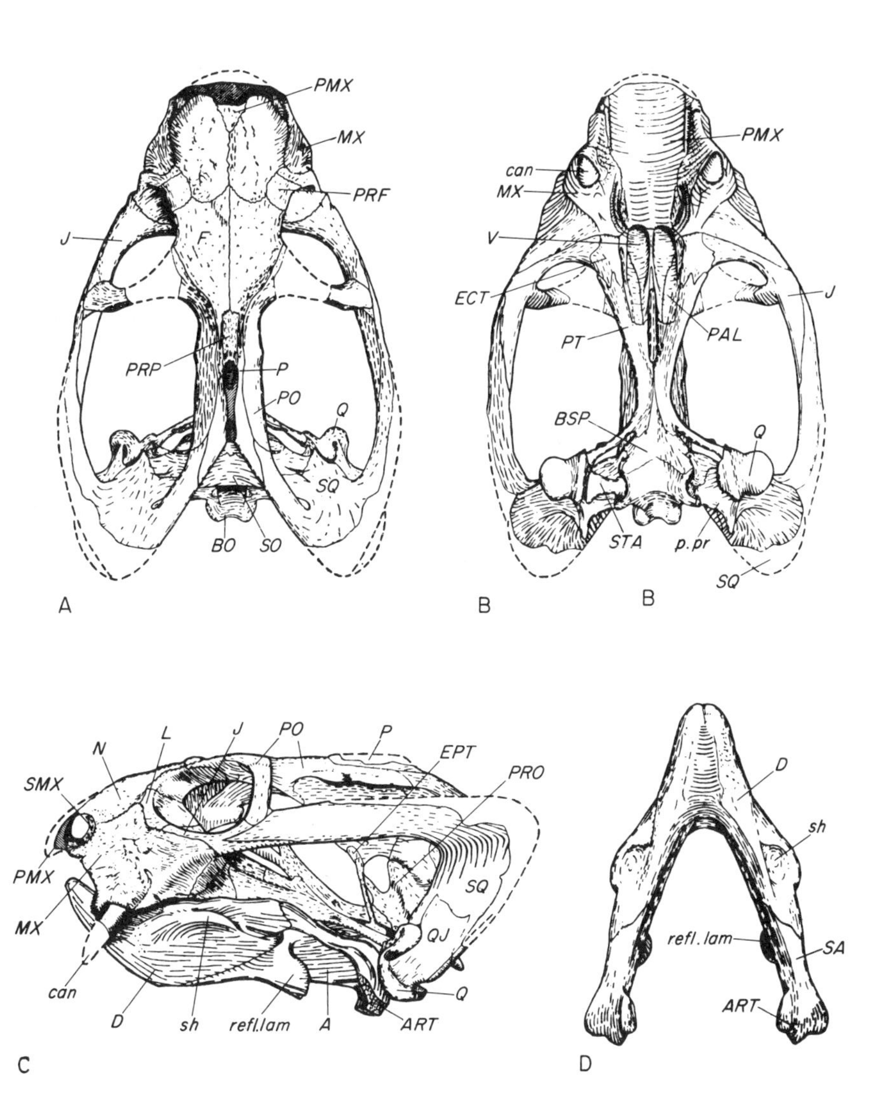


FIG. 49. The skull of *Kingoria*. A, dorsal view. B, ventral view. C, lateral view. D, dorsal view of the lower jaw. (From Cox, 1959.)

*A*, angular; *ART*, articular; *BO*, basioccipital; *BSP*, basisphenoid; *can*, canine; *D*, dentary; *ECT*, ectopterygoid; *EPT*, epipterygoid; *F*, frontal; *J*, jugal; *L*, lachrymal; *MX*, maxilla; *N*, nasal; *P*, parietal; *PAL*, palatine; *PMX*, premaxilla; *PO*, postorbital; *p.pr*, paroccipital process; *PRF*, prefrontal; *PRO*, prootic; *PT*, pterygoid; *Q*, quadrate; *QJ*, quadratojugal; *refl.lam*, reflected lamina of the angular; *sh*, shelf; *SMX*, septomaxilla; *SQ*, squamosal; *SO*, supraoccipital; *STA*, stapes; *V*, vomer.

Magnification *c.* × 0.44.

and flat and the actual horny layer appears to have been restricted to the lateral margins and the adjacent maxillae. Instead of forming the typical sharp cutting edges, these lateral edges are rounded. The corresponding anterior parts of the dentaries are also smoothly rounded (Fig. 49D). Thus *Kingoria* must have possessed rounded horny edges to its jaws in life, adapted for crushing rather than cutting and suggesting a diet of fruits rather than stems, leaves and roots. *Kombuisia* (Hotton, 1974) is closely related to *Kingoria*, and represents one of the few dicynodont lineages to have survived the Permo-Triassic boundary. It is a small, rare member of the Lower Triassic *Cynognathus*-zone fauna of South Africa.

The second group of Permian dicynodonts share with *Kingoria* a reduction of the palatine bone, and these two groups may therefore be related to one another. However, this second group is characterised by the development of a notch in the margin of the maxilla, immediately in front of the tusk (or the caniniform process in those forms lacking the tusk). This modification probably indicates that the anterior part of the feeding apparatus was more important than the posterior part, while the notch itself may be a device to facilitate cutting of slender, but tough stems and roots. Primitive members of this group include *Robertia* (King, 1981a), and *Emydops* which has lost its postcanine teeth, both from the *Tapinocephalus*-zone (Fig. 50A–C). However, much the most widespread form is *Diictodon* (Fig. 50D,E), which occurs from the *Tapinocephalus*-zone right through to the *Daptocephalus*-zone in South Africa, and also in China (Cluver and Hotton, 1981). *Diictodon* is an advanced dicynodont, for as well as loss of the teeth, the intertemporal region of the skull has become very narrow, indicating further development of the jaw musculature. A particularly interesting member of this group is *Cistecephalus* (Fig. 51A–G) and its close allies *Cistecephaloides* (Cluver, 1974a, 1978) and *Kawingasaurus* (Cox, 1972), which were small forms highly adapted for a fossorial mode of life. The skull is short and broad, with a wide intertemporal roof forming a battering-ram like structure. The occiput is wide to provide a large area for the neck musculature, and the neck vertebrae are immobile. The forelimb is distinctly mole-like, with a short, stout humerus, powerful muscle processes, and highly ossified articulatory condyles. There is a well-developed olecranon process on the ulna, and a broad forefoot with enlarged digits. *Myosaurus* (Cluver, 1974b) is another short, broad-headed form (Fig. 51H–I) and is structurally intermediate between a typical member of this group such as *Emydops* and the extremely specialised *Cistecephalus*. *Myosaurus* is another of the few dicynodonts to have persisted into the Triassic, although it is restricted to the *Lystrosaurus*-zone at the base of that period. (See note 2 added in proof, p. 351.)

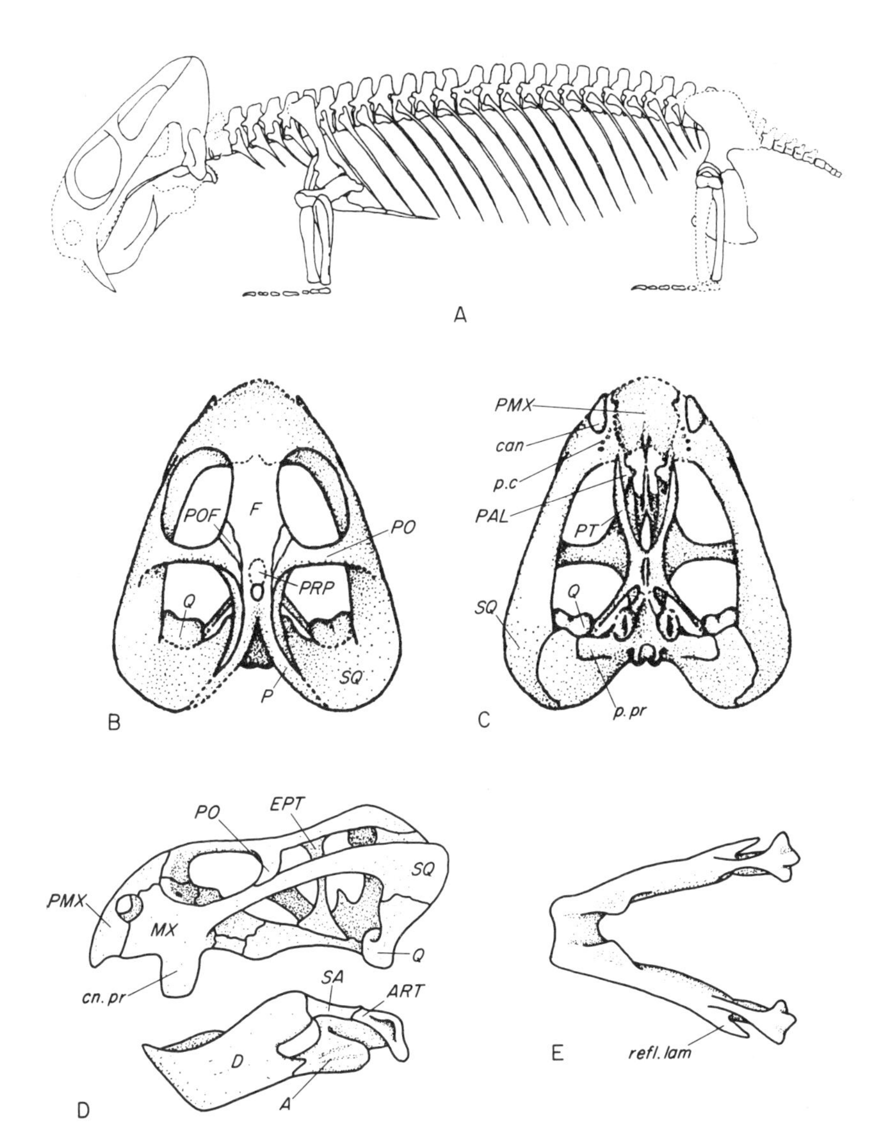


FIG. 50. *Robertia*: A, skeleton. B, skull in dorsal view. C, skull in ventral view. D, lateral view of skull of *Diictodon*. E, dorsal view of lower jaw of *Diictodon*. (A–C from King, 1981a; D and E redrawn from Cluver and Hotton, 1981.)

*A*, angular; *ART*, articular; *can*, canine; *cn.pr*, caniniform process; *D*, dentary; *EPT*, epipterygoid; *F*, frontal; *MX*, maxilla; *P*, parietal; *PAL*, palatine; *p.c*, postcanine. *PMX*, premaxilla; *PO*, postorbital; *POF*, postfrontal; *p.pr*, paroccipital process; *PRP*, preparietal; *PT*, pterygoid; *Q*, quadrate; *refl.lam*, reflected lamina of the angular; *SA*, surangular; *SQ*, squamosal.

Magnifications A, *c.* × 0.27; B and C, *c.* × 0.69; D and E, *c.* × 0.5.

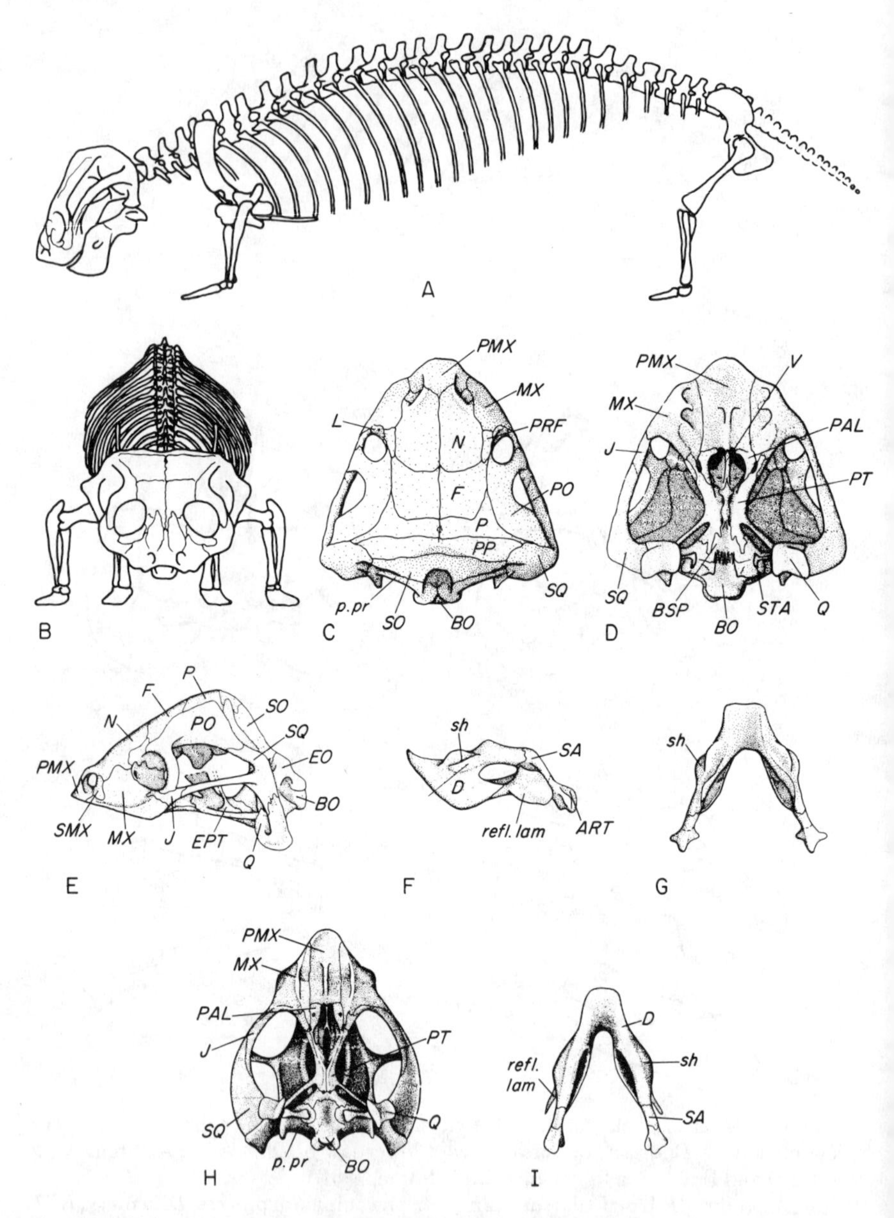


FIG. 51. *Cistecephalus* and its relations. A, skeleton of *Cistecephalus* in lateral view. B, the same in anterior view. Skull of *Cistecephaloides*: C, dorsal view; D, ventral view; E, lateral view; F, lateral view of lower jaw; G, dorsal view of lower jaw. H, ventral view of skull of *Myosaurus*. I, dorsal view of the lower jaw of *Myosaurus*. (A and B from Cluver, 1978; C–G from Cluver, 1974b; H and I from Cluver, 1974a.)

The third of the Permian dicynodont groups produced an even greater radiation than those just discussed, a radiation which included all the larger forms, and which also led ultimately to the second phase of dicynodont evolution during the Triassic. Members of this group are distinguished by a larger, leaf-shaped palatine bone and the elaboration of the dorsal surface of the dentaries, with a longitudinal groove for the support of the horny covering. Thus in this group the more posterior part of the feeding apparatus appears to have become more important than in other dicynodonts.

*Pristerodon* (Fig. 52A–D) is a primitive member of the group from the *Tapinocephalus*-zone. Postcanine teeth are still present, the uppers set at an angle across the jaw. A well-developed dorsal shelf occupies the lateral side of the dentary for insertion of part of the jaw-closing musculature. The adaptive radiation which occurred from a *Pristerodon*-like form includes such familiar types as *Oudenodon* (Keyser, 1975) which has lost the tusks but retained a relatively wide intertemporal region of the skull and narrow snout (Fig. 52E–G). *Rhachiocephalus* included giant dicynodonts, with a skull length of 40–50 cm. They are also tuskless and narrow-snouted, and probably evolved from an *Oudenodon*-like ancestor. They were possibly browsing animals, gaining their food from the higher placed stems and leaves of plants, for in skull shape they are analogous to the Black Rhino. In contrast, there is a wider snouted form, *Aulacocephalodon*, in which the front edge of the snout formed a transverse cutting edge. They were probably the grazers of the time, feeding at ground level, and they too reached a large size, with the skull exceeding 50 cm in some species.

The genus *Dicynodon* has been much misused as a taxonomic "catch-all"; Haughton and Brink (1954) for example listed no less than 111 species. However, many of these are now known to be synonyms of one another, and others appertain to different genera such as *Diictodon* and *Oudenodon*. An acceptable modern definition of the genus, based on the narrow intertemporal region, short basicranial axis of the skull, and the large palatine bones limits it to a much more acceptable status (Cluver and Hotton, 1981). Even so, *Dicynodon* (Fig. 53) remains a very common

*A*, angular; *ART*, articular; *BO*, basioccipital; *BSP*, basisphenoid; *D*, dentary; *ECT*, ectopterygoid; *EO*, exoccipital; *EPT*, epipterygoid; F, frontal; *J*, jugal; *L*, lachrymal; *MX*, maxilla; *N*, nasal; *P*, parietal; *PAL*, palatine; *PMX*, premaxilla; *PO*, postorbital; *PP*, postparietal; *p.pr*, paroccipital process; *PRF*, prefrontal; *PT*, pterygoid; *Q*, quadrate; *QJ*, quadratojugal; *refl.lam*, reflected lamina of the angular; *SA*, surangular; *sh*, dentary shelf; *SMX*, septomaxilla; *SO*, supraoccipital; *SQ*, squamosal; *STA*, stapes; *V*, vomer.

Magnifications: A and B, *c*. × 0.33; C–G, *c*. × 0.5; H and I, *c*. × 0.65.

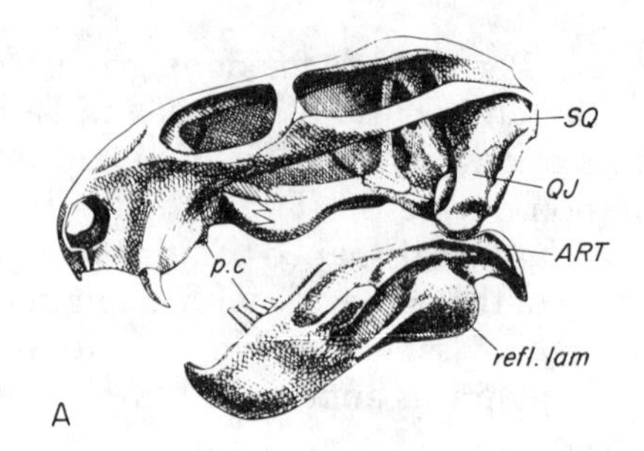


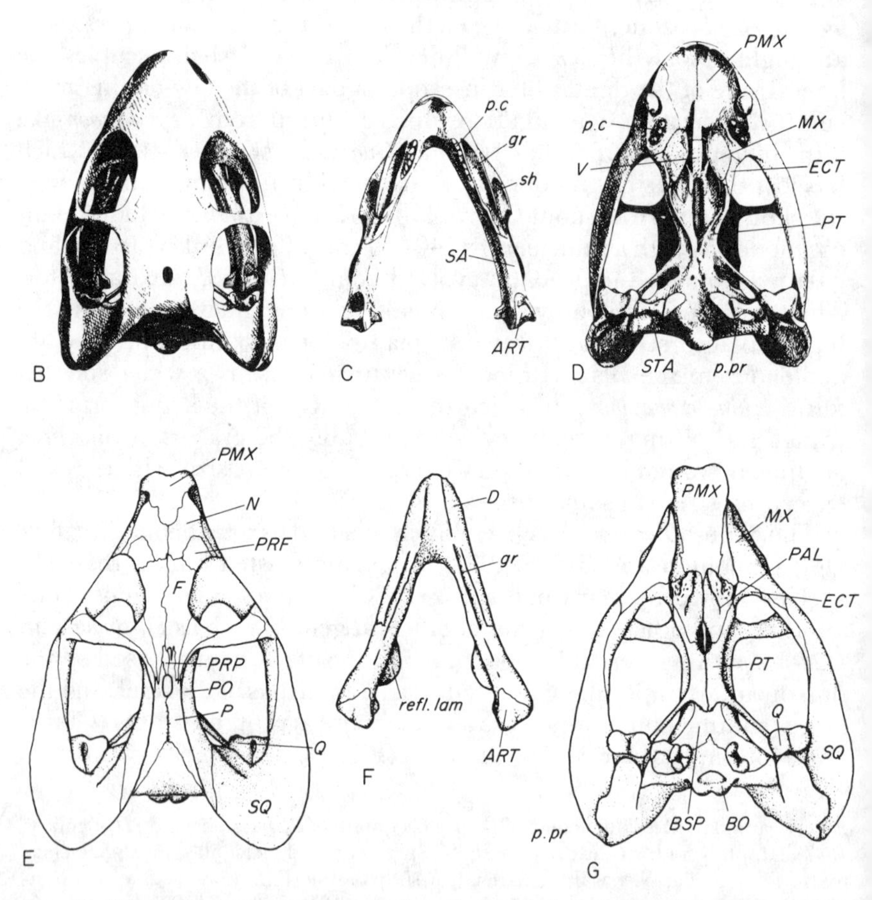


FIG. 52. Skull of *Pristerodon*: A, lateral view; B, dorsal view; C, dorsal view of lower jaw; D, ventral view. Skull of *Oudenodon* in E, dorsal view; F, dorsal view of lower jaw; G, ventral view. (A–D from Crompton and Hotton, 1967, described as *Emydops* (*Postilla* Peabody Mus. Nat. Hist., Yale University); E–F, original drawings by Gillian King of a specimen of *O.baini* in the Oxford University Museum No. TSK 67.)

form towards the end of the Permian, with species of very variable size and detailed structure. *Daptocephalus* (Ewer, 1961) is closely related to *Dicynodon*, and may even be synonymous.

The end of the *Daptocephalus*-zone of the Karroo marks the close of the Permian period, and with very few exceptions such as *Myosaurus* and *Kombuisia* this great adaptive radiation of dicynodonts died out. However, one important line (Fig. 57) persisted because in the lower part of the Triassic a very specialised dicynodont called *Lystrosaurus* was abundant. *Lystrosaurus* had a world-wide distribution; specimens have been recovered not only from regions as far apart as South Africa, India and Antarctica, which were formerly part of the same supercontinent Gondwanaland, but also from China, Indo-China and Russia in the northern hemisphere. Indeed, *Lystrosaurus* may well be regarded as the most successful single mammal-like reptile of all time.

The skull of *Lystrosaurus* (Cluver, 1971) resembled that of a typical dicynodont except for dramatic changes in the proportions (Fig. 54). The snout is greatly deepened and the temporal region shortened. At the same time, the base of the skull shortened, with the result that the snout bends downwards anterior to the orbits. The nostrils have remained at a high level, as have the orbits. The feeding structures, on the other hand, have been carried far ventrally. These features have led to the conclusion that *Lystrosaurus* was an amphibious animal, feeding in shallow water much as modern hippopotami do. The failure of the feet to ossify fully, unknown in other dicynodonts, supports this conclusion. The postcranial skeleton is otherwise little modified, still having the rotund form, short but stout limbs, and reduced tail of the Permian dicynodonts. The origin of the line leading from Permian forms to *Lystrosaurus* lay close to the genus *Dicynodon* and its relatives such as *Daptocephalus*. Comparison of *Lystrosaurus* with, for example *Dicynodon trigonocephalus* shows a series of similarities not generally found in Permian dicynodonts (King, 1981b). Apart from the obvious deepness and downturning of the snout, shortening of the temporal fenestra and reduction of the length of the interpterygoid vacuity, both forms possess a labial fossa and reduced epipterygoid and ectopterygoid. The charac-

*ART*, articular; *BO*, basioccipital; *BSP*, basisphenoid; *can*, canine; *D*, dentary; *ECT*, ectopterygoid; *F*, frontal; *gr*, groove; *L*, lachrymal; *MX*, maxilla; *N*, nasal; *P*, parietal; *PAL*, palatine; *p.c*, postcanine tooth; *PMX*, premaxilla; *PO*, postorbital; *p.pr*, paroccipital process; *PRP*, preparietal; *PT*, pterygoid; *Q*, quadrate; *QJ*, quadratojugal; *refl.lam*, reflected lamina of the angular; *SA*, surangular; *sh*, shelf on dentary; *STA*, stapes.

Magnifications: A–D, *c.* × 0.61; E–G, *c.* × 0.35.

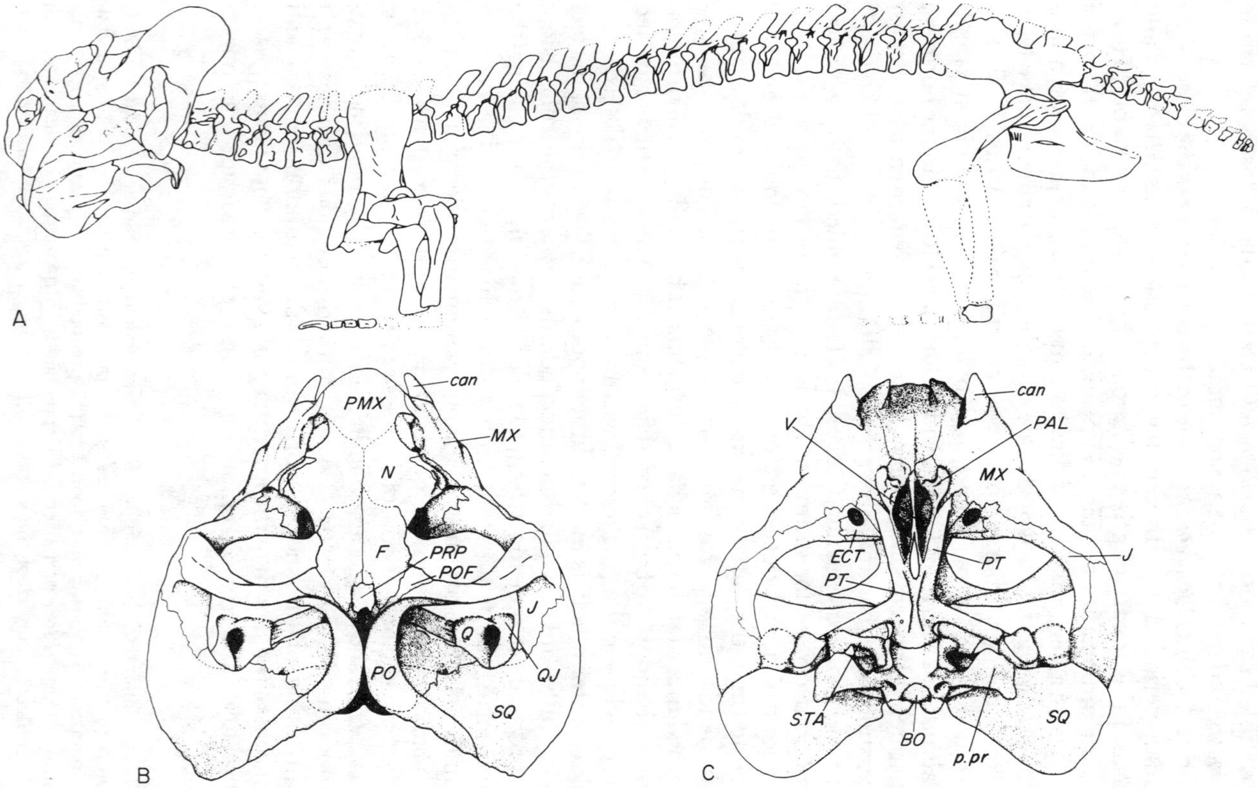

FIG. 53.

teristic contact between the pterygoid and maxilla bones of *Lystosaurus* is not matched in *Dicynodon trigonocephalus*, but is present in *Daptocephalus*, for example.

Whilst *Lystrosaurus* is restricted to this very narrow time zone at the base of the Triassic, the same line from which it diverged continued, producing a range of large Triassic forms, (Keyser and Cruickshank, 1979) which did not finally become extinct until the Upper Triassic. All these forms are broadly similar to one another, and they share several features with *Lystrosaurus*, showing that together they are a monophyletic group. The relatively short basicranial axis and reduction of the interpterygoid vacuity, contact between the pterygoid and maxilla, reduction of the ectopterygoid, and the presence of labial fossae all support this view. However the main Triassic forms were modified in different ways to *Lystrosaurus*. They tended to enlarge, with skull lengths of 25–50 cm, and the snout is relatively long. Various more detailed similarities demonstrate the close relationship between these Triassic forms, such as the posterior extension of the reflected lamina of the angular, which in some cases actually reaches the articular condyle, the development of an anterior process, bearing a dorsal process, of the epipterygoid foot, and the great reduction or even loss of the ectopterygoid and postfrontal bones. The postcranial skeleton is similar to that of the Permian dicynodonts, but is very much more massive, as would be expected of large animals (Fig. 55). The trunk is short and barrel-shaped, and the limb bones stout. A peculiar feature of most of them is the separate ossification of the olecranon process of the ulna. The pelvis is particularly remarkable, for the ilium has a huge anterior expansion, and the pubis is very much reduced. Cruickshank (1978) has made the interesting suggestion that the Triassic radiation of dicynodonts involved adaptation to the habit of feeding on the higher parts of the plants. He sees the animals rearing up to grasp branches with their well-developed claws and pulling them down, or even using their large heads to push small trees over. The structure of the pelvis certainly supports the first proposal, for the ilium would be well placed for the

FIG. 53. *Dicynodon*. A, skeleton in lateral view. B, skull in dorsal view. C, skull in ventral view. (From King, 1981b.)

*BO*, basioccipital; *can*, canine; *ECT*, ectopterygoid; *F*, frontal; *J*, jugal; *MX*, maxilla; *N*, nasal; *PAL*, palatine; *PMX*, premaxilla; *p.pr*, paroccipital process; *PO*, postorbital; *POF*, postfrontal; *PRP*, preparietal; *PT*, pterygoid; *Q*, quadrate; *QJ*, quadratojugal; *SQ*, squamosal; *STA*, stapes.

Magnifications: A, *c.* × 0.14; B and C, *c.* × 0.26.

attachment of muscles tending to raise the front end of the animal off the ground.

All the known forms can be included in a single family Kannemeyeriidae (Keyser and Cruickshank, 1979), although some authors (e.g. Cox,

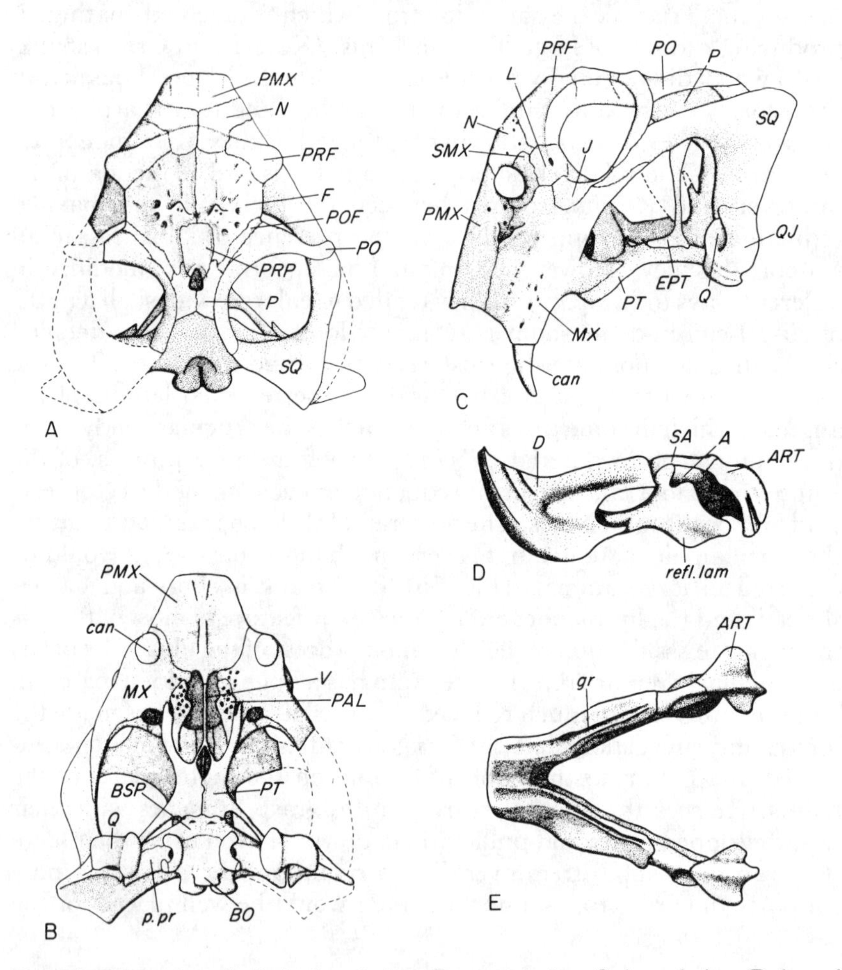


FIG. 54. *Lystrosaurus* skull. A, dorsal view. B, ventral view. C, lateral view. D, lateral view of lower jaw. E, dorsal view of lower jaw. (From Cluver, 1971.)

*A*, angular; *ART*, articular; *BO*, basioccipital; *BSP*, basisphenoid; *can*, canine; *D*, dentary; *EPT*, epipterygoid; *F*, frontal; *gr*. groove; *J*, jugal; *L*, lachrymal; *MX*, maxilla; *N*, nasal; *P*, parietal; *PAL*, palatine; *PMX*, premaxilla; *PO*, postorbital; *POF*, postfrontal; *p.pr*, paroccipital process; *PRF*, prefrontal; *PRP*, preparietal; *PT*, pterygoid; *Q*, quadrate; *QJ*, quadratojugal; *refl.lam*, reflected lamina of the angular; *SA*, surangular; *SMX*, septomaxilla; *SQ*, squamosal; *V*, vomer.

Magnification *c*. × 0.42.

1965) have preferred to separate the more distinctive genera as separate families. *Kannemeyeria* itself (Figs 55 and 56A) is common in the Lower Triassic *Cynognathus*-zone of the South African Karroo, and very similar forms have also been found in the Russian Lower Triassic. The snout is characteristically narrow and pointed, and the intertemporal region is produced as a high, narrow parietal crest. Well-developed tusks are present. A second kind, characterised by a wide, square-cut snout, and lower parietal crest is represented by such forms as *Dinodontosaurus* (Cox, 1965) from the Middle Triassic of South America (Fig. 56B), and *Tetragonias* (Cruickshank, 1967) from the more or less contemporaneous Manda Beds of Tanzania. Other related forms are also known from the Lower Triassic rocks of Shansi in China, and from Russia. During the Middle to Upper Triassic a final group of advanced forms evolved, losing or at least reducing their tusks. *Stahleckeria* and *Ischigualasta* (Fig. 56C) are from South America, while the last surviving of all the dicynodonts is *Placerias* (Camp, 1956), which is found in the Upper Triassic of Arizona (Fig. 56D).

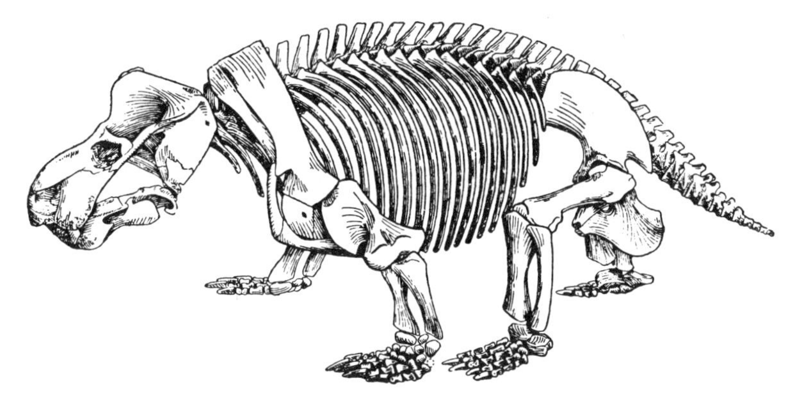

FIG. 55. Skeleton of *Kannemeyeria*. (From Pearson, 1924.) Magnification *c.* × 0.05.

## Functional anatomy

### *Feeding mechanism*

Yet another way of modifying the primitive therapsid feeding mechanism occurred in the anomodonts, as an adaptation for a herbivorous diet, and possibly also for consuming soft invertebrates. The tendency to replace the teeth by heavily keratinised, horny epithelium was

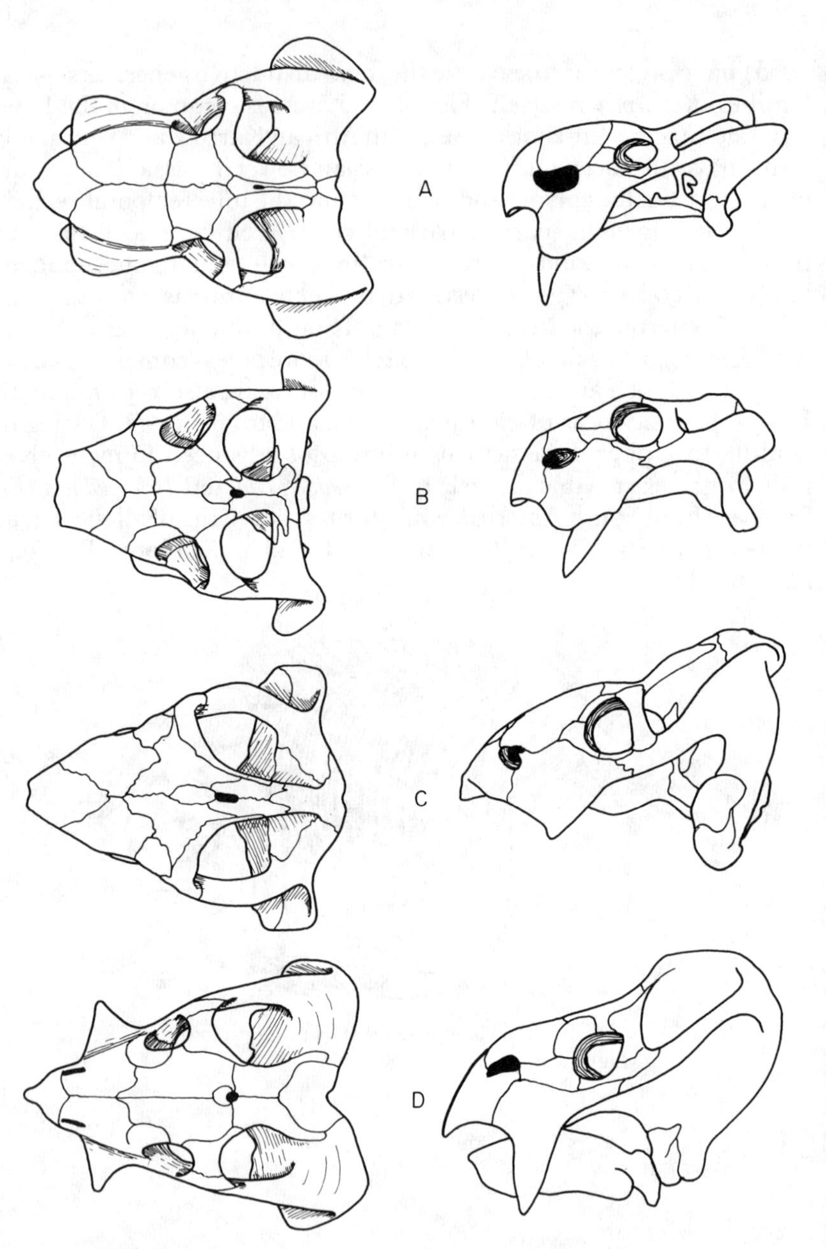


FIG. 56. Triassic dicynodont skulls. A. *Kannemeyeria*. B, *Dinodontosaurus*. C, *Ischigualasto*. D, *Placerias*. (A redrawn after Keyser and Cruickshank, 1980; B–D redrawn after Cox, 1965.)

already manifested in the primitive form *Venjukovia*, where the post-canine teeth must have acted against opposing horny pads on the palate and lateral side of the dentary. In the more highly evolved dicynodonts,

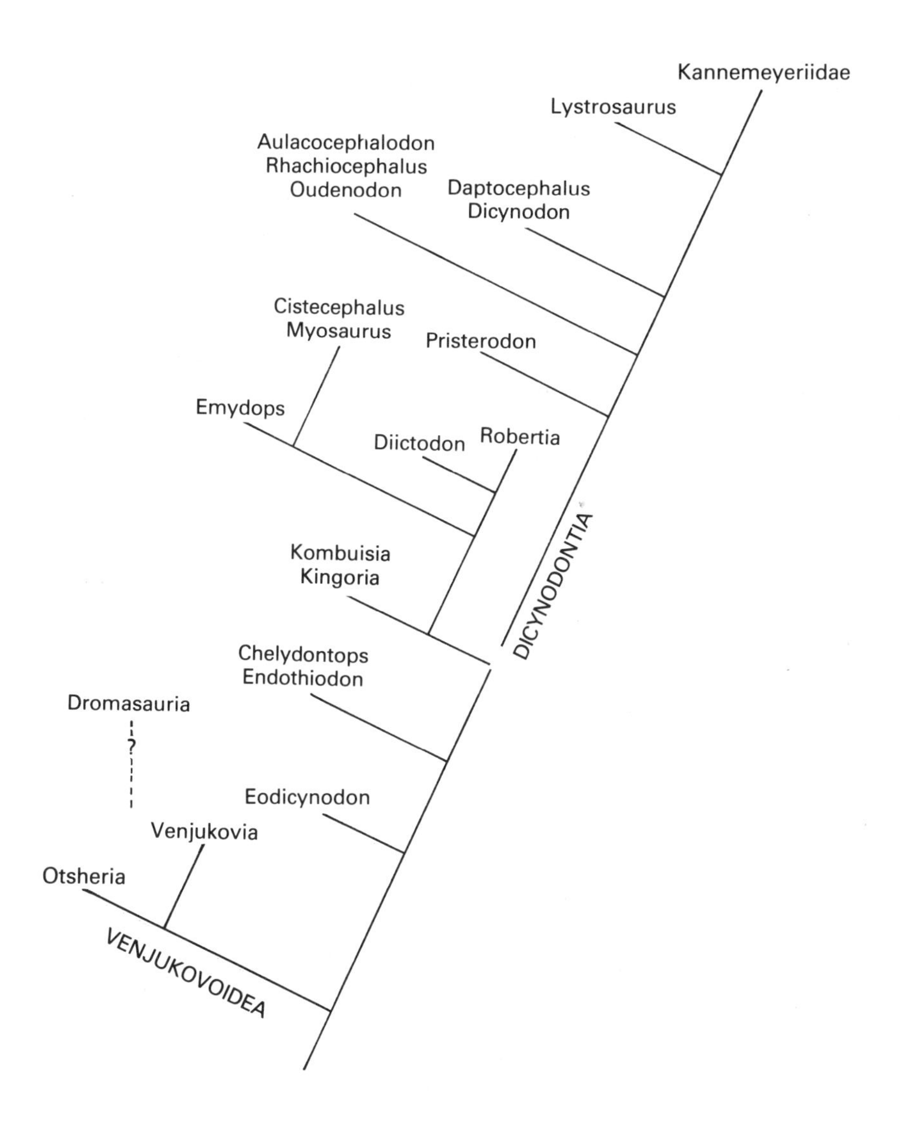


FIG. 57. Phylogeny of the anomodonts.

horn completely replaced the incisors and eventually the postcanines as well in several lines. The evidence for the presence of horn is the nature of the bony surface, where a mass of tiny nutritive foramina and some larger ones indicates that structures requiring a high degree of vascularisation were present. Exactly similar features occur in the turtles and birds, two other vertebrate groups which abandoned teeth in favour of horn. The advantages of horn over teeth are that larger surface areas for grinding, or continuous, sharp edges for cutting are possible, and the structure can be replaced continuously and evenly by growth as it wears away. This avoids the necessary interruption to parts of the triturating surfaces that must occur with a tooth replacement mechanism.

The exact shape of the horny plates cannot be known, but the form of the supporting bone probably matches it approximately. The distinction between a generally food-gathering area anterior to the canines and a food-processing area behind them, which occurs between incisor teeth and postcanine teeth in more primitive animals, has been maintained in the dicynodonts. The front part of the snout and lower jaws of typical dicynodonts indicate that a large beak was present, probably reminiscent of a parrot's beak. The edges were sharp and the upper beak bit against the outer sides of the lower beak. The extent to which the horn of the upper beak spread inwards, over the surface of the secondary palate to form a platform against which the upturned anterior tip of the lower beak could work seems to have varied among the different genera, being extensive in *Dicynodon*, for example, but very limited in *Kingoria*. The anterior beak functioned to cut off suitable sized pieces of vegetation from the plants by a shearing or slicing action. The posterior, triturating surfaces consisted of a more or less flat plate centred around the palatine bone, against which a sharp blade carried on the dentary acted. In primitive forms, the lower blade was still formed from postcanine teeth, which are arranged as a short row of compressed teeth with posterior serrations. In the more advanced forms, these teeth were lost, to be replaced by a blade of horn, presumably of a similar overall shape to the tooth blade. It became more firmly attached in some genera by the appearance of a deep trough on the dorsal surface of the dentary, for example in *Dicynodon* and *Oudenodon*. Functionally, the postcanine triturating surfaces must have acted like a knife working against a wooden board, slicing up the food finely before swallowing.

The manner in which the lower jaw moved in order to achieve these effects is indicated by the specialised nature of the jaw articulation (Fig. 58E,F), which permited extensive antero-posterior movements of the jaw, as well as simple orthal rotation. The quadrate condyle is divided into two parts, lateral and medial, by a deep groove which runs longi-

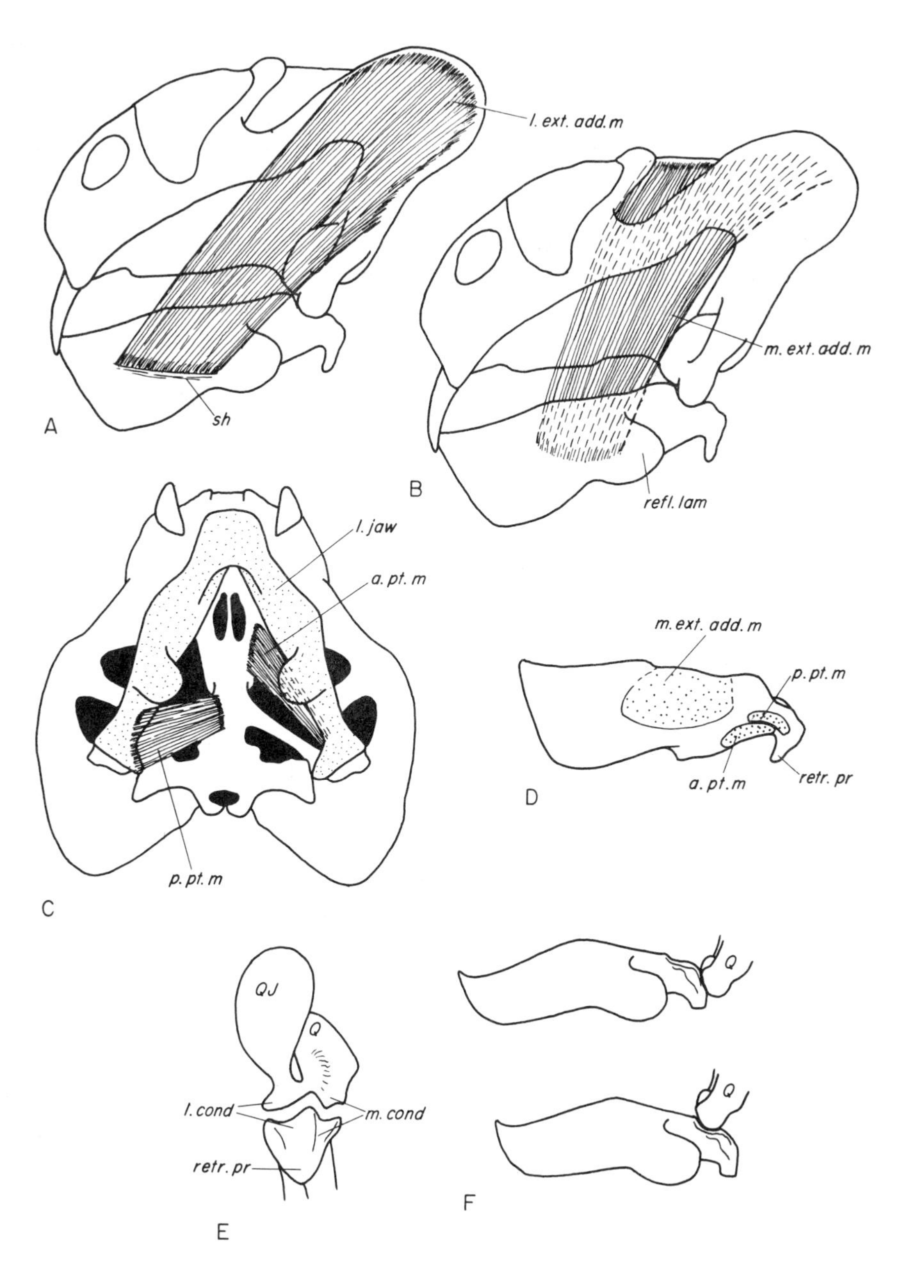


FIG. 58. Jaw functioning in dicynodonts. A, lateral view of the skull of *Dicynodon* showing superficial external adductor musculature. B, the same but showing the deep external adductor musculature. C, ventral view of *Dicynodon* showing internal adductor musculature. D, medial view of lower jaw of *Dicynodon*. E, posterior view of the jaw

tudinally. It corresponds to the articular condyle which is similarly double. The small medial condyle is separated by a high ridge from the large lateral part. Because of this groove and ridge arrangement no medial shift of the articular across the quadrate is possible. The large lateral condyle of the quadrate is gently convex from side to side and from front to back, while the lateral articular condyle of the lower jaw is gently concave from side to side. However, from the front to the back it is strongly convex, so that the anterior part faces more or less dorsally, but the posterior part faces backwards. There is no restraint therefore on the articular and hence the lower jaw moving backwards and forwards, as well as opening and closing upon the quadrate condyle (Watson, 1948; Crompton and Hotton, 1967). The overall action of the lower jaw was protraction, or a forward shift while the jaws were opened, followed by closure of the jaws in the protracted state. The anterior biting surfaces met for cutting off a piece of vegetation. Shredding of the food by the posterior biting surfaces was achieved by retracting the lower jaw while in a closed position, thus drawing the lower blade backwards across the upper plate. A rapid sequence of cutting cycles using the postcanine plates would have produced a shredding up of the food, preparatory to swallowing.

The reason for believing that the main power stroke was retraction rather than protraction is that the greatest bulk of the adductor jaw musculature pulled backwards upon the lower jaw (Crompton and Hotton, 1967). The size of the jaw muscles (Fig. 58A–C) is indicated by the enormous size of the temporal fenestra. There were two major components of the external adductor muscle. A more internal part, the medial external adductor, arose from the medial edge of the temporal fenestra, even invading the external surface of the intertemporal bar in advanced forms such as *Dicynodon* and the kannemeyeriids. This origin spread round to the anterior-facing surface of the squamosal at the back of the skull, and also presumably to the undersurface of a temporal aponeurosis covering the fenestra. This part of the adductor musculature was directly comparable to the whole external adductor of more primitive mammal-like reptiles but a second part, the lateral

articulation. F, lateral view of the lower jaw and quadrate in the protracted (upper) and retracted (lower) positions. (A–E redrawn after King, 1981b; F redrawn after Crompton and Hotton, 1967.)

*a.pt.m*, anterior pterygoideus muscle; *l.ext.add.m*, lateral part of external adductor muscle; *l.cond*, lateral condyle; *l.jaw*, lower jaw; *m.cond*, medial condyle; *m.ext.add.m*, medial part of external adductor muscle; *p.pt.m*, posterior pterygoideus muscle; *Q*, quadrate; *QJ*, quadratojugal; *refl.lam*, reflected lamina of the angular; *retr.pr*, retroarticular process; *sh*, dentary shelf.

external adductor, was a new development. It was associated with the characteristic ventral extension of the squamosal, lateral to and below the temporal fenestra. As well as carrying the jaw articulation ventrally, this process also provided a large new anteriorly facing surface for muscle attachment, and the area was continued anteriorly along the zygomatic arch itself, as far as the postorbital bar. The zygomatic arch is bowed upwards and flattened dorso-ventrally to create a broad surface that faces directly towards the lower jaw. The insertion of the external adductor muscle complex was on the dorsal surface of the dentary behind the level of the dentary tooth plate, and the dorsal surface of the surangular. In certain advanced forms such as *Dicynodon* and *Kingoria*, a definite lateral shelf develops on the dentary for the insertion of the lateral part of the muscle alongside the posterior tooth plate. No coronoid process is present in dicynodonts except the aberrant *Chelydontops* (Cluver, 1975), but a rugose patch of bone at the postero-dorsal tip of the dentary probably marks a point of tendinous attachment of the medial part of the muscle. No doubt the insertion of the medial part also extended onto the medial side of the postdentary bones as well, as in more primitive therapsids (King, 1981b).

Less prominent, but equally important in the functioning of the lower jaw was the internal adductor complex, consisting of an anterior pterygoideus muscle and posterior pterygoideus muscle, equivalent to these muscles in modern reptiles. The pterygoideus complex (Fig. 58C) attached to the palatal surfaces of the posterior part of the maxilla, the ectopterygoid and the pterygoid. It ran backwards, downwards and outwards to an insertion at the back of the lower jaw. Part of it inserted on the inner surface, part to the ventral surface of the median articular condyle, part probably to the downturned retroarticular process, and part wrapped around the ventral edge of the jaw to attach to the lateral surface of the post-dentary bones, behind the reflected lamina of the angular.

The function of the reflected lamina of the angular is as problematical in dicynodonts as in other groups. It differs considerably from the lamina of gorgonopsids, for there is no ridge on the lateral surface, and in well-preserved specimens it is seen to be large, extending ventrally and medially well below the level of the jaw. The outer surface is finely striated, ridges radiating postero-ventrally from the centre of the lamina. The complex masticatory system of dicynodonts makes it certain that there must have been a very well-developed tongue, activated by the various muscles of the animal's throat. The size and complexity of the reflected lamina may be related to this, as King, (1981b) supposes, especially as there is a large hyoid bone preserved between the two jaw rami in certain specimens (e.g. Barry, 1968). Exactly which muscles might be involved and their exact orientations is not known. Possibilities include an intermandibularis, connecting the two rami of the lower jaw, a mylohyoideus running to the ossified hyoid apparatus and a branchiomandibularis also related to the hyoid. Another suggested function for the reflected lamina is the insertion of a muscle which runs antero-dorsally. Parrington (1955) suggested that the forerunner of the characteristic mammalian superficial masseter muscle originally attached to the lamina in all therapsids. This view is no longer widely accepted in the case of therapsids generally, but may be true of certain advanced dicynodonts (King, 1981b), where a possible attachment of the muscle to the skull occurs on the maxilla, just behind the position of the canine tusk. Finally, the probability that jaw opening musculature also attached to part of the reflected lamina, as already suggested for other therapsids (p. 91), exists in the case of dicynodonts as well.

The way in which the jaw muscles acted in order to move the jaws in the manner required for effective use of the tooth plates is clear. Protraction was the role of the pterygoideus complex, which attached to the jaw in such a way that it pulled forwards as well as upwards. Conversely, the enormous external adductor complex produced a powerful backward force on the jaw, as well as closing it. Therefore, these two muscle complexes between them were capable of producing the antero-posterior oscillations of the jaw, synchronously with the opening and closing. Detailed variations of the system no doubt occurred in different forms. Cluver (1974a) for example, has shown that in the case of the specialised wide-skulled form *Cistecephalus*, lateral movement of the lower jaw was also possible, and King (1981b) believes that in at least one species of *Dicynodon*, active biting could occur during the protraction stroke as well as the retraction stroke of the jaw.

The evolution of this specialised feeding mechanism had a profound effect on the overall architecture of the skull (King, 1981b). The advanced carnivorous therapsids increased the moment arm of the external adductor muscle by developing a coronoid process of the dentary. In the dicynodonts, however, a quite different method of increasing the moment arm was adopted. The area of insertion of the muscle on the jaw shifted forwards, which would normally be expected to interfere with the orbits and posterior part of the palate. To avoid this, the orbits moved relatively forwards themselves, resulting in the short snout and long postorbital regions, whilst the palate became very narrow, involving eventually the loss of the lateral pterygoid processes. Forward shift of the insertion of the muscles would also tend to reduce the angle at which they ran from the jaw, but this angle is maintained by lowering the position of the jaw articulation. Thus the great ventral extension of the squamosal below the temporal fenestra occurred, incidentally providing a new area for the origin of external adductor muscle fibres. The muscle retained its length because of the long temporal fenestra, and it did not restrict the gape of the jaws. It also maintained a large posterior component of force, thereby creating the condition necessary for the development of the characteristic retraction power stroke of feeding.

In terms of these evolutionary changes, *Venjukovia* and *Otsheria* had already achieved the essential dicynodont structures. In the case of *Venjukovia* (Barghusen, 1976), the snout is short and the temporal fenestra long and relatively high up. The new, lateral division of the external adductor muscle was already differentiated, originating from the central extension of the squamosal and at least the posterior half of the zygomatic arch. The pterygoideus muscle complex was also well developed and associated with a narrowing of the posterior part of the palate and reduction of the lateral pterygoid processes. Despite its dental specialisations, there is little doubt that a form such as *Venjukovia* represents an early manifestation of the characteristic remodelling of the skull in the dicynodonts.

## *Locomotion*

The structure of the dicynodont postcranial skeleton resembles that of the gorgonopsids (p. 53) in general, although there are a few specialisations indicating that the mode of locomotion had been slightly modified from a basic therapsid type. The limbs are relatively short, with the lower leg still shorter than the humerus/femur, at least in the few forms which have been adequately described, such as *Kannemeyeria* (Pearson, 1924), *Dicynodon* (King, 1981b), and *Cistecephalus* (Cluver, 1978).

The shoulder joint is anatomically and functionally as described for gorgonopsids, and the main difference in the forelimb concerns the arrangement of the supracoracoideous muscle, the principal limb protractor (Fig. 59C). The acromion process of the scapula to which the outer end of the clavicle attaches is turned outwards, so creating a large notch beneath it. This notch leads from the outer surface of the procoracoid and lower part of the scapula, in an antero-dorsal direction, to a fossa occupying the anterior part of the inner face of the scapula. In primitive therapsids, including the gorgonopsids, the supracoracoideus is restricted to the lateral surface of the scapulo-coracoid, but in dicynodonts it has expanded via the notch to occupy the inner face of the scapula as well (Cox, 1959; King, 1981b). As will be seen (p. 244) a similar modification occurred in the cynodonts, giving rise to a muscle equivalent to the supraspinatus of the mammals. Functionally, the extension of the supracoracoideus in this way probably increased the extent to which the humerus could protract and retract, thus increasing the length of the stride. Otherwise, the forelimb functioned in a sprawling manner similar to that of other therapsids.

The hindlimb is even more distinctive in dicynodonts. The ilium is expanded forwards as a great plate of bone, and the posterior part is

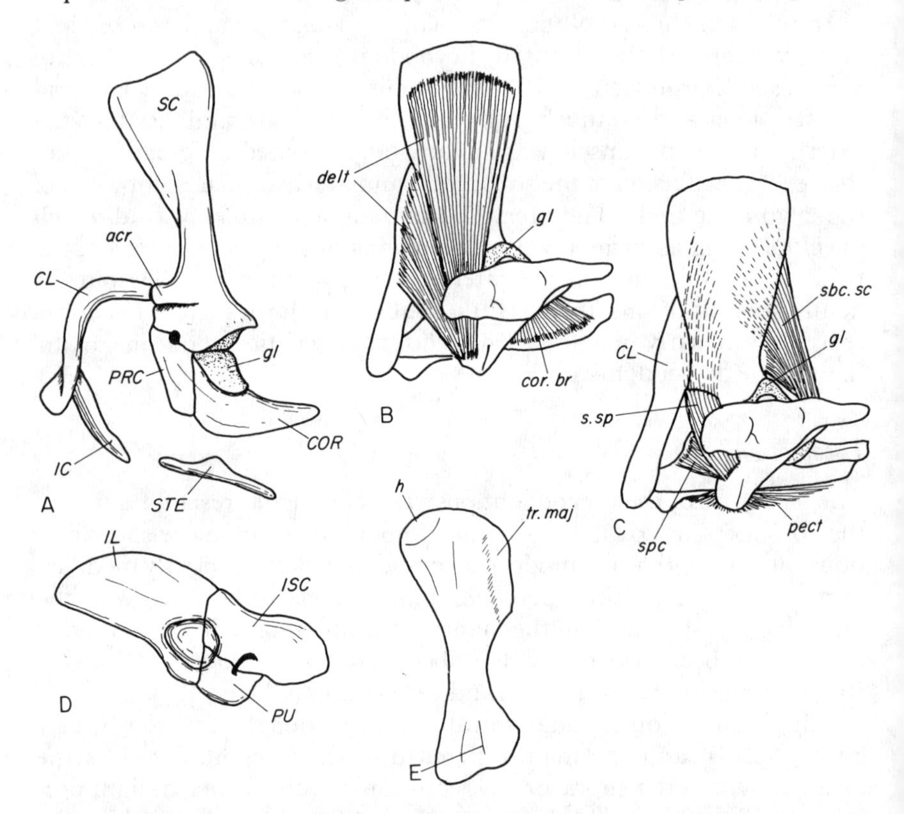


Fig. 59. A, shoulder girdle of the dicynodont *Kingoria* in lateral view. B, lateral view of the pectoral girdle and humerus of *Dicynodon* showing the superficial musculature. C, the same showing the deeper shoulder musculature. D, the pelvis of *Kingoria* in lateral view. E, dorsal view of the left femur of *Dicynodon*. (A and D redrawn after Cox, 1959; B, C and E redrawn after King, 1981b.)

*acr*, acromion; *cor.br*, coraco-brachialis muscle; *CL*, clavicle; *COR*, coracoid; *delt*, deltoideus muscle; *gl*, glenoid; *h*, head of femur; *IC*, interclavicle; *IL*, ilium; *ISC*, ischium; *pect*, pectoralis muscle; *PRC*, procoracoid; *PU*, pubis; *sbc.sc*, subcoracoscapularis muscle; *SC*, scapula; *spc*, supracoracoideus muscle; *s.sp*, muscle equivalent to supraspinatus; *STE*, sternum; *tr.maj*, trochanter major.

reduced (Fig. 53). The pubis is also much more reduced than in any other therapsids except advanced cynodonts, and there is no symphysis, or at most only a weak ventral connection between the pelvis of each side. The femur is very characteristic for the trochanter major is a massive, flattened process extending almost half-way down the shaft, and quite disguising the S-shaped curvature of the bone (Fig. 59E). Surprisingly there is no internal trochanter, which is the prominent feature of the ventral surface of all other therapsids. One of the most peculiar features of all is the considerable variation in the structure of the pelvis and hindlimb in dicynodonts. The pelvis of some forms, particularly the Triassic genera, is very high but that of, for example, *Kingoria* (Cox, 1959) is long, low and very mammalian in superficial appearance (Fig. 59D). The head of the femur also varies from barely distinguishable from the shaft to a highly ossified spherical structure set off at a marked angle from the shaft, in *Cistecephalus* (Cluver, 1978) and the South American Triassic forms (Cox, 1965) for example. By analogy with advanced cynodonts and mammals, the changes in the pelvis suggest that a more or less erect gait had evolved, as concluded by Watson (1960), Cox (1965) and Cluver (1978). On the other hand, Pearson (1924) believed that *Kannemeyeria* had a more or less sprawling gait, and King (1981b) suggested a complex mode of operation of the hindlimb of *Dicynodon* which was closer to sprawling than erect, although not strictly either.

In view of the relatively short limbs, and the shortness of the tibia and fibula compared to the femur, a permanent erect gait seems improbable. Also, the absence of the internal trochanter, and the distal extension of the trochanter major suggest that the insertion of the obturator externus (i.e. the posterior part of the pubo-ischio-femoralis externus from the ischium), which in other therapsids inserts on the internal trochanter, has shifted posteriorly to insert on the underside of the trochanter major in dicynodonts. This change would be appropriate only if the hindlimb gait was sprawling. Unfortunately, no dicynodont ankle joint has been studied sufficiently to demonstrate whether it had the complex action, appropriate to a dual-gaited hindlimb, of other therapsids (p. 121).

## *Middle ear*

The controversy about whether or not a tympanic membrane was present applies as much to dicynodonts as to the other mammal-like reptiles. Parrington (1945) believed that a tympanum attached to the squamosal and quadrate bones, and Cox (1959), described a small process in the occiput of *Kingoria* which he termed the tympanic process (Fig. 49B) believing that it marked the attachment of the top of a tympanum. On the other hand, the same problems to this interpretation of the dicynodont ear arise as occur in other groups of mammal-like reptiles. The stapes is massive and, except in the most primitive form, even lacks a foramen to reduce its weight; and it abuts strongly against the quadrate bone. While neither of these features can be said to prohibit the possibility of a tympanically operated ear, neither is expected in an ear that is claimed to have worked in the manner of modern reptiles. It seems unlikely that high-frequency hearing occurred in dicynodonts.

### *General biology*

The early anomodonts, both the primitive venjukovoids of Russia and the dicynodonts of the *Tapinocephalus*-zone of South Africa were remarkable for their small size compared to the other great herbivore group of the time, the tapinocephalid dinocephalians. This contrast is particularly striking in the *Tapinocephalus*-zone fauna, where the dicynodonts were around 25–50 cm in length and the dinocephalians two metres and more. To survive in the high latitudes within which South Africa lay at that time, these small dicynodonts must surely have either developed a high metabolic rate and body insulation, or else the habit of hibernation during the cold season. There is no clear evidence for endothermy, although some hints that it had evolved may be detected. The histological structure of the bone of dicynodonts is identical to that of other therapsids (p. 96), but as has been seen, this is not necessarily an indication of a high metabolic rate. On the other hand, being so small, the probability of the body temperature being maintained largely by an elevated metabolic rate is higher than in the case of dinocephalians. The presence of a secondary palate in dicynodonts is ambiguous because in this case the structure is manifestly part of the feeding mechanism, providing a platform for the lower jaw to work against. It does not prove that a higher rate of oxygen uptake was necessary. The more efficient feeding mechanism itself suggests that the rate of food uptake had increased, although it might alternatively indicate that the food source was of a low quality, or sparsely available. As with all therapsids, there is no good evidence for the development of hair for insulation. However, there is even less evidence bearing on the possibility of hibernation. All that can be said is that the dicynodonts may have been endotherms. One adaptation to temperature physiology, whether they were endotherms or ectotherms, concerns the overall body shape. The body is compact, the tail greatly reduced and the limbs short. These modifications may have been to reduce the rate of heat loss by reducing the surface area (Geist, 1971).

The sense organs of anomodonts probably changed their emphasis compared to other therapsids. The sense of olfaction was reduced when the snout shortened. Ridges like those of gorgonopsids for the support of turbinal cartilages (p. 126) are present in some forms, including *Kingoria* (Kemp, 1969a) and *Lystrosaurus* (Cluver, 1971), but in most they are absent or only weakly developed (King, 1981b). The eyes, however, probably became the dominant sense organs. They were always relatively large, and in several forms came to face more forwards than sideways suggesting that high resolution images of distant objects were formed.

The dicynodont brain, to judge from the nature of endocranial casts, was not developed to any greater extent than in typical therapsids (Cluver, 1971; Hopson, 1979).

The phenomenal success of the dicynodonts must be attributed to their highly specialised feeding mechanism. Not only was it efficient in general as a means of feeding on plant material, but also it was capable of a wide variety of slight modifications suitable for a series of restricted food types. A large number of species was possible, each adapted to a different part of the continuing *Glossopteris* flora (p. 99) for its food. A curious problem, however, is why the dicynodonts remained relatively small during the *Tapinocephalus*-zone. The large herbivore niches were occupied by the dinocephalians, and pareiasaur reptiles, but there is a complete absence of herbivores of an intervening size range. That larger dicynodonts are theoretically possible is indicated by the abundance of such forms in the later zones of the Karroo. The answer may be that the intermediate-sized herbivore niches were occupied by juvenile dinocephalians, or possibly that dicynodonts did in fact hibernate and had to remain small enough to be able to find suitable sites for the winter.

At any event, the close of the *Tapinocephalus*-zone saw the extinction of the dinocephalians, and the full potential of dicynodont adaptive radiation was realised during the succeeding *Cistecephalus* and *Daptocephalus*-zone times. It is not known whether this radiation was due simply to the absence of competing herbivore groups, or to a general warming of the climate of southern Africa which is believed to have occurred.

The biology of the highly successful, cosmopolitan *Lystrosaurus* in the lowest part of the Triassic is particularly interesting in view of its apparent specialisations for a semi-aquatic life. The cause of the extinction of the great mass of dicynodonts, and many other groups, at the close of the Permian is unknown at present, but the subsequent radiation of *Lystrosaurus* is probably related to a world-wide transgression of the sea, raising its level and resulting in a great increase in the areas of low-lying marshy, deltaic regions (Anderson and Cruickshank, 1978). The rest of the fauna of the *Lystrosaurus*-zone supports this interpretation, for it consists mainly of a variety of small therapsids such as the cynodonts (p. 180) and therocephalians (p. 160), several genera of labyrinthodont amphibians of which *Lydekkerina* is the best known, and the earliest representatives of the archosaur reptiles, the proterosuchids, which were semi-aquatic. Thus the fauna contains only small terrestrial animals, plus semi-aquatic forms, which is appropriate to a habitat of small areas of dry land surrounded by extensive wetlands (Parrington, 1948).

# 9 Therocephalians

THE THEROCEPHALIANS were another group of carnivorous therapsids, some of which had a superficial resemblance to gorgonopsids, but others were smaller, sometimes very small indeed, and must have been insectivorous. One late group actually adopted a herbivorous diet, but in a manner quite different to the dicynodonts. The earliest appearance of the group is in the *Tapinocephalus*-zone of the South African Karroo, and a very fragmentary specimen from the Zone II of the Russian Late Permian is also known. Unlike the groups dealt with so far, no very primitive members of the Therocephalia are known; all the distinctive characters are present at the first appearance of the group in the record, when they were already fairly diverse. A few forms survived the Permo-Triassic boundary and occur in deposits of Lower Triassic age.

## Systematics

The Therocephalia is yet another group which combined standard therapsid characters with a number of specialisations absent from other therapsid groups. Derived characters shared with any of the individual kinds of therapsids so far discussed have not been discovered and therefore the relationships to these other groups are obscure. (As will be seen, there is a relationship to the final therapsid group, the cynodonts, p. 180.) Prevalent classifications place the therocephalians with the gorgonopsids and eotitanosuchians as a taxon Theriodonta, implying a monophyletic relationship between these forms. However, the characters which unite these forms include a large canine tooth, relatively high jaw articulation, well-developed lateral pterygoid processes of the palate and so on, all of which are ancestral therapsid features and of no value in determining phylogenetic relationships. One derived character shared between therocephalians and gorgonopsids is the presence of a coronoid process of the dentary, but it was clearly evolved independently in the two groups because its detailed structure differs considerably. The Therocephalia must be regarded at present as yet another line of therapsid evolution whose relationships cannot be resolved at the moment, a view taken by Boonstra (1972).

Like in all the other therapsid groups, enlargement of the temporal fenestra has occurred, but in yet another manner. The fenestra has expanded medially to such an extent that only a narrow intertemporal region remains, which often develops as a high crest in larger forms. In the process, the postorbital bone was reduced, so that the parietal bone forms a large part of the lateral face of the intertemporal bar, between the postorbital in front and the squamosal behind. In this it differs from the condition seen in those dinocephalians and dicynodonts which also form narrow intertemporal crests. The therocephalian fenestra is relatively elongated, but the squamosal bone does not flare backwards above the quadrate region as it does in gorgonopsids. The fenestra also remains at a relatively high level, but there is no development of the squamosal ventro-lateral to the fenestra as occurs so characteristically in the dicynodonts. Because of the particular way in which the fenestra has evolved, the therocephalian occiput is wide but low. The supraoccipital bone is very broad across the occipital surface.

A functional correlate of the enlarged fenestra is the coronoid process of the dentary, superficially resembling that of the gorgonopsids, but in fact rather different in structure. It is flat and broad, rather than triangular in section as in gorgonopsids, and quite clearly evolved in parallel. The reflected lamina is also characteristic, being exceptionally large, strongly fluted, and quite free from the jaw along its dorsal edge.

The dentition consists of up to seven incisors, and a well-developed canine. In addition there can be up to three precanine teeth, that is, small incisor-like teeth implanted not in the premaxilla but in the maxilla anterior to the canine tooth. The postcanines are very variable. Larger forms tend to have reduced postcanines in the manner of the gorgonopsids, but in the smaller forms they are typically retained as an important part of the feeding mechanism. More complex, multi-cusped postcanine teeth appear in a few of the late, highly specialised members of the group.

The snout of most forms is relatively long and often heavily built, a feature also reflected in the relatively long palate and short basicranial region of the skull. One of the most distinctive characters of all is a large fenestra on either side of the hind part of the palate, the suborbital vacuity. Another highly distinctive feature is the presence of a medially directed process of the squamosal, which contacts the braincase, thereby cutting off a pterygo-paroccipital foramen (Fig. 69). The epipterygoid tends to expand as a broad sheet lateral to the middle region of the braincase, although in most forms this is not as marked as in the cynodonts.

The therocephalian postcranial skeleton is basically therapsid in structure, but with a few unique specialisations. The ilium is particularly well developed, and has a small, but constant anterior process on the front edge. The femur has developed an extra trochanter on the medial side, probably equivalent to the mammalian trochanter minor. The scapula blade is very narrow and lacks an out-turned acromion process. The limbs in general were relatively long and slender, and the lower leg approached the length of the respective humerus and femur. Finally, even in the early therocephalians, the digital formula had achieved the mammalian condition of 2.3.3.3.3, in both front and hind feet, paralleling the dicynodonts.

There has been a tendency in the past to separate a variety of small, variously advanced forms from the rest, as the Scaloposauria (Brink, 1965) or Bauriamorpha (Watson and Romer, 1956). These do not, however, share much more than smallness with one another, and they possess all the basic characters of the rest of the therocephalians. It is doubtful whether they constitute a monophyletic group in any case, and there is certainly no justification for excluding them from the same taxon as other therocephalians.

The interrelationships and classification of the Therocephalia are not yet fully worked out. Probably the most primitive member known is represented by the single, poorly preserved skull of *Crapartinella* (Mendrez, 1975) from the *Tapinocephalus*-zone. This small form (Fig. 60A) possesses six upper incisors, a single canine, and twelve simple conical postcanines, which is probably close to the primitive dentition for the group. There is no development of a secondary palate, so that the internal nares are continuous with the recesses in the palate which received the lower canines when the jaws closed. Other primitive features are the presence of palatal teeth on both the vomer and the pterygoid, and the paired nature of the vomers.

The most familiar of the *Tapinocephalus*-zone forms are the pristerognathids (Fig. 60B–D), some of which tend to become quite large with a skull length of over 30 cm. The snout is long and heavy, and the intertemporal region is in the form of a high, sharp crest. Up to seven incisor teeth are present in the upper jaw, and about six in the lower. All are relatively long and sharp, and are followed by the single large canine. The postcanines are, however, reduced, both in size, and also in number for only three to nine are present. The palate, like that of

*Crapartinella* is primitive, with no secondary palate, but palatal teeth are no longer present. *Trochosaurus* and certain similar forms have been placed in a separate family to the pristerognathids, on the basis of the presence of two functional upper canines in each jaw. However, it appears that in such cases one of the canines is actually in the process of replacing the other and that normally only a single one was present. Thus these genera are now regarded as members of the Pristerognathidae (Van den Heever, 1980).

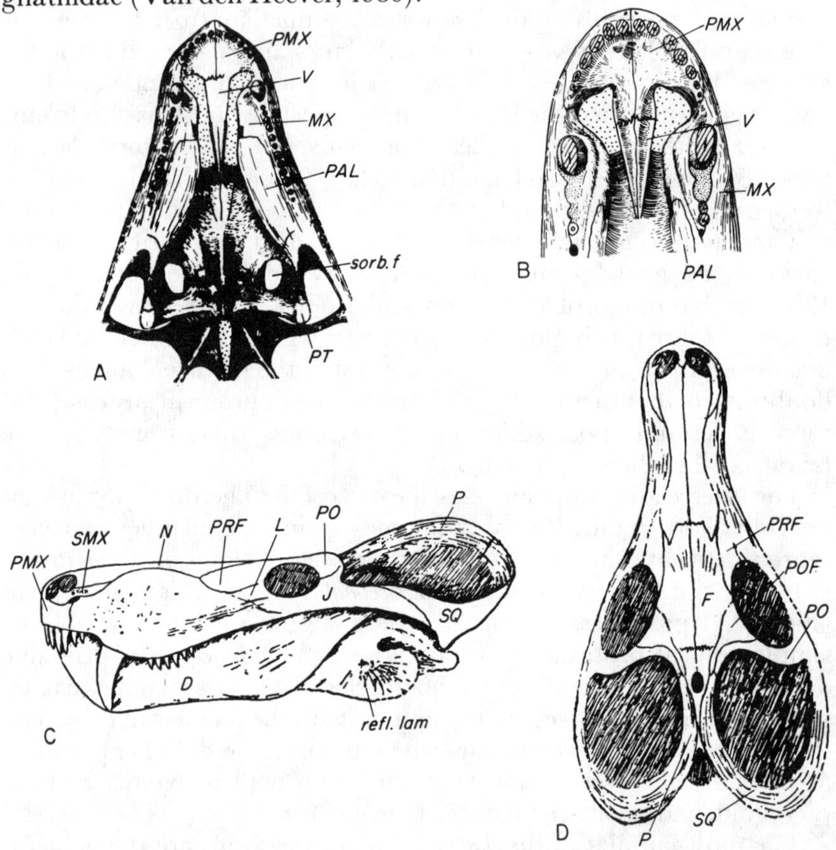

FIG. 60. Primitive therocephalians. A, palate of *Crapartinella* in ventral view. B, palate of *Pristerognathus* in ventral view. C, skull of *Pristerognathus* in lateral view. D, skull of *Pristerognathus* in dorsal view. (A and B from Mendrez, 1975; C and D from Broom, 1932.)

*D*, dentary; *F*, frontal; *J*, jugal; *L*, lachrymal; *MX*, maxilla; *N*, nasal; *P*, parietal; *PAL*, palatine; *PMX*, premaxilla; *PO*, postorbital; *POF*, postfrontal; *PRF*, prefrontal; *PT*, pterygoid; *refl.lam*, reflected lamina of the angular; *SMX*, septomaxilla; *sorb.f*, suborbital fenestra: *SQ*, squamosal; *V*, vomer.

Magnifications: A, *c.* × 0.85; B, *c.* × 0.4; C and D, *c.* × 0.25.

A series of rather short, wide-snouted forms constitute the family Moschorhinidae (Annatherapsidae). Again no secondary palate is developed, but the vomers are specialised by a wide anterior expansion. Only five upper incisors are present, along with a precanine and single canine. The most primitive member is *Annatherapsidus* (Tatarinov, 1974), which has retained a fairly prominent postcanine dentition of six well-developed teeth, and also has a postfrontal bone, and palatal teeth on the pterygoid. It is a Russian Late Permian form from Zone IV. *Moschorhinus* (Mendrez, 1974a) is slightly later occurring in both the latest Permian *Daptocephalus*-zone and the basal Triassic *Lystrosaurus*-zone of South Africa (Fig. 61A–E). The skull is particularly short and wide, and the postcanine dentition of the upper jaw is reduced to three small teeth immediately behind the canine. Both the postfrontal bone and the palatal teeth have been lost. *Euchambersia* (Fig. 61F) is a highly specialised offshoot of which but two specimens are presently known. There is a very deep recess in the side of the snout, between the single canine tooth and the orbit. The canine has a groove down its outer side, and the bone surounding both the base of the tooth and the walls of the recess is covered in fine foramina. *Euchambersia* appears to have evolved a poison gland associated with a snake-like fang for administering a venomous bite. The snout is very short and wide, and the post-orbital bar and zygomatic arch slender, but probably complete (Mendrez, 1974a).

Another specialised group is the family Whaitsiidae (Fig. 62), in which a rudimentary secondary palate has appeared in the form of a short medial process of the maxilla meeting the vomer, just behind the canine. The internal choana is therefore separated from the recess for the lower canine. The snout is constricted behind the upper canines, so that the dentary directly opposes the maxilla simultaneously on both sides of the jaw. The dorsal surface of the dentary has a very characteristic concave profile as seen from the side. The postcanine dentition tends to be reduced, and in the advanced whaitsiids like *Theriognathus* (=*Whaitsia*) no postcanine teeth are present at all. The opposing dentary and maxilla surfaces are heavily rugose and it is likely that they were covered in heavily keratinised epithelium, forming dicynodont-like horny tooth plates. Other specialisations include the tendency to close the suborbital vacuity and the interpterygoid vacuity, and the broad expansion of the epipterygoid (Kemp, 1972a). *Moschowhaitsia* is a primitive whaitsiid from Russia, which still possesses postcanine teeth, a precanine, and a suborbital vacuity (Tatarinov, 1963).

The rest of the therocephalians fall into a series of families of fairly small forms, which retain a significant postcanine dentition, and tend to

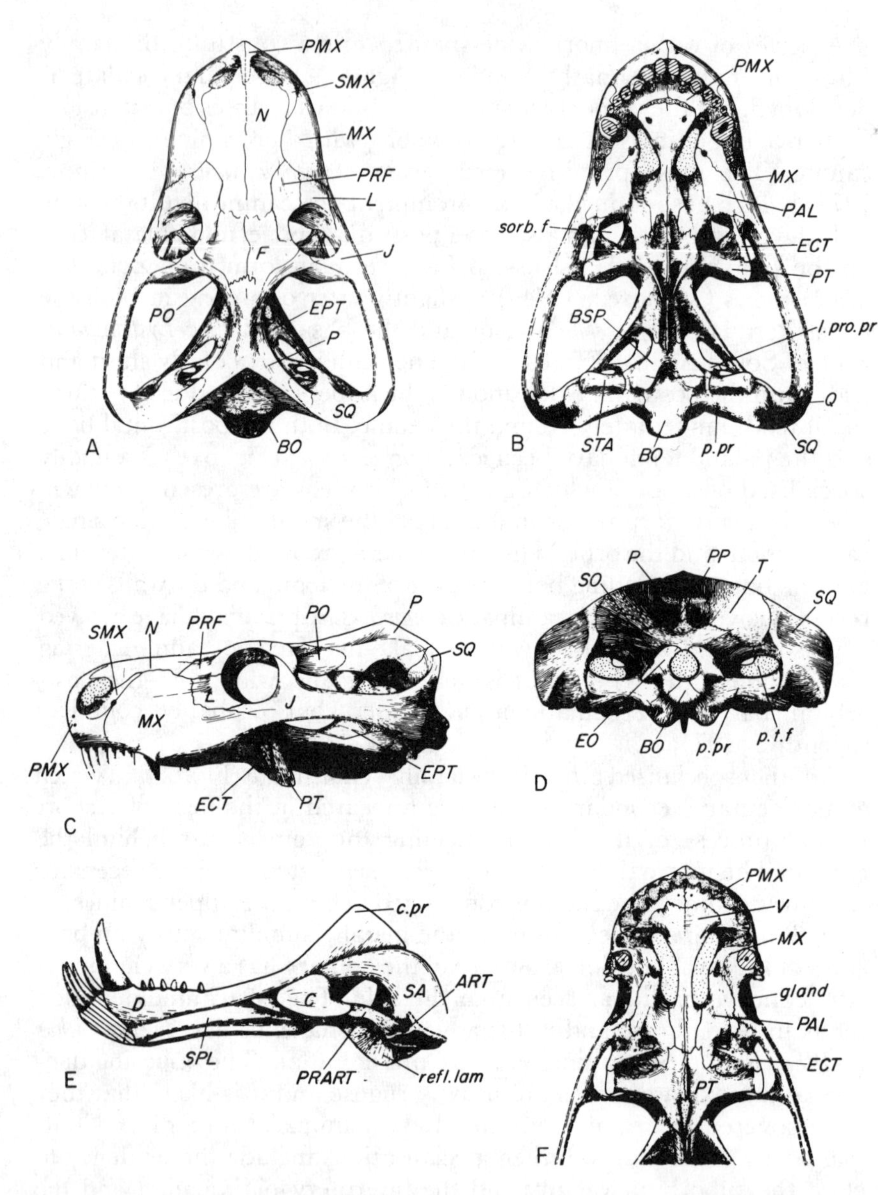


FIG. 61. Skull of *Moschorhinus*. A, dorsal view. B, ventral view. C, lateral view. D, posterior view. E, medial view of the lower jaw. F, anterior part of the skull of *Euchambersia* in ventral view. (A–D from Mendrez, 1974a; E from Mendrez, 1974b; F from Mendrez, 1975.)

*ART*, articular; *BO*, basioccipital; *BSP*, basisphenoid; *C*, coronoid; *c.pr*, cornoid process; *D*, dentary; *ECT*, ectopterygoid; *EO*, exoccipital; *EPT*, epipterygoid; *F*, form a secondary palate, at least to some extent. The postorbital bar and zygomatic arch are slender and the former is sometimes incomplete. It is these which have sometimes been separated as the Scaloposauria or Bauriamorpha. Their taxonomy has yet to be fully worked out, although the valuable work of Christianne Mendrez, interrupted by her sad and untimely death, has gone a considerable way in this direction.

The most primitive forms are the ictidosuchids, represented by *Ictidosuchoides* (Fig. 63A) which has not developed a secondary palate and retains two precanine teeth in the maxilla (Crompton, 1955a; Mendrez, 1975). It is from the *Cistecephalus*-zone of South Africa. *Regisaurus* (Fig. 63B,C) is a later form from the *Lystrosaurus*-zone (Mendrez, 1972), in which the maxilla has expanded medially to contact the vomer for a short distance, thereby creating a short secondary palate. The precanine teeth have been lost, leaving only the six incisors anterior to the canine. Another form, *Lycideops* (Fig. 63D), is clearly related to *Regisaurus*, but has a much longer secondary palate, and has retained two precanines.

The family Scaloposauridae is composed of very small animals (some at least of which may prove to be juvenile ictidosuchids). The genus *Scaloposaurus* (Fig. 62E–H) occurs in the *Daptocephalus*-zone and *Lystrosaurus*-zone. The secondary palate is well developed but, like that of *Lycideops*, consists of the maxilla meeting the vomer and leaving the latter fully exposed in ventral view. Two precanine teeth are still present, the canine is reduced in size, and the postcanines are remarkable for the appearance of small accessory cuspules on at least some of the teeth (Crompton, 1955). These take the form of a tiny posterior and sometimes a tiny anterior cuspule at the base of the main cusp. Overall, the scaloposaurids are quite the smallest of the therocephalians, and must be judged to have been exclusively feeders on small invertebrates, particularly insects.

*Ericiolacerta* (Fig. 64), known from a single skeleton of *Lystrosaurus*-zone age (Watson, 1931; Mendrez, 1975), is further advanced over the ictidosuchids and scaloposaurs. The secondary palate is completed,

frontal; *J*, jugal; *L*, lachrymal; *l.pro.pr*, lateral process of the prootic; *MX*, maxilla; *N*, nasal; *P*, parietal; *PAL*, palatine; *PMX*, premaxilla; *PO*, postorbital; *PP*, postparietal; *p.pr*, paroccipital process; *PRART*, prearticular; *PRF*, prefrontal; *PT*, pterygoid; *p.t.f*, post-temporal fenestra; *Q*, quadrate; *q.ra.pt*, quadrate ramus of the pterygoid; *refl.lam*, reflected lamina of the angular; *SA*, surangular; *SMX*, septomaxilla; *SQ*, squamosal; *SO*, supraoccipital; *sorb.f*, suborbital fenestra; *SPL*, splenial; *STA*, stapes; *T*, tabular; *V*, vomer.

Magnifications: A–D, *c.* × 0.33; E, *c.* × 0.44; F, *c.* × 0.22.

and for the first time the paired maxillae meet one another in the midline, obscuring most of the vomer from ventral view. Six incisors, a precanine, a small canine and six postcanines were present in the upper

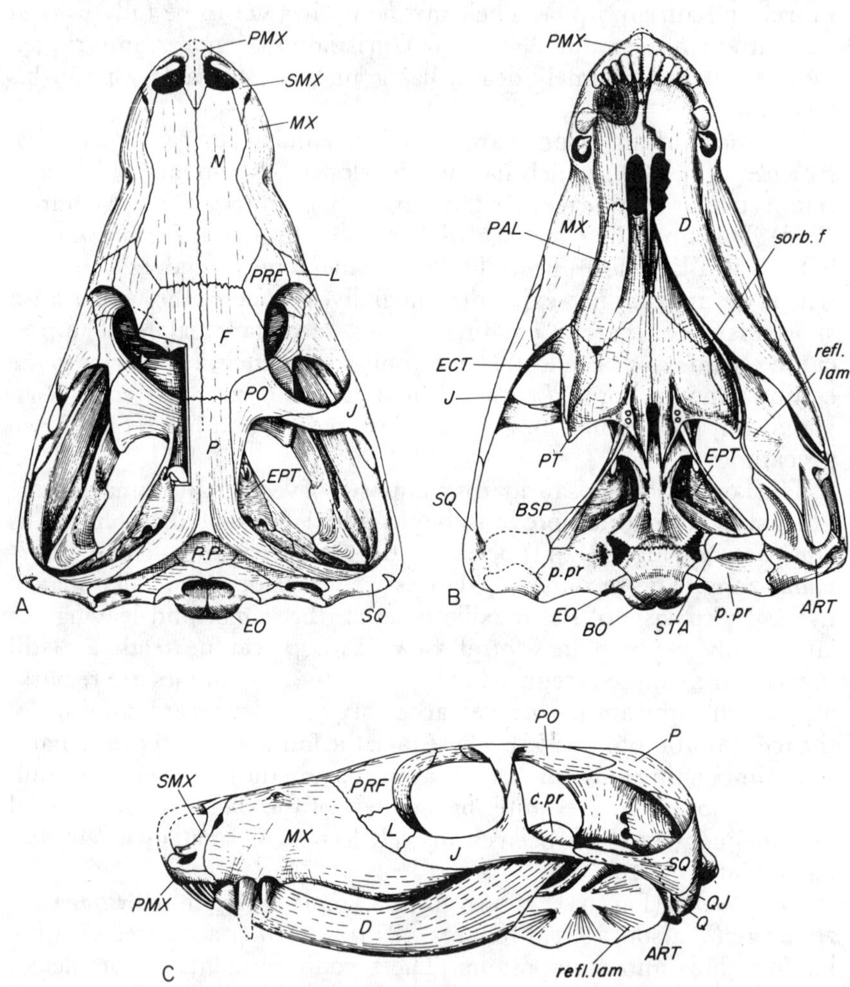


FIG. 62. The whaitsiid therocephalian *Theriognathus*. Skull of a juvenile specimen in A, dorsal view; B, ventral view; C, lateral view. (From Brink, 1956b, as *Aneugomphius*.)

*ART*, articular; *BO*, basioccipital; *BSP*, basisphenoid; *c.pr*, coronoid process; *D*, dentary; *ECT*, ectopterygoid; *EO*, exoccipital; *EPT*, epipterygoid; *F*, frontal; *J*, jugal; *L*, lachrymal; *MX*, maxilla; *N*, nasal; *P*, parietal; *PAL*, palatine; *PMX*, premaxilla; *PO*, postorbital; *PP*, postparietal; *p.pr*, paroccipital process; *PRF*, prefrontal; *PRO*, prootic; *PT*, pterygoid; *Q*, quadrate; *QJ*, quadratojugal; *refl.lam*, reflected lamina of the angular; *SMX*, septomaxilla; *SQ*, squamosal; *sorb.f*, suborbital fenestra; *STA*, stapes; *V*, vomer.

Magnification *c.* × 1.0.

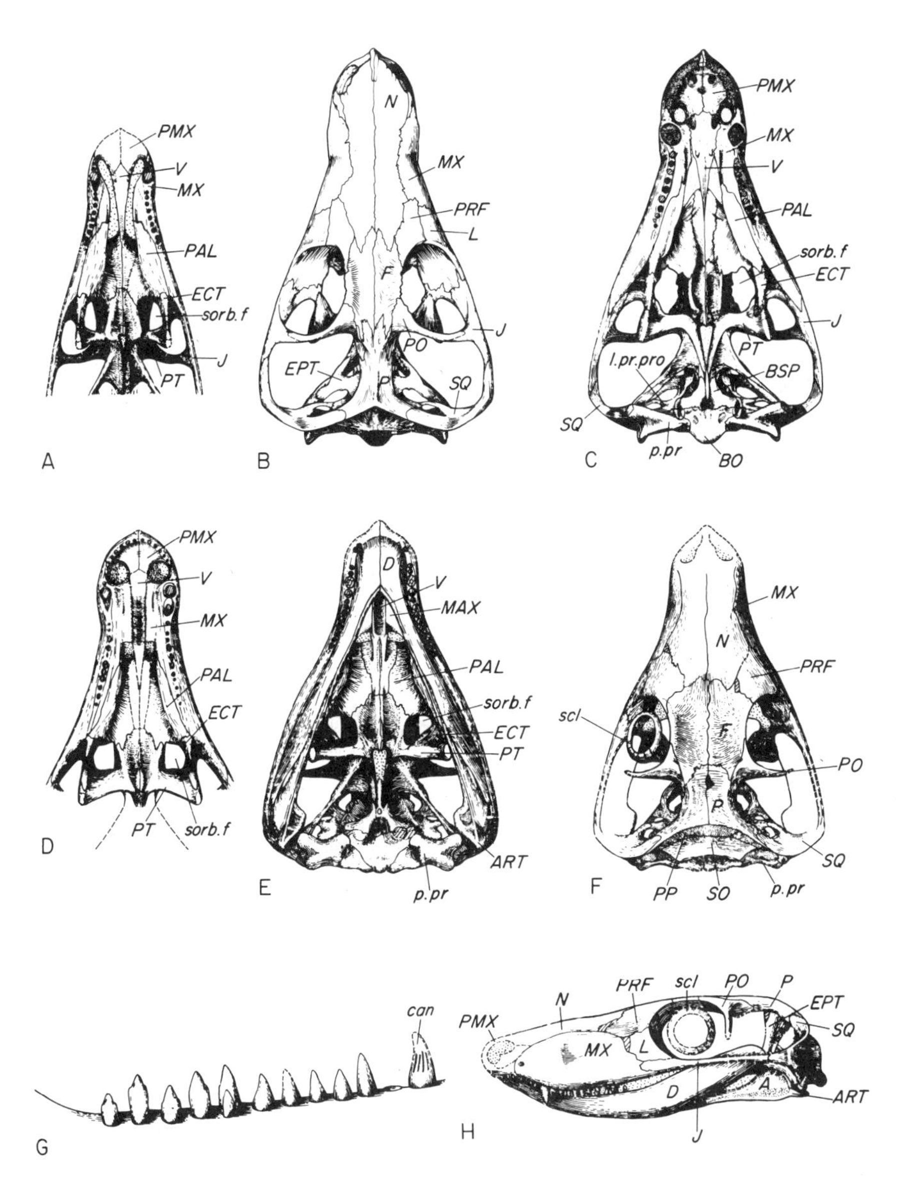


FIG. 63. A, palate of *Ictidosuchoides* in ventral view. B, skull of *Regisaurus* in dorsal view. C, skull of *Regisaurus* in ventral view. D, palate of *Lycideops*. E, skull of *Scaloposaurus* in ventral view. F, the same in dorsal view. G, internal view of the left postcanine teeth of *Scaloposaurus*. H, skull of *Scaloposaurus* in lateral view. (A and D from Mendrez, 1975; B and C from Mendrez, 1972; E–H from Mendrez-Carroll, 1979.)

*scl*, sclerotic ring. For other abbreviations see Fig. 61.

Magnifications: A, *c.* × 0.3; B and C, *c.* × 0.45; D, *c.* × 0.23; E, F and H, *c.* × 0.90; G, × 3.0.

jaw, and the latter show signs of being tricuspid, like the scaloposaurs. The lower jaw is extremely shallow and delicate, as befits such a small animal, but appears to have the same basic structure as all other therocephalians. It too possesses tricuspid postcanine teeth.

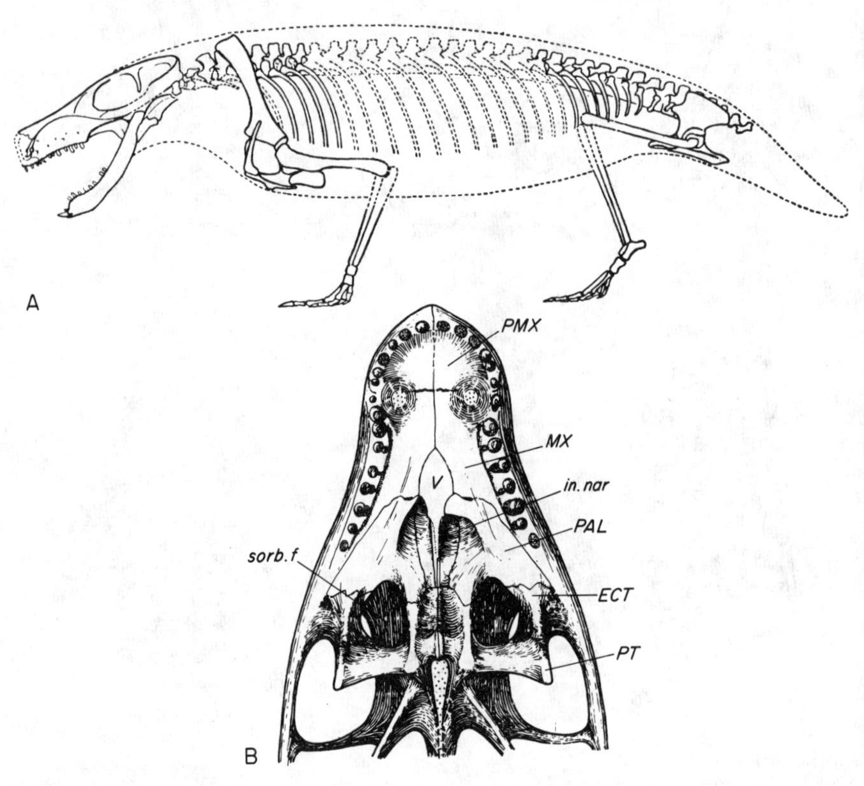


Fig. 64. *Ericiolacerta*. A, skeleton. B, palate. (A from Watson, 1931; B from Mendrez, 1975.)

*ECT*, ectopterygoid; *in.nar*, internal nares; *MX*, maxilla; *PAL*, palatine; *PMX*, premaxilla; *PT*, pterygoid; *sorb.f*, suborbital fenestra.

Magnifications: A, *c*. × 0.6; B, *c*. × 2.0.

The most advanced of all therocephalians were the bauriids (Fig. 65), which occurred latest of all, in the *Cynognathus*-zone of the Lower Triassic. In *Bauria* (Brink, 1963), a complete secondary palate is again present, and the paired maxillae continue their medial connection further posterior than in *Ericiolacerta*, obscuring even more of the vomer (Fig. 65C). The most remarkable feature of *Bauria* is its highly complex dentition (Gow, 1978). There are only four upper incisors, followed by a

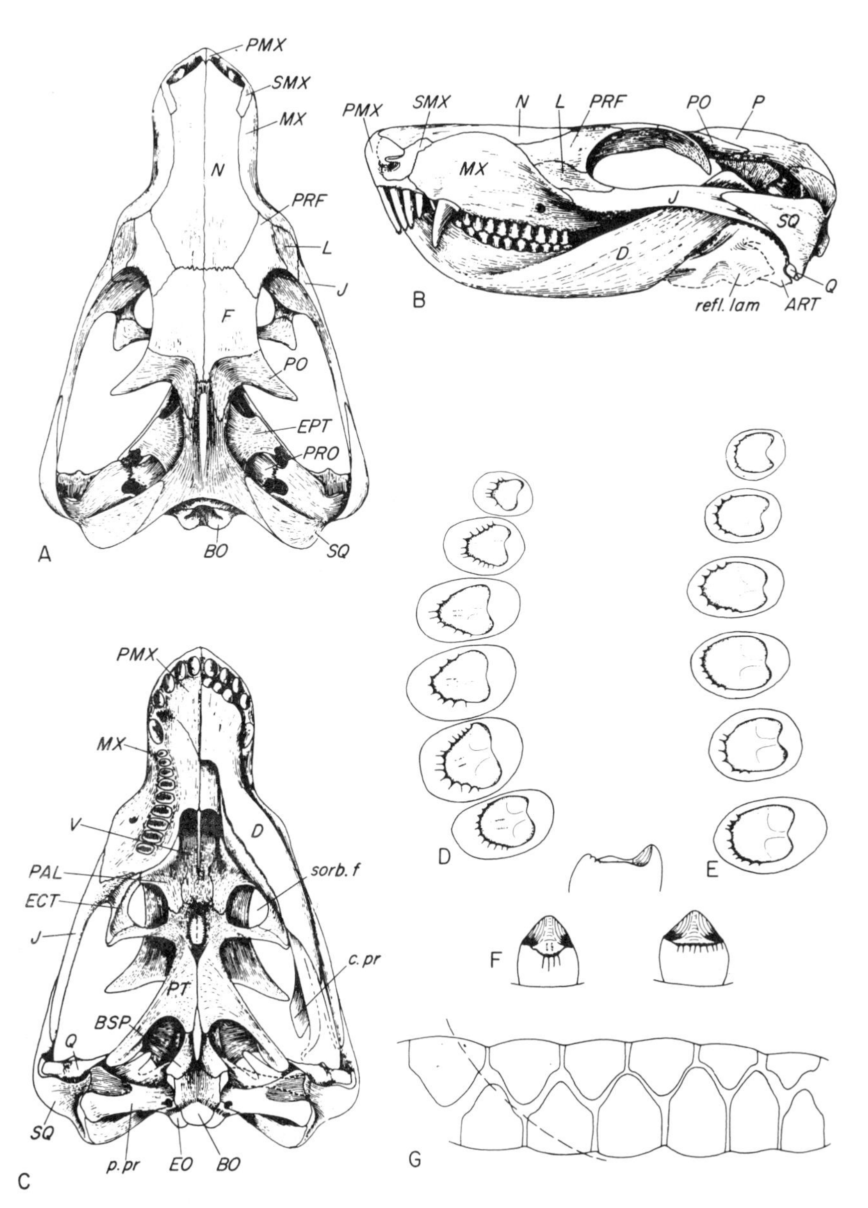


FIG. 65. The skull of *Bauria*. A, dorsal view. B, lateral view. C, ventral view. D, crown view of the lower right postcanine teeth. E, crown view the upper left postcanine teeth. F, anterior view of lower postcanine (above), internal view of lower postcanine (left) and internal view of upper postcanine (right). G, lateral view of the occluding right postcanines, with the position of the coronoid process of the dentary shown by the dashed line. (A–C from Brink, 1963a; D–G from Gow, 1978.)

modest canine. The postcanine teeth are expanded from side to side. Each one, upper and lower, consists of a large main cusp on the labial (outer) side, and a row of extra cuspules on the lingual (inner) side. Upper and lower teeth intermesh, forming a series of shearing surfaces, suitable for cutting up fibrous material. It seems clear that *Bauria* was adapted for a herbivorous diet. A further specialisation, reminiscent of the whaitsiids and also some of the cynodonts, is a constriction of the snout immediately behind the canine region. This allows both the left and the right dentitions to be active simultaneously. As in the scaloposaurids the postorbital bar is incomplete, giving a superficially mammalian appearance to the skull. The epipterygoid is widened.

## Functional anatomy

### *Feeding mechanism*

The therocephalian dentition is very variable, indicating that a number of different feeding strategies were developed in different groups. The reduction of the postcanine teeth in pristerognathids (Fig. 60B,C) and moschorhinids (Fig. 61) parallels the similar occurrence in gorgonopsids, and indicates that the jaws were used in a kinetic fashion. The lower jaw was accelerated from the open position, and the kinetic energy so generated was dissipated when the large dagger-like canines closed upon the prey. It is doubtful whether the incisor teeth of any therocephalians could interdigitate as in the gorgonopsids. In specimens where the lower jaw is preserved in place, the lower incisors have closed internal to the upper incisors without intermeshing, and there is no apparent mechanism of the jaw articulation region to permit an anterior shift of the lower jaw. Use of the incisors was probably much cruder than in gorgonopsids, consisting simply of embedding in the prey and tearing the flesh by movements of the whole skull. The complete loss of the postcanines in whaitsiids (Fig. 62) was different, for in them horny tooth plates probably developed in place of the teeth. Possibly they had adopted an omnivorous or scavenging role. Such a habit must have

*ART*, articular; *BO*, basioccipital; *BSP*, basisphenoid; *c.pr*, coronoid process; *D*, dentary; *ECT*, ectopterygoid; *EO*, exoccipital; *EPT*, epipterygoid; *F*, frontal; *J*, jugal; *L*, lachrymal; *MX*, maxilla; *N*, nasal; *P*, parietal; *PAL*, palatine; *PMX*, premaxilla; *PO*, postorbital; *p.pr*, paroccipital process; *PRF*, prefrontal; *PRO*, prootic; *PT*, pterygoid; *Q*, quadrate; *refl.lam*, reflected lamina of the angular; *SMX*, septomaxilla; *sorb.f*, suborbital fenestra; *SQ*, squamosal; *V*, vomer.

Magnifications: A–C, *c.* × 0.5; D–G, *c.* × 3.0.

been available and there are no other therapsids obviously adapted for it.

The smaller therocephalians which retained a full postcanine dentition (Fig. 63) were the only therapsids before the cynodonts which adopted an insectivorous mode of life. Besides relatively small size, insectivory requires a good series of small but sharp teeth for catching the prey and piercing the cuticle. Relative agility is also likely to be required, and these forms have the longest limbs and most gracile skeletons of all therapsids, again prior to cynodonts.

Only the whaitsiid jaw articulation has been studied from a functional point of view (Kemp, 1972b), although the structure of this region of other therocephalians is similar as far as is known, and it differs markedly from any other therapsids except cynodonts. The articular facet of the lower jaw (Fig. 66A,B) faces backwards and inwards, and bears against an antero-laterally facing quadrate condyle (Fig. 66F,G). The purpose of this particular arrangement was to provide maximum resistance to adductor jaw muscles which pulled in a postero-medial direction from their insertion on the lower jaw (Fig. 66C). Thus the back end of the jaw could not be forced off the quadrate in either a medial or a posterior direction. However, a subtle difficulty existed because the axis of each hinge joint ran forwards and inwards, rather than being aligned transversely. Yet it would have been impossible to open and close the jaws unless the left hinge axis and the right hinge axis both coincided with the same transverse line from left to right through the skull. The resolution to the problem seems to have been the development of a quadrate that could move relative to the rest of the skull. As the jaws opened and closed, the quadrate made suitable adjustments to prevent disarticulation of the articular bone from the quadrate bone (Fig. 66D,E). This highly complex system serves to maintain the resistance to the jaw muscles, while at the same time allowing a large gape. It is interesting to compare it with the equally complex gorgonopsid jaw articulation, which serves a similar end but by quite a different mechanism.

The greatest bulk of the external adductor muscle (Fig. 67) arose from the medial and posterior edges of the temporal fenestra. The intertemporal region of the larger forms tended to develop as a high, sharp-crested edge, the sagittal crest, with a broad lateral facing surface continuous posteriorly with the anterior-facing surface of the squamosal. These two surfaces are bounded ventrally by a sharp edge above the braincase, which marks the ventral limit of the muscle origin. The muscle, along with a presumed temporal aponeurosis, descended to an insertion on the coronoid process of the dentary, and the dorsal and medial faces of the postdentary bones. This relatively high level of insertion on the jaw, well above the level of the jaw articulation, means that the muscle was inclined strongly postero-medially, yet at the same time generated a large torque about the jaw hinge.

The external adductor muscle extended onto two other parts of the skull. The first was the process of the squamosal which descends medial to the quadrate, and then turns inwards to meet a process of the prootic bone of the braincase. The anterior surface of this squamosal process is a smooth continuation of the main muscle-bearing face of the squamosal above. The insertion of the part of the muscle attached here must have been on the inner face of the lower jaw, in front of the articular.

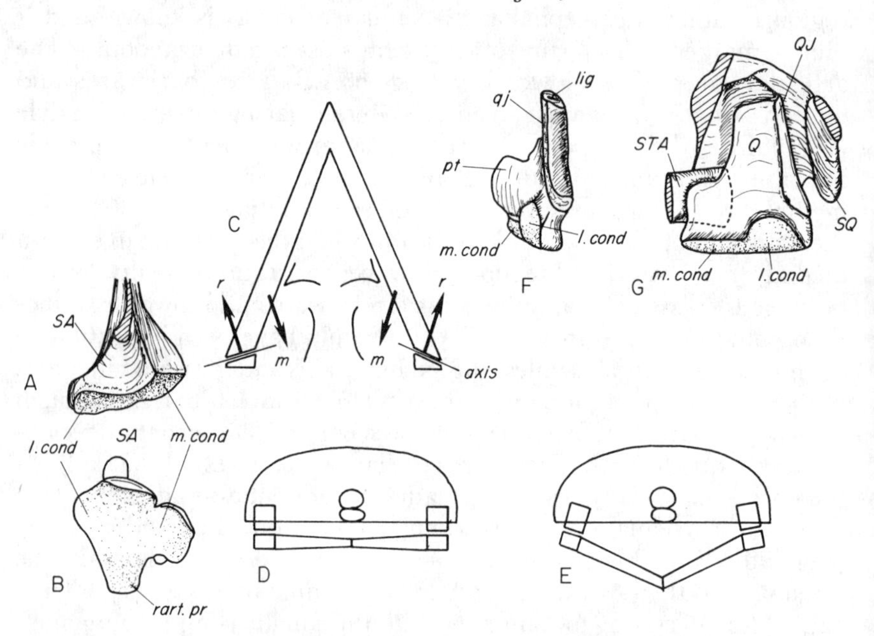


Fig. 66. Therocephalian jaw articulation. A, hind end of left lower jaw in dorsal view. B, the same in posterior view. C, dorsal view of the lower jaws and jaw hinge diagrammatically, showing the axis of the hinge and the direction of the muscle force and hinge reaction. D, diagrammatic posterior view of the skull, lower jaws and jaw articulation with the jaws closed. E, the same but with the jaws open, indicating the compensatory movement of the quadrate. F, lateral view of the left quadrate. G, anterior view of the left quadrate–quadratojugal complex, *in situ*. (Redrawn after Kemp, 1972b.)

*axis*, axis of the jaw articulation; *l.cond*, lateral condyle; *lig*, attachment of ligament controlling the quadrate; *m*, muscle force; *m.cond*, medial condyle; *pt*, wing of the quadrate attaching to the pterygoid; *Q*, quadrate; *QJ*, quadratojugal; *qj*, area of attachment to the quadratojugal; *r*, reaction force at hinge; *ratr.pr*, retroarticular process of the articular; *SA*, surangular; *SQ*, squamosal; *STA*, stapes.

The second extension of the external adductor muscle was laterally. The posterior part of the zygomatic arch is widened and forms a smooth fossa facing inwards, just lateral to the quadrate complex. Muscle fibres from here ran forwards and downwards, to insert on the hind edge of a powerful ridge of the lateral surface of the angular, above and in front of the reflected lamina of the angular. At least to a small extent, the muscle probably also inserted between the lamina and the main body of the angular (Kemp, 1972b).

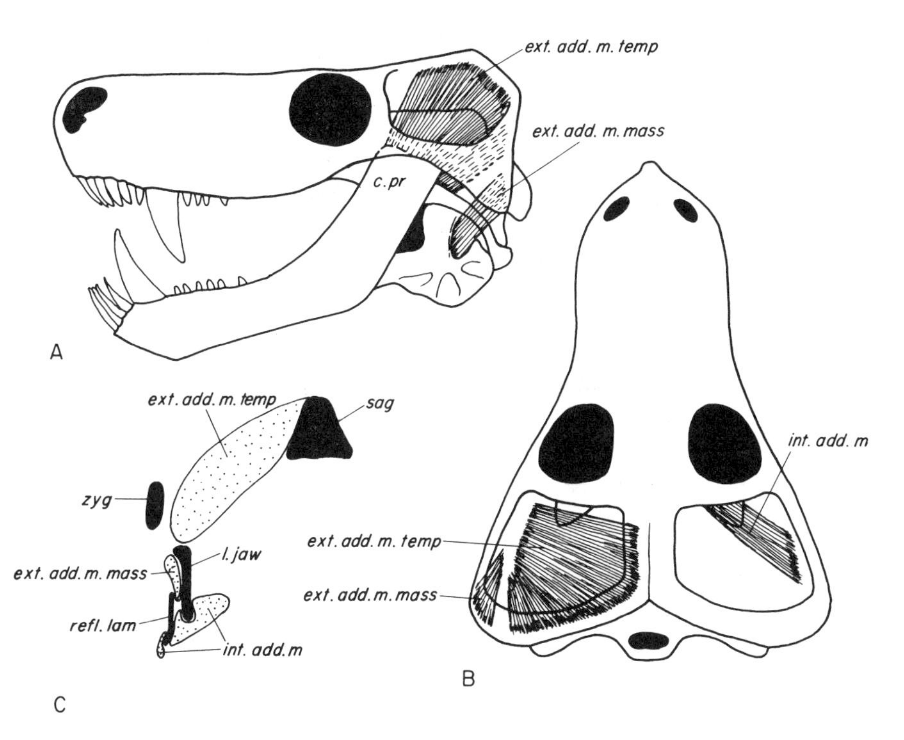


FIG. 67. Therocephalian jaw musculature. A, lateral view of skull. B, dorsal view of skull. C, transverse section through the temporal region of the skull. (Shape of the skull based on Mendrez, 1974a.)

*c.pr*, coronoid process; *ext.add.m.mass*, superficial part of the external adductor muscle, equivalent to the mammalian masseter; *ext.add.m.temp*, deep part of the external adductor muscle, equivalent to the mammalian temporalis; *int.add.m*, internal adductor muscle (pterygoideus); *l.jaw*, lower jaw; *refl.lam*, reflected lamina of the angular; *sag*, sagittal (intertemporal) crest; *zyg*, zygomatic arch.

The internal adductor muscle was as well developed as in any other therapsids. The pterygoideus musculature originated from the posterior face of the large lateral pterygoid process, and also from the dorsal surface of the posterior parts of the palate. The characteristic suborbital vacuity of the palate may be analogous to the temporal fenestra, with respect to the pterygoid musculature. Muscle fibres attached to the anterior edge and sides of the fenestra, and no doubt to an aponeurotic sheet across it, giving them a stronger anchorage. The evident need for the pterygoideus muscle fibres to extend so far onto the palate in therocephalians may relate to the relatively short post-palatal skull length. The insertion of the pterygoideus muscle was probably the inner face of the keel of the angular, and its posterior continuation, the reflected lamina. The relatively posterior position of the lamina further indicates the need for reasonably long pterygoideus muscle fibres. It is possible too that some part of the muscle wrapped around the ventral edge of the lower jaw, to insert within the recess between the reflected lamina and the body of the angular, as indicated by a widening of the ventralmost part of the recess, at least in the whaitsiids (Kemp, 1972b).

Other functions of the reflected lamina, which is exceptionally well developed in this group, probably included the origin of musculature related to the hyoid apparatus and a muscular tongue, as has been argued in dicynodonts. At least part of the musculature for opening the jaw may also have attached to it. As in other therapsids, a large, ventrally directed retroarticular process of the articular bone occurs behind the jaw articulation, for the insertion of a depressor mandibuli jaw opening muscle, and possibly also for part of the pterygoideus musculature.

Looking at the whole jaw musculature of therocephalians, it can be seen to have differed in its organisation from all the previous therapsid groups discussed, and to have developed certain new properties not so far met. The main pull is postero-medially by means of the large adductor externus, although to some extent the posterior component would tend to be balanced by the anterior component of the pterygoideus muscle force. However, any tendency for the latter muscle to disarticulate the jaw by pulling it forwards off the quadrate, as might occur at wide gapes, was countered by the action of the two specialised parts of the external adductor on either side of the quadrate. These exerted a largely posterior force on the jaw, and because they inserted close to the jaw hinge, they did not change length much during the cycle of jaw movement. They therefore maintained a large, virtually isometric contraction, and functioned to guarantee stability of the jaw hinge. Such an arrangement is particularly suited to the needs of an active carnivore, and represents an alternative solution to that found by the gorgonopsids.

## *Locomotion*

The structure of the postcranial skeleton of therocephalians is sufficiently similar to that of the gorgonopsids (p. 115) for the conclusions about locomotion in that group to apply equally well here. Thus the forelimb functioned in a sprawling manner, whilst the hindlimb was capable of both sprawling and a more erect gait. There are certain unique features of therocephalians which suggest that the more erect gait was of greater significance than in gorgonopsids and other primitive therapsids (Kemp, 1978). The tibia and fibula are about the same length as the femur for example. There is also evidence that the protractor muscle, the pubo-ischio-femoralis internus had migrated antero-dorsally to a greater extent. The ilium has a broad concavity on the lower part of its anterior region, and a small anterior process, which may indicate an extensive area of origin of the muscle. The insertion on the femur is marked by the appearance of a new trochanter, roughly equivalent to the mammalian trochanter minor, on the anterior region of the bone (Fig. 68A). As the more dorsal parts of this muscle were best placed for the more erect gait, the development implies a more energetic use of this particular mode of locomotion.

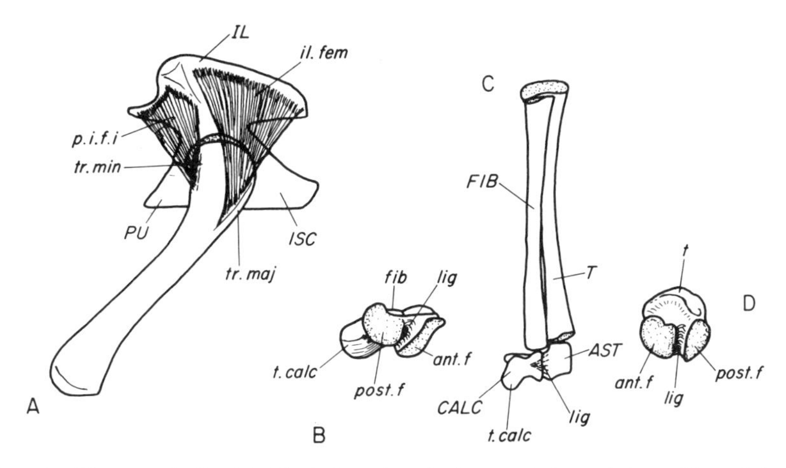


FIG. 68. Therocephalian hindlimb, based on *Regisaurus*. A, lateral view of the pelvis and femur in the erect gait. B, medial view of the left calcaneum. C, posterior view of the lower leg. D, lateral view of the left astragalus. (Redrawn from Kemp, 1978.)

*ant.f*, anterior facet; *AST*, astragalus; *CALC*, calcaneum; *F*, femur; *FIB*, fibula; *fib*, articulation for the fibula; *il.fem*, ilio-femoralis muscle; *IL*, ilium; *ISC*, ischium; *lig*, astragalo-calcaneal ligament as if bones transparent; *p.i.f.i*, pubo-ischio-femoralis internus muscle; *post.f*, posterior facet; *PU*, pubis; *T*, tibia; *t*, articulation for the tibia; *t.calc*, tuber calcis; *tr.maj*, trochanter major; *tr.min*, trochanter equivalent to the mammalian trochanter minor.

*Middle ear*

The stapes (Fig. 69) is a short, robust bone lacking a stapedial foramen and therefore resembling in general the dicynodont stapes. However, it does not abut directly against the quadrate as in those forms, although the gap may have been filled by cartilage. As in all the therapsids considered so far, the stapes does not have the kind of features, particularly weight reduction, that would be expected if it was capable of the transmission of high-frequency vibrations via a tympanic membrane, and there is no obvious site for a tympanic membrane anywhere near the distal end of the stapes. The likely conclusion must be that therocephalians were still without a tympanum, and could only detect low-frequency, relatively high-energy sound waves impinging on the lower jaw and transmitted via the hyoid apparatus or quadrate to the stapes and fenestra ovalis.

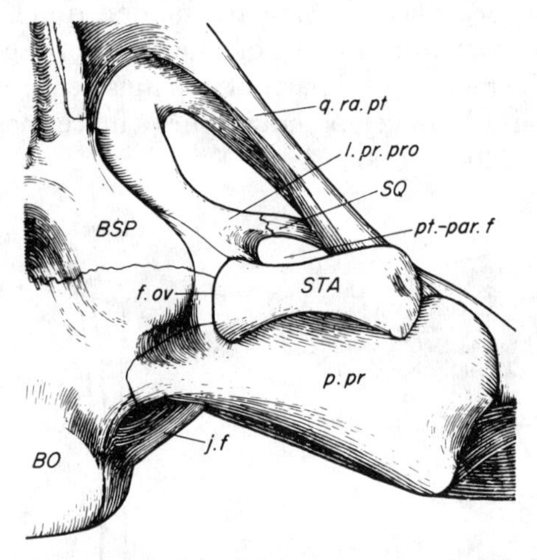


FIG. 69. Ventral view of the left middle ear region of the skull of *Moschorhinus*. (From Mendrez, 1974a.)

*BO*, basioccipital; *BSP*, basisphenoid; *f.ov*, fenestra ovalis; *j.f*, jugular foramen; *l.pr.pro*, lateral process of the prootic; *p.pr*, paroccipital process; *pt.par.f*, pterygo-paroccipital foramen; *q.ra.pt*, quadate ramus of the pterygoid; *SQ*, squamosal; *STA*, stapes.

Magnification *c.* × 1.2.

## *General biology*

The Therocephalia are interesting because of their diversity of structure and presumed habits. It is the only lineage of the therapsids during the Permian which solved the problems of insectivory, while both large carnivores, and ultimately herbivores, were also included. Indeed, in a microcosmic way the relatively insignificant therocephalian radiation mimicked the vastly more successful cynodont radiation of the Triassic Period, for as well as diversity, the therocephalians also show hints of an increase in general biological complexity through time. Thus the later members tend to develop a secondary palate and more complex post-canine teeth, trends which culminate in the bauriids which were superficially rather cynodont-like. Inevitably, the reasons why both diversity and progress were evidently more possible in this group than in, say, gorgonopsids or dicynodonts are obscure. There is no evidence from the structure of the primitive, pristerognathid therocephalians that their temperature regulation was any more developed than in other Permian therapsids. The advanced bone histology, large size, and their existence in the *Tapinocephalus*-zone temperate deposits can best be interpreted as signs that they were inertial homiotherms like dinocephalians (p. 97). However, the simple fact that much smaller forms evolved may indicate that the metabolic rate was higher, although it is, of course, possible that these smaller individuals hibernated. At any rate, the incipiently multicusped teeth of forms such as Scaloposaurids and complex dentition of the bauriids clearly suggest that the rate of food processing had increased in them.

Coupled with the fully formed secondary palate separating the air passage from the masticatory part of the buccal cavity, this is good evidence for a substantially increased metabolic rate by the end of the Permian.

Therocephalians possessed well-developed ridges in the snout, presumably for the support of turbinal cartilages as in gorgonopsids (p. 126) and so in this group too, olfaction was probably an important sense. The brain is not adequately known. What little evidence there is from posterior braincase casts (Olson, 1944) indicates that the brain was no larger or better differentiated than in contemporary therapsids such as gorgonopsids and dicynodonts.

# 10 Cynodonts

THE CYNODONTS were the last of the main therapsid groups to appear in the fossil record, being unknown prior to the last part of the Late Permian *Daptocephalus* or possibly the underlying *Cistecephalus*-zone of South Africa, and the Zone IV of Russia (Fig. 1). Specimens of the same age are also known from Karroo exposures in Tanzania and Zambia. Although these earliest forms are in several respects more primitive than Triassic cynodonts, they are nevertheless unmistakably at the cynodont level of evolution. The group persisted throughout the Triassic period, in the course of which many of their characters changed to a very mammalian condition. Indeed, the cynodonts are the therapsids most closely related to mammals. Two groups which are particularly mammalian in organisation are the tritylodontids and the trithelodontids (=ictidosaurs). They have been placed in groups separate from the cynodonts in the past, but are generally regarded as very advanced members of the Cynodontia nowadays.

## Systematics

The temporal fenestra of the cynodonts has enlarged in a manner similar to that seen in the therocephalians, but carried to greater lengths. It has expanded medially, producing a narrow intertemporal or sagittal crest which is much deeper than in typical therocephalians, even in small cynodonts. The postorbital bone is again reduced and restricted to the anteriormost part of the fenestra, and most of the crest is formed by the parietal bone. The fenestra has also expanded posteriorly, tending to cause the development of a backwardly reflected squamosal like that of gorgonopsids, and laterally to give a bowed lower temporal bar, or zygomatic arch. This arch also bows upwards, well above the level of the jaw hinge. The lower jaw is characterised by a relatively large dentary with a broad coronoid process rising above the level of the postdentary bones. There is a depression on the lateral surface of the coronoid process indicating that the adductor musculature had invaded the lateral surface of the jaw. The reflected lamina of the angular is reduced to a small, thin sheet of bone even in the most primitive forms.

The cynodont dentition is remarkable for the development of complex multi-cusped postcanine teeth behind the invariably fairly prominent canine. No doubt in association with the dental elaboration, a secondary palate is present, which is incomplete in the primitive forms, but in which the palatine plays a prominent role in contrast to the equivalent structures in therocephalians and dicynodonts. Several other skull features of cynodonts are equally characteristic, although of less obvious functional significance. Thus the nasal bones are expanded posteriorly, so that they meet the lachrymals. On the occipital surface, the supraoccipital is very narrow while the tabulars are broad and completely surround the post-temporal fenestra. The occipital condyle is at least incipiently double. The floor of the braincase has become thinner, and the basisphenoidal tubera present in other therapsids are lost. The epipterygoid is broadly expanded to form a large, thin sheet of bone lateral to the side wall of the braincase.

The postcranial skeleton of the most primitive cynodonts (Fig. 72) has few important differences from that of other therapsids such as gorgonopsids and therocephalians, although by the Triassic a number of profound modifications had occurred. Early characteristic features include a reduction of the atlas centrum and its fusion to the axis vertebra, promoting greater flexibility of the head on the vertebral column. The dorsal vertebrae tend to be differentiated into anterior thoracic and posterior lumbar vertebrae, and the ribs of the latter are reduced and immovably fixed to the vertebrae. The two heads of each rib become confluent at least in the more posterior vertebrae. The ilium is relatively elongated anteriorly and the pubis reduced.

An early view of the origin of cynodonts was that they evolved from gorgonopsids (Watson, 1921), but it is now clear that the similarities between these two groups are no more than retained ancestral therapsid features. Romer (1969a) believed that none of the other therapsid groups were closely related to the cynodonts, and that the ancestor of the latter lay at the very primitive therapsid level, exemplified by the eotitanosuchians. If this is true, then the cynodonts were a fifth independently evolved lineage from the remote therapsid ancestor. However, there are several features in which primitive cynodonts and therocephalians are alike, features clearly derived rather than ancestral in nature (Kemp, 1972a). The most immediate one is the form of the temporal fenestra and lower jaw, with the marked medial expansion of the former and associated coronoid process of the latter. The sagittal crest of both groups is very similar in construction, with reduction of the postorbital and dominanace of the parietal. The arrangement of the bones of the jaw is also very similar, for both have a flattened coronoid process, with the surangular overlapping its lower part over the medial surface of the jaw. An intramandibular fenestra is present in both, between the surangular above and the prearticular below. The nature of the bones of the jaw articulation is almost identical, with the condyles set at an angle to the transverse line, and articular facets of the articular facing largely backwards. Even the manner by which the quadrate complex is attached to the squamosal is comparable, with the quadratojugal extending into a slit in the squamosal. Other derived characters possessed by both groups include the structure of the basicranial region of the skull and possession of a connecting bridge between the quadrate ramus of the pterygoid and the prootic region of the side wall of the braincase. However, the bridge differs between the two, for in therocephalians it consists of a connection between the squamosal and the prootic, while in cynodonts the squamosal is not involved, and the connection is simply between a lateral prootic process and the quadrate ramus. This may therefore be a case of parallel evolution.

Among the known therocephalians, the whaitsiids are possibly the closest to the cynodonts for they have a very broad, cynodont-like epipterygoid involved in the side wall of the braincase, and reduction of the suborbital vacuity. However, more primitive whaitsiids such as *Moschowhaitsia* (Tatarinov, 1963, 1964) still have a suborbital vacuity, and therefore this character at least may have changed independently in cynodonts and advanced whaitsiids. Certainly, the whaitsiids have a number of specialisations indicating that a common ancestor between them and cynodonts could not have advanced much beyond a fairly generalised therocephalian-like form.

### *Primitive cynodonts*

The cynodont with the greatest number of primitive, that is unmodified, ancestral therapsid characters is the Russian Late Permian *Dvinia* (Fig. 70), known only from two skulls and a few other fragments (Tatarinov, 1968). It has six upper incisors and a small precanine tooth, and six lower incisors, thereby resembling primitive therocephalians. The dentary is relatively small and the postdentary bones correspondingly well developed, and although an adductor fossa is present on the

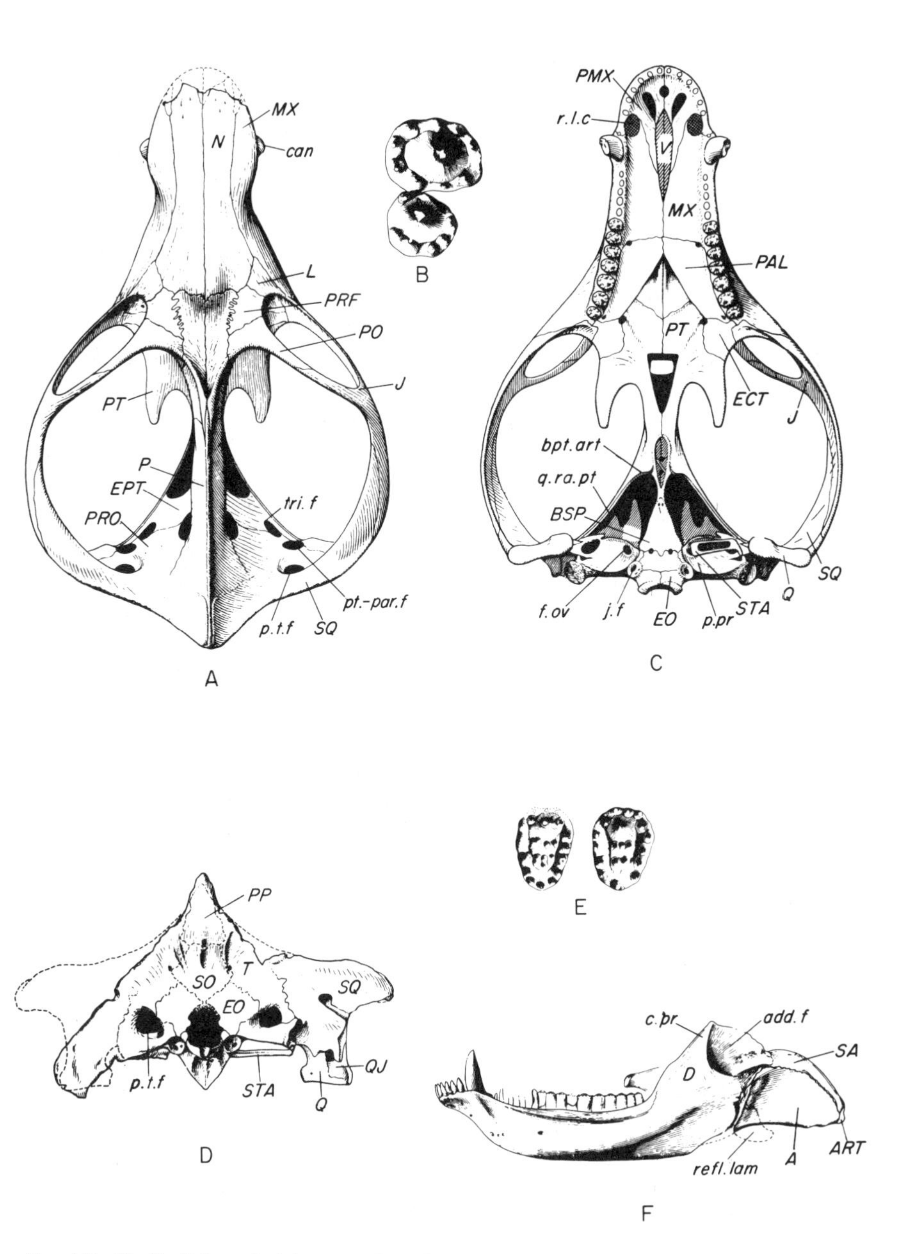


FIG. 70. Skull of the primitive cynodont *Dvinia*. A, dorsal view. B, enlarged crown view of last two upper right postcanines. C, ventral view. D, posterior view. E, enlarged crown view of two lower right postcanines. F, lateral view of lower jaw. (From Tatarinov, 1968: *Postilla*, Peabody Mus. Nat. Hist., Yale University.)

lateral surface of the coronoid process, it is restricted to a dorsal position (Fig. 70F). The reflected lamina is not completely known, but is larger than in later cynodonts, although still smaller and more delicate than in other therapsids. The secondary palate of *Dvinia* is actually completed medially for much of its length, in contrast to other Late Permian cynodonts. There is, however, a large anterior cleft exposing the vomer in ventral view. The vomers themselves are paired, which contrasts with all the other cynodonts known, where they are fused to form a median bone. Other primitive features seen in ventral view are the relatively large ectopterygoid, the persistent interpterygoid vacuity, and the width between the paired quadrate rami of the pterygoids. From what little is known about the postcranial skeleton, the ribs appear to be unexpanded and the ilium relatively little expanded, compared to Triassic cynodonts.

Set against these primitive features, however, *Dvinia* has several unique specialisations, indicating that it had evolved some considerable distance from the hypothetical common ancestor that it shared with all other cynodonts. The most remarkable feature is the enormous expansion of the temporal fenestra, both anteriorly between the orbits, and postero-medially so that the hind end of the sagittal crest is actually the posteriormost point of the skull. The increase in adductor musculature indicated by the temporal region is correlated with specialisation of the postcanine dentition (Fig. 70 B,E). The anterior postcanine teeth are slender and relatively simple, but the posterior seven are transversely widened and have very complex crowns. The uppers have a large central main cusp surrounded by up to about nine accessory cusps around the crown margin. In the case of the lower teeth, the main central cusp is placed towards the external side of the tooth and there are about fourteen accessory cusps around the margin. A further cluster of about six cuspules occupies the area between the main cusp and the internal crown margin. These complex teeth do not compare closely

*A*, angular; *add.f*, adductor fossa; *ART*, articular; *bpt.art*, basipterygoid articulation; *BSP*, basisphenoid; *c.pr*, coronoid process; *D*, dentary; *ECT*, ectopterygoid; *EO*, exoccipital; *EPT*, epipterygoid; *F*, frontal; *f.ov*, fenestra ovalis; *J*, jugal; *j.f*, jugular foramen; *L*, lachrymal; *MX*, maxilla; *N*, nasal; *P*, parietal; *PAL*, palatine; *PMX*, premaxilla; *PO*, postorbital; *PP*, postparietal; *p.pr*, paroccipital process; *PRF*, prefrontal; *PRO*, prootic; *PT*, pterygoid; *p.t.f*, post-temporal fenestra; *pt.par.f*, pterygo-paroccipital foramen; *Q*, quadrate; *QJ*, quadratojugal; *q.ra.pt*, quadrate ramus of the pterygoid; *refl.lam*, reflected lamina of the angular; *r.l.c*, recess for the lower canine; *SA*, surangular; *SO*, supraoccipital; *SQ*, squamosal; *STA*, stapes; *tri.f*, trigeminal foramen; *V*, vomer.

Magnification *c.* × 0.9.

with any of the later cynodont teeth, and were used in a quite different manner. Instead of the upper teeth occluding with the lower teeth, they bit against the dorsal shelf formed by the dentary. The lower teeth bit against the secondary palate, internal to the upper teeth. These bony surfaces were presumably covered in heavily keratinised epithelium, after the manner of dicynodonts. The diet of the animal probably included at least some herbaceous material, although the continued presence of well-developed canines suggests that animal food was also eaten. Possibly insects constituted much of their food (Tatarinov, 1968).

*Procynosuchus* (Fig. 71) is a rather less specialised primitive cynodont, from the *Daptocephalus*-zone and possibly also the *Cistecephalus*-zone of southern Africa (Kemp, 1979, 1980b). Several specimens of primitive cynodonts from South Africa have been placed in other genera, but at least most of these are probably *Procynosuchus* in various states of maturity and preservation (Hopson and Kitching, 1972). *Procynosuchus* possesses all the primitive characters mentioned for *Dvinia*, except that the vomers are fused, and the lower incisor teeth are reduced to four. These indicate that it has a relationship with all the later cynodonts (Fig. 87). The secondary palate has failed to meet medially at all, however, suggesting that the closure of the secondary palate in *Dvinia* occurred independently of the closure in Triassic forms.

The postcanine teeth are much less specialised than those of *Dvinia* and are of a form that was probably ancestral to the teeth of all the other cynodonts. The more anterior postcanines are fairly simple, with a large, slightly recurved, sharp crown and a poorly developed ridge or cingulum on the inner side. The more posterior postcanines still have the single dominant cusp, but the internal cingulum is much more prominent, bearing a series of around five cuspules (Fig. 74A). There is no trace of an external cingulum on either upper or lower teeth. The tooth rows are relatively long, and they diverge posteriorly, rather than being almost parallel as in *Dvinia*. Also there is some evidence in the form of wear facets on the teeth that the lowers made at least a crude contact with the uppers as they passed one another. *Procynosuchus* appears to have been adapted to eating insects (Kemp, 1979). The postcranial skeleton (Fig. 72) is well known (Kemp, 1980b) and consists of a fairly generalised therapsid skeleton with a number of specialisations superimposed, some of which are to be found in later cynodonts, but others are unique to *Procynosuchus*. The vertebrae lack the accessory zygapophyseal articulations found in the typical Triassic forms, and the ribs show no signs of expanded costal plates. The scapula is broad and flat, the corocoids relatively large and the glenoid widely open. The

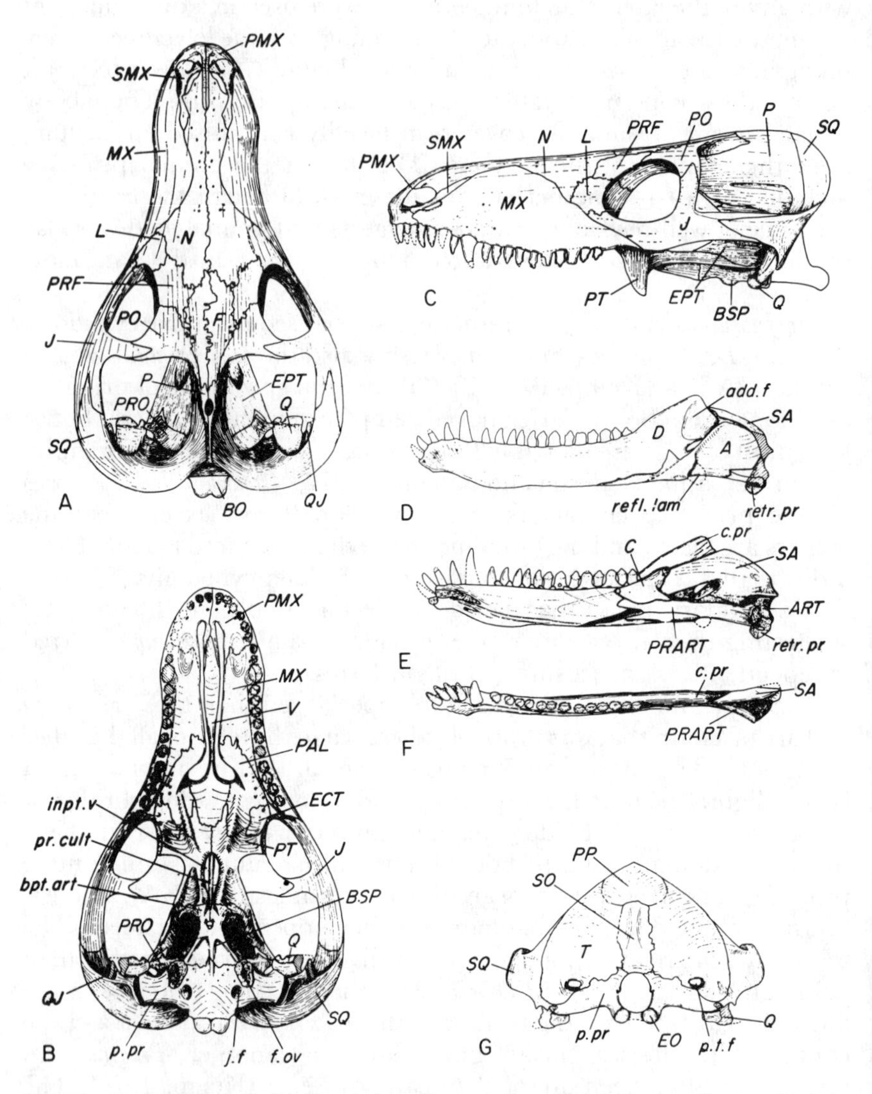


FIG. 71. Skull of the primitive cynodont *Procynosuchus*. A, dorsal view. B, ventral view. C, lateral view. D, lateral view of lower jaw. E, medial view of lower jaw. F, dorsal view of lower jaw. G, posterior view. (From Kemp, 1979.)

*A*, angular; *add.f*, adductor fossa; *ART*, articular; *BO*, basioccipital; *bpt.art*, basipterygoid articulation; *BSP*, basisphenoid; *C*, coronoid; *c.pr*, coronoid process; *D*, dentary; *ECT*, ectopterygoid; *EO*, exoccipital; *EPT*, epipterygoid; *F*, frontal; *f.ov*, fenestra ovalis; *inpt.v*, interpterygoid vacuity; *J*, jugal; *j.f*, jugular foramen; *L*, lachrymal; *MX*, maxilla; *N*, nasal; *P*, parietal; *PAL*, palatine; *PMX*, premaxilla; *PO*, postorbital; *PP*, postparietal; *p.pr*, paroccipital process; *PRART*, prearticular; *pr.cult*, processus ilium is only slightly expanded forwards. However, in common with later cynodonts, the vertebral column is differentiated into thoracic and lumbar regions by a slight increase in massiveness of the more posterior vertebrae, and a reduction in the length of their ribs. The more posterior lumbar ribs become immoveably attached to the transverse processes of the vertebrae. Particular specialisations of *Procynosuchus* are probably adaptations for swimming, for this animal appears to have been as adept in water as on land. Thus the zygapophyses of the lumbar vertebrae are horizontally oriented, permitting extensive lateral undulations of the posterior part of the body and tail. The limb bones and feet are extraordinarly flat, suggesting the use of the limbs as paddles.

Apart from these postcranial specialisations, *Procynosuchus* must be judged structurally close to the ancestor of all later cynodonts. The next most primitive group is the family Galesauridae (=Thrinaxodontidae). Its best known member is the relatively abundant *Thrinaxodon* from the basal Triassic *Lystrosaurus*-zone of South Africa, and it has also been recorded from the contemporaneous Fremouw Formation of Antarctica (Colbert and Kitching, 1977). However, a more primitive member of the family, *Cynosaurus* comes from the latest Permian *Daptocephalus*-zone of South Africa, and therefore this is one of the very few therapsid families to have succeeding in crossing the Permo-Triassic boundary. The rather specialised *Tribolodon* is the only certain galesaurid to have been found in the younger Lower Triassic *Cynognathus*-zone.

*Thrinaxodon* (Fig. 73) is often assumed to represent the stem from which all later cynodonts descended. Certainly it combines several advanced cynodont features with a number of primitive ones, but it does appear to have some specialisations as well. All the galesaurids have reduced the number of incisors to four uppers and three lowers, and the precanines are lost. The number of postdentary teeth is reduced to between seven and nine, which is probably one of the specialisations of the group. The actual structure of the postcanines differs between the different constituent genera, but all tend to enlarge the anterior and posterior cingulum cusps in line with the main, large cusp. Compared to *Procynosuchus*, the rest of the cingulum tends to be reduced or even lost (Fig. 74). In *Thrinaxodon* (Fig. 74F), the upper teeth have more or less

cultriformis of the parasphenoid; *PRF*, prefrontal; *PRO*, prootic; *PT*, pterygoid; *p.t.f*, post-temporal fenestra; *Q*, quadrate; *QJ*, quadratojugal; *refl.lam*, reflected lamina of the angular; *retr.pr*, retroarticular process; *SA*, surangular; *SMX*, septomaxilla; *SO*, supraoccipital; *SQ*, squamosal; *T*, tabular; *V*, vomer.

Magnification *c.* × 0.55.

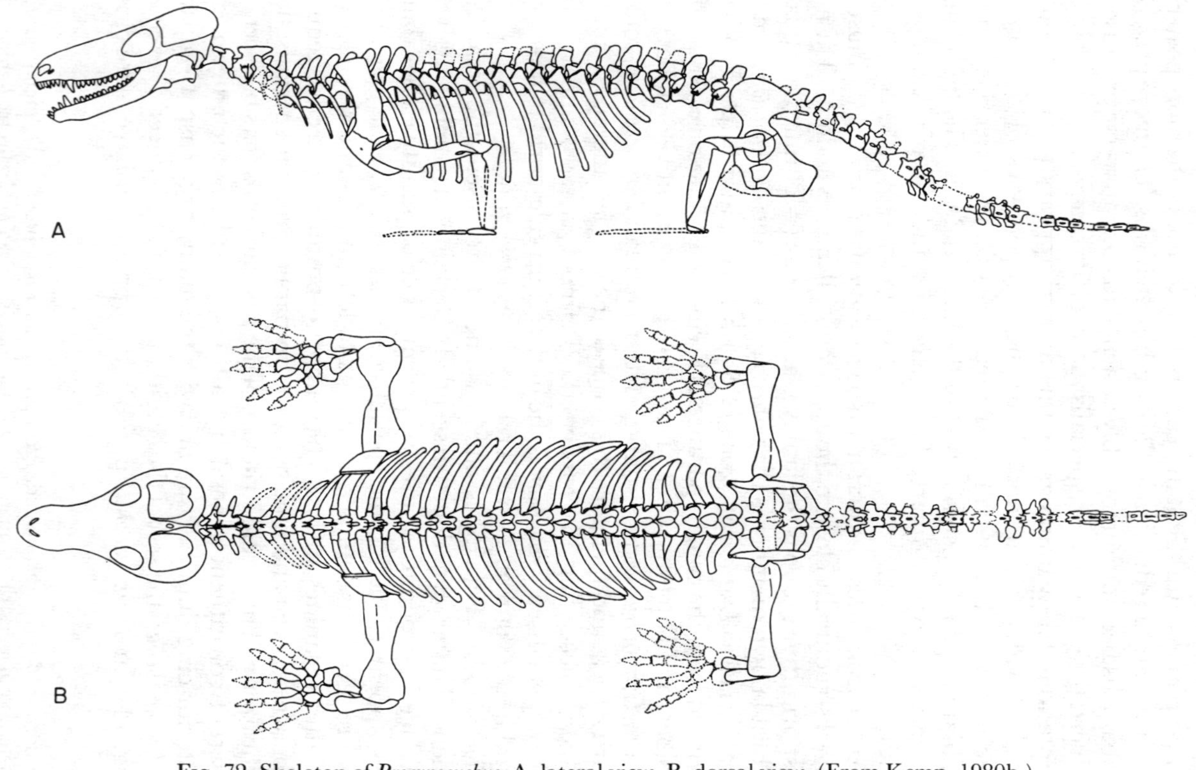


FIG. 72. Skeleton of *Procynosuchus*. A, lateral view. B, dorsal view. (From Kemp, 1980b.) Magnification *c.* × 0.26.

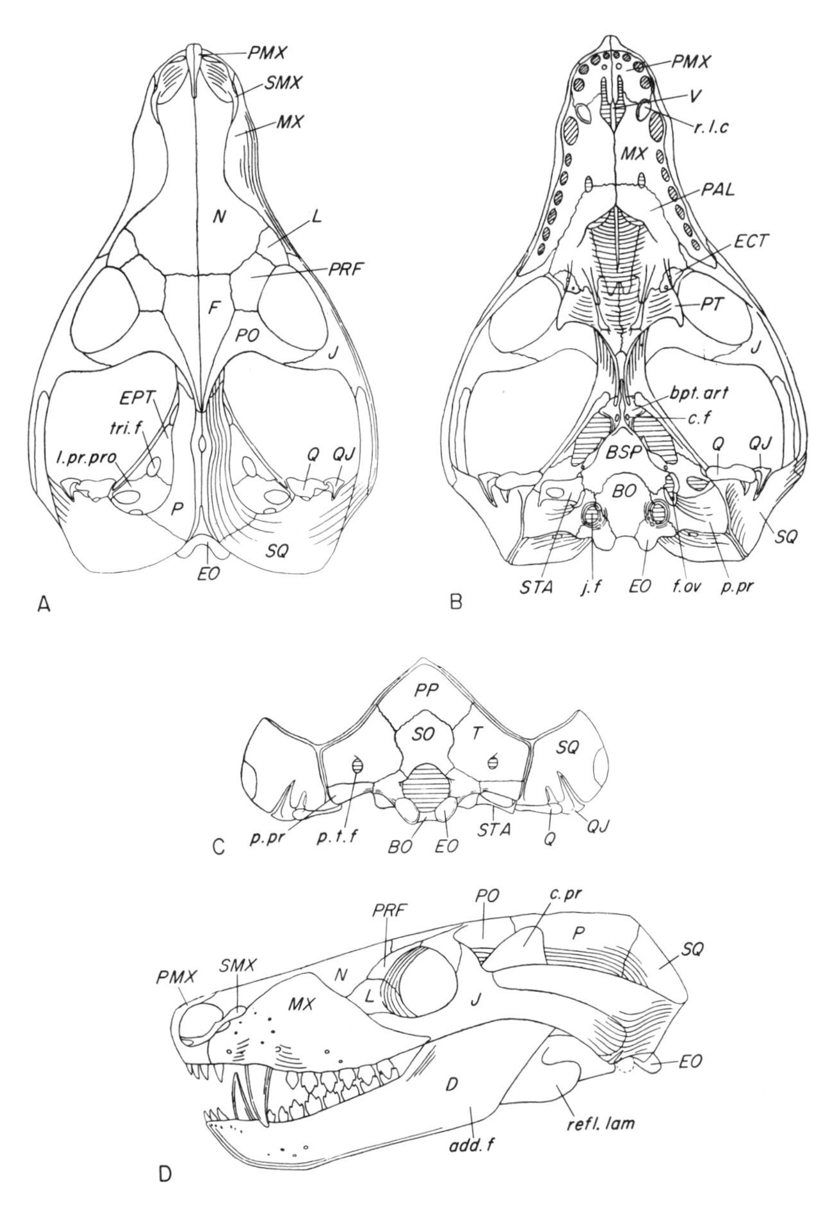


FIG. 73. The skull of the galesaurid cynodont *Thrinaxodon*. A, dorsal view. B, ventral view. C, posterior view. D, lateral view. (From Parrington, 1946.)

*add.f*, adductor fossa; *BO*, basioccipital; *bpt.art*, basipterygoid articulation;. *BSP*, basisphenoid; *c.f*, carotid foramen; *c.pr*, coronoid process; *D*, dentary; *ECT*, ectopterygoid; *EO*, exoccipital; *EPT*, epipterygoid; *F*, frontal; *f.ov*, fenestra ovalis; *J*, lost the cingulum, but that of the lower teeth is retained and a further accessory cusp is added at the back of the tooth, in the more complex, posterior teeth. *Cynosaurus* (Fig. 74B) has teeth rather similar to *Thrinaxodon* but with no cingulum at all (Van Heerden, 1976). The postcanines of *Galesaurus* (Fig. 74D) (Parrington, 1934) are very odd, for the large main cusp curves backwards to overhang a prominent posterior accessory cusp, but neither anterior accessory cusp nor cingulum is present. *Tribolodon* (Broili and Schröder, 1934a) has tall, fairly slender postcanines (Fig. 74E). Prominent anterior and posterior accessory cusps are present, but separated from the main cusp by shallow grooves, while large anterior and posterior cingulum cusps are also present, on the inner side. All these different types of postcanine teeth must reflect different feeding habits, although details have not been worked out. In general all the galesaurids were probably small carnivores or insectivores.

The adductor jaw musculature of galesaurids is advanced over *Procynosuchus*, as indicated by the larger temporal fenestra, and particularly the broader, more robust zygomatic arch (Fig. 73). In connection with the latter, the occipital surface is wide and relatively low, and is deeply

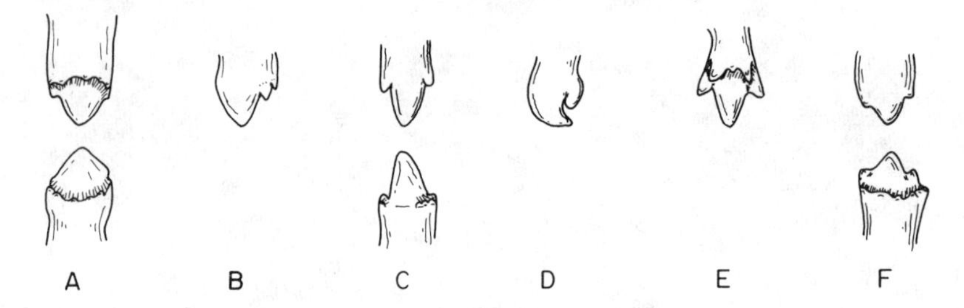


FIG. 74. Postcanine teeth of *Procynosuchus* and galesaurids. A, upper and lower postcanines of *Procynosuchus* in internal view. B, upper postcanine of *Cynosaurus* in external view. C, upper postcanine in external view and lower postcanine in internal view of *Nanictosaurus*. D, upper postcanine of *Galesaurus* in external view. E, upper postcanine of *Tribolodon* in internal view. F, upper and lower postcanines of *Thrinaxodon* in external view. (A redrawn after Kemp, 1979; B and C after Van Heerden, 1976; D after Broom, 1932; E after Broili and Schröder, 1934a; F after Crompton, 1972b.)

jugal; *j.f*, jugular foramen; *L*, lachrymal; *l.pr.pro*, lateral process of the prootic; *MX*, maxilla; *N*, nasal; *P*, parietal; *PAL*, palatine; *PMX*, premaxilla; *PO*, postorbital; *PP*, postparietal; *p.pr*, paroccipital process; *PRF*, prefrontal; *PRO*, prootic; *PT*, pterygoid; *p.t.f*, post-temporal fenestra; *Q*, quadrate; *QJ*, quadratojugal; *refl.lam*, reflected lamina of the angular; *r.l.c*, recess for the lower canine; *SMX*, septomaxilla; *SO*, supraoccipital; *SQ*, squamosal; *STA*, stapes; *T*, tabular; *tri.f*, trigeminal foramen; *V*, vomer.

Magnification *c.* × 1.0.

incised dorsally betwen the occipital plate proper and the laterally flaring zygomatic arch. The dentary has enlarged and the postdentary bones are correspondingly smaller; most immediately noticeable is the now huge coronoid process of the dentary, which extends high up into the temporal fenestra, and bears an adductor fossa over the whole lateral surface down to the ventral edge. Nevertheless, the general construction of the lower jaw is similar to that of *Procynosuchus*. The quadrate complex is smaller and less exposed below the squamosal. It has developed a more complex attachment to the squamosal, whereby the slit into which the quadratojugal passes is deeper, and the quadrate has a posterior flange which fits into a second slit in the squamosal (Parrington, 1946; Crompton, 1972a).

The secondary palate is complete except for a small anterior fissure in most galesaurids, although *Cynosaurus* is more primitive in retaining an incomplete, *Procynosuchus*-like palate. The ectopterygoid bone is reduced, and the lateral pterygoid processes of the back of the palate are smaller and more delicately built than in the more primitive cynodonts. The interpterygoid vacuity has closed (although Estes, 1961, reported a vacuity in a juvenile *Thrinaxodon*) and the processus cultriformis of the parasphenoid is reduced to a small splint above the closed vacuity.

*Thrinaxodon* (Fig. 75) and *Galesaurus*, of which the postcranial skeleton

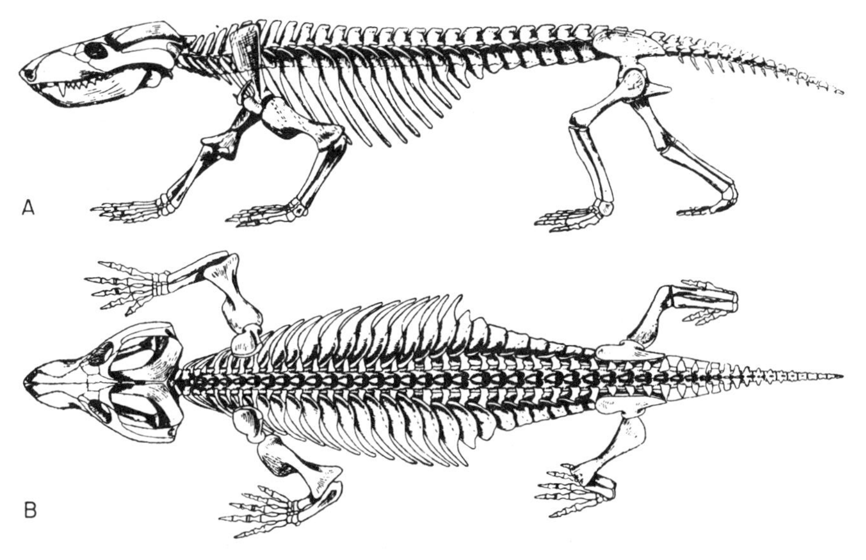


Fig. 75. The postcranial skeleton of *Thrinaxodon*. A, lateral view. B, dorsal view. (From Brink, 1956a.)
Magnification *c.* × 0.3.

is known, have a number of quite remarkable features, particularly of the vertebrae and ribs (Jenkins, 1971a). Pairs of accessory articulations have developed between a peg below the posterior zygapophysis and a groove below the anterior zygapophysis of the next vertebra behind. The ventral head of each rib, the capitulum, articulates with a parapophyseal facet that is shared equally between the centra of adjacent vertebrae. The proximal part of the rib shaft has expanded to form a broad plate, which is overlapped by the plate of the next anterior rib, and in turn overlaps the rib-plate behind. The plates are approximately horizontal in the thoracic region of the vertebral column, and the rib is continued distally as a slender shaft. More posteriorly, in the lumbar region, the plates become almost vertically disposed, and no distal shaft is developed. The rib-cage is therefore much shorter than in *Procynosuchus*. Changes in the limb girdles include the development of a deeply concave scapula blade with a narrowing base, and a reduced coracoid plate. The upper half of the glenoid, formed from the scapula, is an almost flat, horizontal surface which restricted the movements of the humerus more than in *Procynosuchus*. The pelvic girdle has a greater expansion forwards of the ilium, and further reduction of the pubis, and the trochanter major of the femur is somewhat better developed.

### *Advanced cynodonts*

An important radiation of cynodonts occurred in the Triassic subsequent to *Lystrosaurus*-zone times, with about half a dozen families represented (Fig. 87). The radiation appears to have been world-wide, for specimens are known from southern Africa, South America, Russia, China and India. Very late advanced forms, tritylodontids, have also been discovered in North America and western Europe. The relationship between all these advanced cynodonts on the one hand and the galesaurids on the other is indicated by the common possession of numerous derived characters, for example the reduction of the incisor teeth to four uppers and three lowers, and loss of the precanines. The enlarged dentary bone and extensive adductor fossa of the lateral surface of galesaurids are necessary precursors of the even more advanced lower jaw of the later forms, and the nature of the quadrate complex and its mode of attachment is remarkably similar. The structure of the palate, with reduced ectopterygoid and lateral pterygoid processes, and closure of the interpterygoid vacuity is similar in all the Triassic cynodonts including galesaurids. Equally strong evidence for the relationship comes from the postcranial skeleton, where the expanded ribs, concave scapula and anteriorly extended ilium and reduced pubis of galesaurids is matched in the advanced forms. The only exception is that in certain very advanced Middle and Upper Triassic cynodonts the ribs are unexpanded. Bonaparte (1963a) suggested that this indicates a separate origin of these forms from a pre-galesaurid level, but close relatives of these which still possess expanded ribs are now known (Jenkins, 1970; Kemp, 1980c). It must be assumed therefore that a few terminal members of the Triassic cynodont radiation secondarily simplified the rib structure.

The question of whether all the later, advanced cynodonts constitute a monophyletic group, or whether more than one line evolved independently from a galesaurid level is controversial. A commonly held view (e.g. Hopson and Crompton, 1969; Crompton and Jenkins, 1973) is that at least two and possibly more of the advanced cynodont groups evolved in parallel. The main argument here is the diverse nature of the postcanine teeth, particularly between the herbivores and the carnivores. However, all the advanced cynodonts possess a spectacular array of features which were clearly derived from the galesaurid condition, and all of which must have been acquired independently more than once if the polyphyletic view is correct. Most obviously, the dentary has enlarged even more at the expense of the postdentary bones, which are reduced to little more than a compound rod set in a trough on the inner side of the dentary and only extend for a short distance below and behind. The jaw articulation is more complex because there is a secondary contact between the surangular bone and the squamosal, alongside the primary articular-quadrate joint. The postcanine teeth of the various families vary in form, but in all cases they have evolved a mechanism of true occlusion between the uppers and the lowers, which involves a posterior movement of the lower jaw, and implies a common neurological basis of mastication. Compared to galesaurids, the base of the braincase is much narrower and the epipterygoid bones have moved closer to the midline, while the occiput has developed a prominent groove on its posterior surface, the external auditory meatus, which is believed to have housed the air-filled tube leading to the middle ear. At the most, the meatus is incipient in galesaurids. The postcranial skeleton also possesses a number of advances, common to at least many of the advanced groups. In view of all these shared, derived features, it is more likely that the advanced cynodont condition arose only once from the galesaurid-like ancestor, and only subsequently diverged into the various separate lines (Fig. 87).

There are two herbivorous groups of advanced cynodonts which are closely related to one another. They appear first in the Lower Triassic *Cynognathus*-zone of the Karroo of South Africa, where the best known

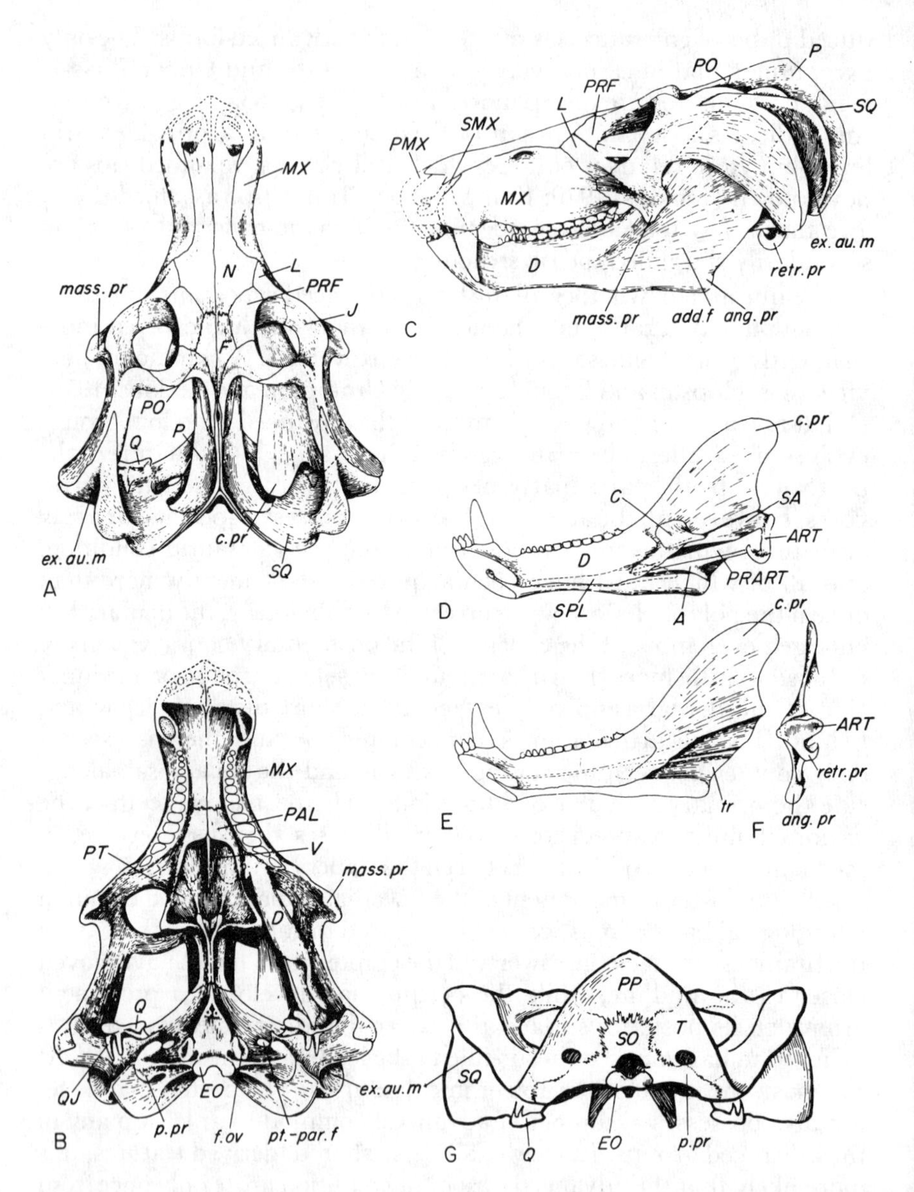


FIG. 76. The skull of *Diademodon*. A, dorsal view. B, ventral view. C, lateral view. D, medial view of the lower jaw. E, medial view of the lower jaw without the postdentary bones. F, posterior view of the lower jaw. G, posterior view. (A–F from Brink, 1963b; G from Broili and Schröder, 1935b.)

*A*, angular; *add. f*, adductor fossa; *ang. pr*, angular process; *ART*, articular; *BSP*, basisphenoid; *C*, coronoid; *c.pr*, coronoid process; *D*, dentary; *EO*, exoccipital; *ex.au.m*, form, *Diademodon*, is relatively abundant (Fig. 76). It was a fairly large animal, with a skull length up to about 40 cm. The shape of the skull is characteristic, for the snout is narrow while the orbital region widens so that the eyes faced forwards. The postorbital region is dominated by a pair of huge temporal fenestrae, separated from one another by the narrow, high sagittal crest. The zygomatic arch bounding each fenestra laterally is a massive bar of bone. A distinct ventral process at its anterior end, immediately below the orbit is termed the masseter process, and indicates that the masseter muscle had spread forwards along the full length of the arch. The temporal fenestra has also expanded posteriorly to a more marked degree than in the primitive cynodonts. An important consequence is that the upper part of the base of the zygomatic arch has turned posteriorly, creating a deep groove beneath it. This groove, which is only incipiently present in a form such as *Thrinaxodon*, is believed to have housed the external auditory meatus, the air-filled tube which led from the outside to the hearing structures of the middle ear. Another characteristic of this part of the skull of *Diademodon* is the deep notch in the dorsal border of the occipital surface, between the occiput proper medially and the root of the zygomatic arch laterally (Fig. 76G).

The increase in the adductor musculature of the jaw indicated by the enlargement of the temporal fenestra is reflected in the structure of the lower jaw. The dentary bone has grown at the expense of the post-dentary bones and dominates the jaw (Fig. 76C–F). There is a huge coronoid process which rises high into the temporal fenestra when the jaw is closed, and also a postero-ventral extension, the angular process, which occupies the approximate area of the jaw formerly occupied by the reflected lamina of the angular. The whole of the posterior region of the dentary was muscle-bearing. The postdentary bones are reduced to little more than a compound rod, lying within a hollowing of the medial surface of the dentary and only extending posteriorly beyond the dentary for a comparatively short distance to the jaw articulation. The articulation itself has become more complex, for in addition to the normal joint formed between the articular of the lower jaw and the

external auditory meatus; *F*, frontal; *f.ov*, fenestra ovalis; *J*, jugal; *L*, lachrymal; *mass.pr*, masseter process; *MX*, maxilla; *N*, nasal; *P*, parietal; *PAL*, palatine; *PMX*, premaxilla; *PO*, postorbital; *PP*, postparietal; *p.pr*, paroccipital process; *PRART*, prearticular; *PRF*, prefrontal; *PT*, pterygoid; *pt.par.f*, pterygo-paroccipital foramen; *Q*, quadrate; *QJ*, quadratojugal; *SA*, surangular; *SMX*, septomaxilla; *SO*, supraoccipital; *SPL*, splenial; *SQ*, squamosal; *rart.pr*, retroarticular process; *T*, tabular; *tr*, trough of dentary for the postdentary bones; *V*, vomer.

Magnifications: A–F, *c.* × 0.26; G, *c.* × 0.35.

reduced quadrate, there is a secondary contact (Fig. 92D,E). This is between the posterior end of the surangular bone and a ventral extension of the squamosal, and lies lateral to the quadrate-articular joint (Crompton, 1972a).

The dentition of *Diademodon* consists of four upper and three lower incisors, as in the galesaurids. The canines are quite prominent. The postcanines are, however, highly modified for dealing with herbivorous material, and for the first time in cynodont evolution, true, accurate occlusion between the upper and lower teeth occurred. The anterior four or so postcanine teeth are simple, conical crowns. These are followed by up to nine (depending on age) transversely widened, multicusped "gomphodont" teeth. The posteriormost two to five postcanine teeth are of yet another type, described as sectorial, in which there is a dominant, somewhat recurved main cusp that is flattened laterally. A number of smaller accessory cusps lie behind the main cusp, in line with it. An anterior accessory cusp is also present, lying in front of and slightly internal to the main cusp. The pattern of the cusps of the gomphodont teeth is important in assessing the relationships of *Diademodon* to other herbivorous cynodonts (Fig. 77A). The upper teeth are very widened across the jaw and are oval in section. The largest cusp lies in the middle of the outer edge of the crown, and sharp, crenulated ridges run both forwards and backwards from this cusp. The anterior, inner and posterior edges of the crown all bear small cusps, and an ill-defined transverse ridge runs across the middle of the crown from the main cusp to the inner edge of the tooth. The lower gomphodont postcanines are less expanded, being approximately circular in crown view. They have a cusp arrangement basically similar to that of the upper teeth, with a large external cusp marking the high point of a longitudinal external ridge, and again a poorly defined transverse crest between the main cusp and the inner edge of the crown. In fact, these various features of the teeth are only displayed in newly erupted teeth. The enamel is extremely thin and becomes rapidly worn away, destroying the features and reducing the crown to no more than a featureless peg of dentine. Occlusion of the teeth occurred by each lower tooth working in the valley formed by two adjacent upper teeth. (Crompton, 1972b).

An evolutionary change that occurred between primitive cynodonts and *Diademodon* concerns the basicranial axis. The epipterygoids, which lie alongside the axis, have moved inwards closer to the basisphenoid, presumably to increase the size of the space through which the adductor jaw muscles pass from the sagittal crest to the lower jaw. In the process, the basipterygoid articulation between the basisphenoid and the pterygoids has been reduced, the carotid foramina which in primitive cynodonts pierce the basisphenoid have been lost, and the lateral process of the prootic which contacts the quadrate ramus of the pterygoid has enlarged.

The postcranial skeleton of *Diademodon* is surprisingly poorly represented in view of the abundance of skulls (Jenkins, 1971a). In most respects it resembles that of *Thrinaxodon*, and the main difference concerns the form of the costal plates of the ribs. These are absent from the

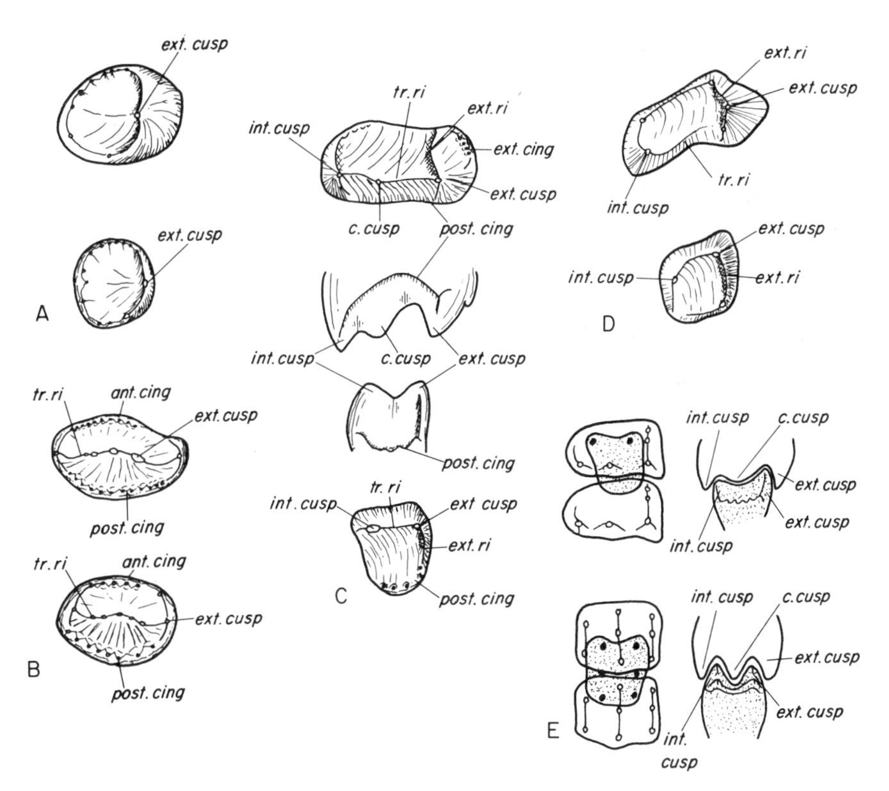


FIG. 77. Postcanine teeth of herbivorous cynodonts. A, upper and lower postcanines of *Diademodon* in crown view. B, upper and lower postcanines of the trirachodont *Cricodon*. C, upper postcanine in crown and posterior views, lower postcanine in posterior and crown views of the traversodontid *Scalenodon*. D, upper and lower postcanines in crown view of the advanced traversodontid *Exaeretodon*. E, comparison of the postcanines of the traversodontid *Massetognathus* with a tritylodontid. Occlusal and posterior views of both are shown, upper teeth clear, lower teeth stippled. (Redrawn after Crompton, 1972b.)

*ant.cing*, anterior cingulum; *c.cusp*, central cusp; *ext. cing*, external cingulum; *ext.cusp*, external cusp; *ext.ri*, external ridge; *int.cing*, internal cingulum; *int.cusp*, internal cusp; *tr.ri*, transverse ridge.

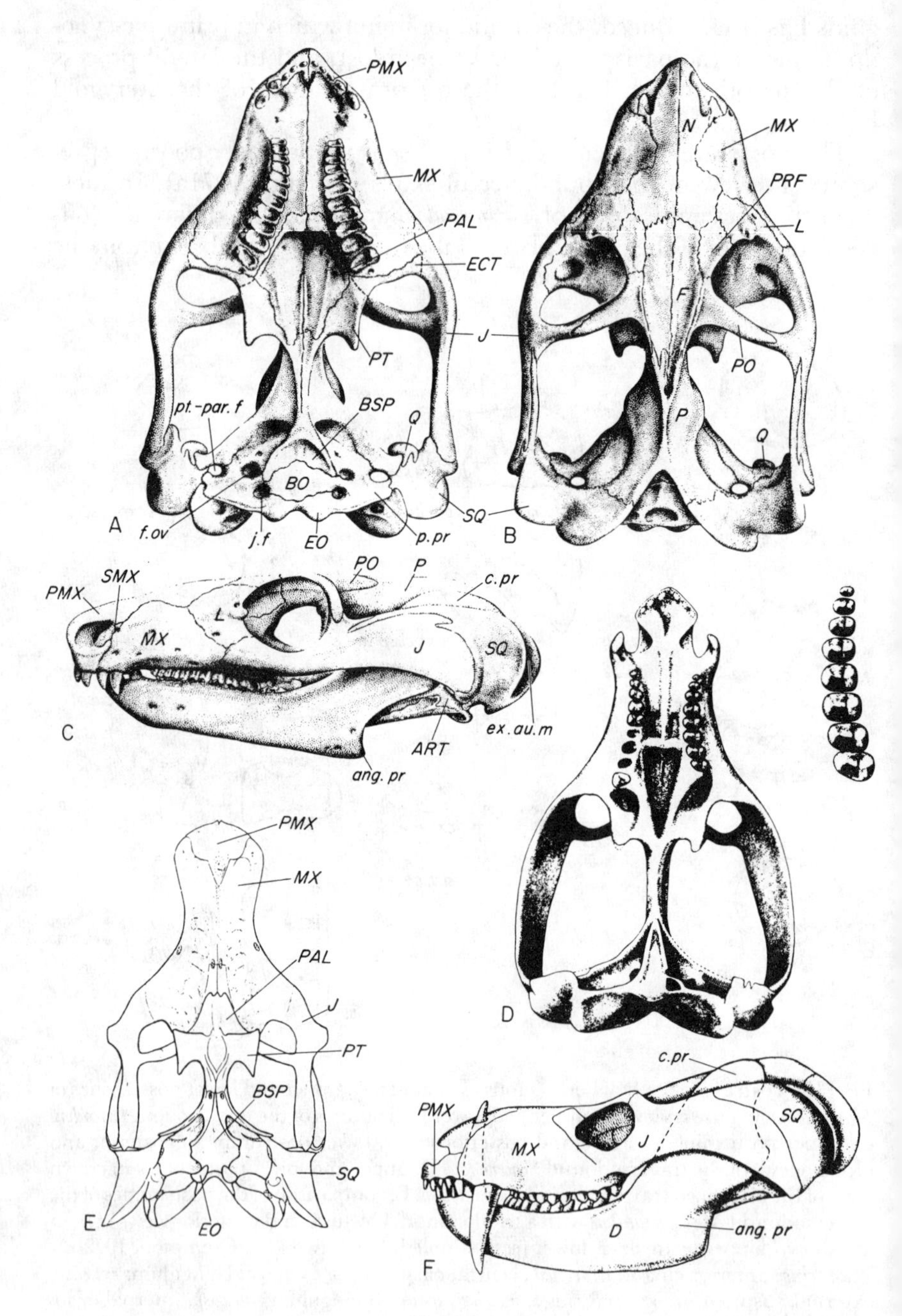
PMX
MX
PAL
ECT
J
PT
pt.-par. f
BSP
Q
BO
A
f.ov
j.f
EO
p.pr
SQ
N
PRF
L
F
PO
P
B
SMX
C
c.pr
ex.au.m
ART
ang. pr
D
E
F

FIG. 78.

cervical and anterior thoracic ribs, and differ in shape on the posterior thoracic and lumbar ribs (Fig. 94F). Each costal plate is relatively smaller than the corresponding structure in *Thrinaxodon*, and is a characteristic diamond shape, restricted more to the proximal end of the rib shaft. A small, but potentially important development of the shoulder girdle is the appearance of a distinct acromion process on the front edge of the scapula blade. It is turned outwards and the distal end of the clavicle attaches to it, creating a space between the scapula and coracoids medially and the clavicle laterally. This was the route by which muscles of the lateral surface of the coracoid plate could invade the medial surface of the scapula (Fig. 96B).

*Trirachodon* is a form closely related to *Diademodon* and also found in the *Cynognathus*-zone of South Africa. It differs from *Diademodon* only in the structure of the postcanine dentition, for the simple type of anterior postcanines are not present, and the cusp arrangement of the gomphodont teeth is not the same. Both the uppers and the lowers have a transverse row of three cusps across the middle of the crown, of which the centre cusp is the largest. A series of much smaller cusps lie on both the anterior and posterior edges. Posterior sectorial teeth almost identical to those of *Diademodon* are present. *Cricodon* (Fig. 77B) from the slightly younger Manda Formation of Tanzania (Crompton, 1955b) is a late surviving relative of *Trirachodon*.

The most diverse and successful of all the herbivorous cynodonts were the family Traversodontidae, in which the skull structure closely resembled that of *Diademodon*, but in which the postcanine teeth had become more effective by elaboration of the transverse crest. Certain primitive traversodontids from the Lower Triassic Puesto Viejo and Rio Mendozo Formations of Argentina, *Pascualgnathus* (Fig. 78F), *Andescynodon* and *Rusconiodon* (Fig. 78D) (Bonaparte, 1970) have teeth in which the transverse ridge is little better defined than in *Diademodon*, and lies in the middle of the crown. The intermediate nature of these

FIG. 78. Skulls of traversodontid cynodonts. A, *Massetognathus* in ventral view. B, *Massetognathus* in dorsal view. C, *Massetognathus* in lateral view. D, *Rusconiodon* in ventral view. E, *Exaeretodon* in ventral view. F, *Pascualgnathus* in lateral view. (A–C from Romer, 1967; D and F from Bonaparte, 1970; E from Bonaparte, 1962.)

*ang.pr*, angular process; *ART*, articular; *BO*, basioccipital; *BSP*, basisphenoid; *c.pr*, coronoid process; *D*, dentary; *ECT*, ectopterygoid; *EO*, exoccipital; *F*, frontal; *f.ov*, fenestra ovalis; *J*, jugal; *j.f*, jugular foramen; *L*, lachrymal; *MX*, maxilla; *N*, nasal; *P*, parietal; *PAL*, palatine; *PMX*, premaxilla; *PO*, postorbital; *p.pr*, paroccipital process; *PRF*, prefrontal; *PT*, pterygoid; *pt.par.f*, pterygo-paroccipital foramen; *Q*, quadrate; *SMX*, septomaxilla; *SQ*, squamosal.

Magnifications: A–C, *c.* × 0.65; D, *c.* × 0.57; E, *c.* × 0.17.

forms offers the best evidence that the traversodontids were in fact closely related to the diademodontids, and that the traversodontid type of gomphodont tooth can be derived from a typical diademodontid tooth.

In typical traversodontids (Fig. 78A–C) of the Middle Triassic, such as *Scalenodon* (Crompton, 1955b; 1972b) from the Manda Formation of Tanzania, *Luangwa* from the Ntawere Formation of Zambia (Kemp, 1980c), and *Massetognathus* from the Chañares Formation of Argentina (Romer, 1967), neither the simple anterior type nor the sectorial posterior type of postcanines are present, leaving only gomphodont teeth (Fig. 77C). The transverse ridge of these is a dominant feature, and has shifted towards the posterior edge of the upper teeth. It is bounded by high internal and external cusps, and the middle of the crest is in the form of a large, low cusp. A longitudinal ridge is also well developed, running along the external edge of the crown forwards form the main external cusp. A deep occlusal basin lies bounded by these two crests, and by a variety of patterns of small cusps and cingulum cuspules in the different forms. As in *Diademodon*, the lower teeth are less widened transversely. The transverse crest of these has shifted anteriorly, and runs between prominent external and internal cusps. Thus the lower teeth too have a wide occlusal basin in the middle of the crown. The effect of these changes in the tooth morphology is that high, sharp transverse crests on both the upper and the lower teeth constitute pairs of opposing shearing edges, facilitating the cutting of fibrous vegetation. The opposing basins, on the other hand, increase the ability of the teeth to crush or pound food material. As in *Diademodon*, the teeth wore rapidly and therefore lost their characteristic features.

The skull structure of typical traversodontids (Fig 78A–C) is basically similar to that of *Diademodon*, the most obvious difference being a tendency to widen the snout but keep the left and right postcanine tooth rows close together. The maxilla therefore comes to overhang the dentition laterally, which may be a method of increasing the volume of the nasal cavity for elaboration of the sense of olfaction (Kemp, 1980c). The temporal fenestra is even larger, particularly antero-laterally, and the zygomatic arch is therefore more or less parallel to the midline of the skull. The postero-ventral region of the dentary enlarged to give a more prominent angular process (Fig. 78C), but the dentary has not grown backwards to invade the jaw articulation at this stage.

The postcranial skeleton of *Luangwa* (Kemp, 1980c) differs in certain small but significant respects from the diademodontids. The costal plates of the thoracic region are similar, but those of the lumbar region are simpler, although equally prominent in size (Fig. 94H). The acromion process of the scapula is larger and carried further laterally, indicating a more extensive invasion of the inner surface of the scapula blade by muscle fibres. The most marked difference, however, concerns the pelvic girdle (Fig. 95C), which has evolved a much more mammalian structure. The ilium extends further forwards and the posterior process is reduced compared to *Diademodon*, the ischium is almost horizontally oriented, and the pubis has turned backwards. Similar changes to the limb girdles are seen in *Massetognathus* (Fig. 79A), while evolution of the ribs has gone even further towards simplification (Fig. 94G). There are no costal plates present in the thoracic region at all, while those of the lumbar region are reduced to no more than delicate anterior and posterior processes at the distal ends of the ribs, which overlap one another (Jenkins, 1970).

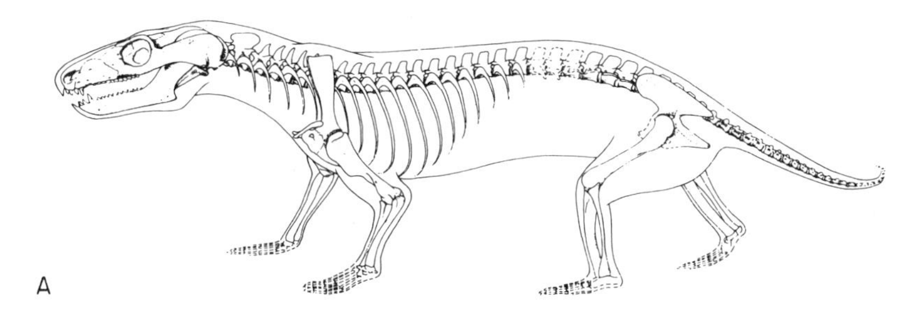


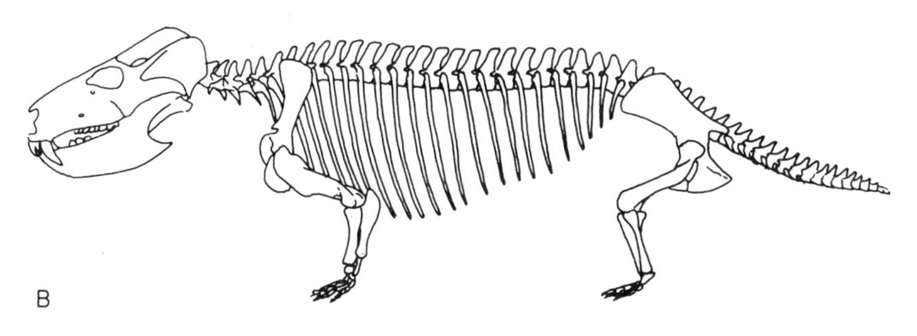


Fig. 79. Skeleton of traversodontid cynodonts. A, *Massetognathus*. B, *Exaeretodon*. (A from Jenkins, 1970; B redrawn after Bonaparte 1963b.)
Magnifications: A, *c*. × 0.21; B, *c*. × 0.068.

The latest and most specialised of the traversodontids occurred in the early part of the Upper Triassic. *Exaeretodon* (Bonaparte, 1962, 1963b) and *Ischignathus* (Bonaparte, 1963c) are from the Ischigualasto Formation of Argentina (Fig. 78E, 79B) and *Scalenodontoides* (Crompton and

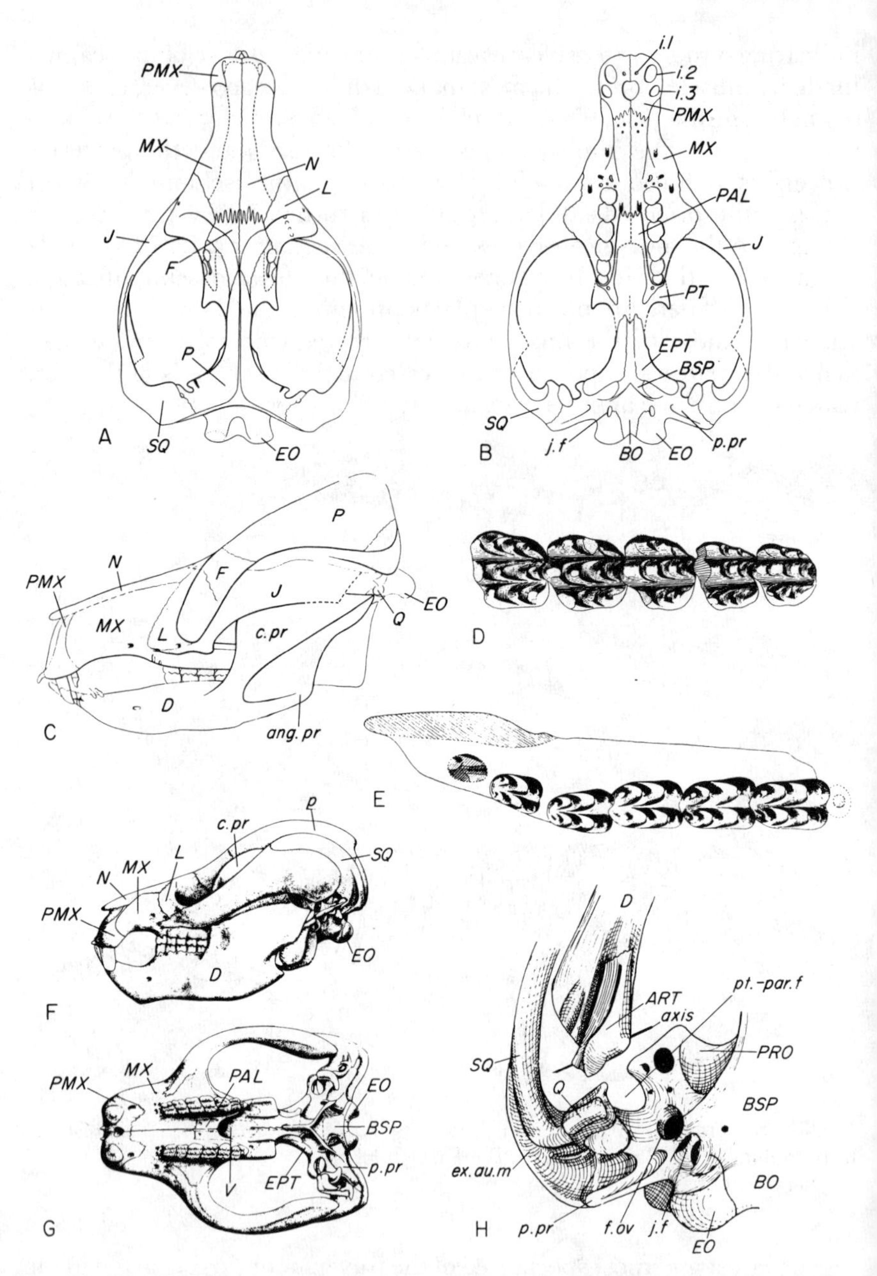


FIG. 80. The skull of tritylodontids. Skull of *Oligokyphus* in A, dorsal view; B, ventral view; C, lateral view. D, upper left postcanines. E, lower left postcanines. Skull of *Bienotherium* in F, lateral view; G, ventral view. H, enlarged ventral view of the postero-lateral part of the skull of *Oligokyphus*. (A–E from Kühne, 1956; F and G from

Ellenberger, 1957) is known from a lower jaw from the Red Beds of the South African Karroo Stormberg Series. In all these, the gomphodont teeth have the same basic structure as the Middle Triassic traversodontids, but the tendency for the adjacent teeth to interlock with one another has become more marked, and each tooth crown has a strongly concave posterior edge which fits against the convex anterior edge of the next tooth back (Fig. 77D). A large diastema separates the first postcanine tooth from the canine and the whole postcanine tooth row is straight and parallel to the skull midline. The postcranial skeleton of the very large *Exaeretodon* (Bonaparte, 1963b) is remarkable for the complete absence of costal plates on any of the ribs, and of accessory zygapophyseal articulations between adjacent vertebrae (Fig. 79B).

The family Tritylodontidae consists of a well-defined group of advanced herbivores. They possess certain mammalian characters such as the loss of the prefrontal and postorbital bones, and multi-rooted postcanine teeth, not found in any other cynodonts, and were indeed regarded as mammals at one time. Their relationships are still not certain, but since the most widely held current view is that they merely represent a further step along the trends noted in the other herbivorous cynodonts, they are discussed at this point. Tritylodontids appear in the Upper Triassic and had a world-wide distribution, since specimens have been found in southern Africa, China, Western Europe, North America and South America. This group also has the distinction of being the only therapsids which survived the Triassic–Jurassic boundary. *Oligokyphus* occurs in abundance as fragments preserved in fissure fills of Rhaetic age in England (Kühne, 1956) while the single lower jaw of *Stereognathus* is from the Middle Jurassic Stonesfield Slate of Oxfordshire.

Tritylodontids varied in size between very small, such as *Oligokyphus* with a skull length around 8 cm (Kühne, 1956) and forms such as *Tritylodon maximus* (Fourie, 1962, 1968), where the skull was about 22 cm in length. With the loss of the postorbital bone (Fig. 80A), the orbit and temporal fenestra have become confluent as in primitive mammals. The

Romer, 1966, after Hopson, reprinted from "Vertebrate Paleontology" by A. S. Romer by permission of the University of Chicago Press; H from Crompton, 1964.)

*ang.pr*, angular process; *ART*, articular; *BO*, basioccipital; *BSP*, basisphenoid; *c.pr*, coronoid process; *D*, dentary; *EO*, exoccipital; *EPT*, epipterygoid; *ex.au.m*, external auditory meatus; *F*, frontal; *f.ov*, fenestra ovalis; *i.1–3 incisors; J*, jugal; *j.f*, jugular foramen; *L*, lachrymal; *MX*, maxilla; *N*, nasal; *P*, parietal; *PAL*, palatine; *PMX*, premaxilla; *PRO*, prootic; *p.pr*, paroccipital process; *PT*, pterygoid; *pt.par.f*, pterygoparoccipital foramen; *Q*, quadrate; *SQ*, squamosal; *V*, vomer.

Magnifications: A–C, *c.* × 0.6; F and G, *c.* × 0.28.

temporal fenestrae themselves are large and separated from one another by a high, narrow sagittal crest. The dentary bone is relatively huge and mammal-like, with very well-developed coronoid and angular processes, and a posterior extension, the articular process, which almost but not quite reaches the jaw articulation. Fourie (1968) claimed that a secondary, mammalian jaw articulation between the dentary and the squamosal exists in *Tritylodon maximus*. However, his specimen is not well preserved, and this structure has not been found in other members of the family (Crompton, 1972a). The extremely reduced postdentary bones lie as a compound rod within a deep groove on the medial side of the dentary. The dentition is perhaps the most remarkable feature of the tritylodontids. The canines are absent, but have been replaced functionally by an enlarged pair of incisors. This is the second of three pairs of incisors in *Oligokyphus* (Fig. 80B), but the sole remaining pair in *Bienotherium* (Fig. 80F,G). A very long diastema separates the incisors from the 6–8 postcanines, each of which is enlarged. The crown of the upper postcanine (Fig. 80D) has three longitudinal rows of crescentic cusps, three cusps in the inner and middle rows and two cusps in the outer row. The lower teeth (Fig. 80E) are similar except that there are only two rows of three cusps each, present. When the teeth met, the two lower rows of cusps occluded with the valleys respectively formed between the three upper rows. Each upper cusp has its concave edge facing forwards, while the concave edges of the lower cusps face backwards. As the lower teeth were drawn backwards along the upper teeth, the opposing cusps formed a great battery of pairs of cutting edges, suitable for the fine mastication of foodstuff (Fig. 77E).

The postcranial skeleton, which is adequately known only in *Oligokyphus* (Fig. 81), is relatively long, slender and short limbed, being rather reminiscent of a modern mustelid mammal. It has evolved a number of mammalian features which are not present in the other herbivorous cynodont families. The ribs have no trace of costal plates. The shoulder girdle is modified by reduction of the coracoid plate and even more extensive reflection of the acromion process away from the main plane of the scapulo-coracoid (Fig. 96D). The humerus (Fig. 96E) is much

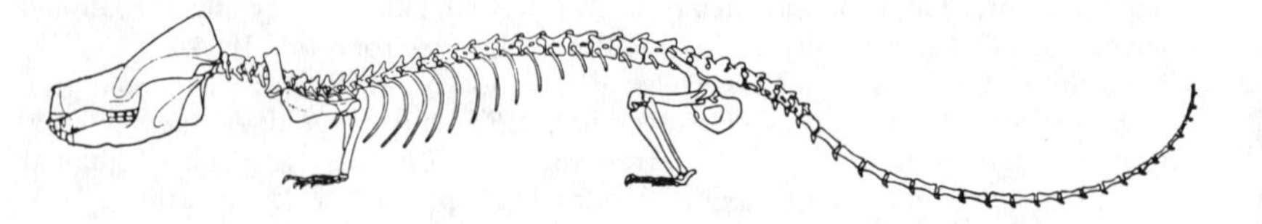

FIG. 81. The skeleton of the tritylodontid *Oligokyphus*. (From Kühne, 1956.) Magnification *c.* × 0.22.

more slender and has less prominent processes than any other cynodonts. The pelvis (Fig. 95D) is particularly mammalian, for the posterior process of the ilium is virtually absent and the anterior process is divided by a ridge into upper and lower parts; for the first time, the femur (Fig. 95E) has developed a mammal-like trochanter minor in addition to a very prominent trochanter major. A point of great functional significance is that the radius and ulna are of about the same length as the humerus, and similarly the lower part of the hindlimb equals the femur in length.

The most important feature of the tritylodontids suggesting a relationship to the herbivorous cynodonts, more specifically to the traversodontids, is the nature of the dentition (Crompton, 1972b). The postcanines of a traversodontid such as *Massetognathus* can be compared with tritylodontid postcanines (Fig. 77E). In this particular traversodontid, extra cusps are present in front of the three main cusps that define the transverse crest of the upper teeth. Similarly, extra cusps lie behind the two cusps forming the transverse crest of the lower teeth. This particular arrangement can be interpreted as an incipient stage in the development of longitudinal rows of cusps. The tendency to form tooth rows parallel to the long axis of the skull is also present in *Massetognathus*, although it is better expressed in the Upper Triassic traversodontids such as *Ischignathus* (Fig. 78E). Also, the mode of operation of the postcanine dentition is essentially similar in tritylodontids and traversodontids, in so far as a posterior movement of the lower jaw is involved during the occlusion of the lower teeth with the uppers. In addition to dental structure, several detailed features of the braincase are similar in the two families; and these tend to support the hypothesis of their relationship (Kemp 1980c), although the braincases of other advanced cynodonts are not sufficiently well known yet for it to be certain that these particular characters are not simply ancestral features of all such groups.

Certain other characters of the tritylodontids pose a problem, however, for the tritylodontids possess several extremely mammal-like characters which are absent from traversodontids. The loss of the postorbital bar and evolution of multi-rooted postcanines are perhaps fairly trivial. However, the whole build of the postcranial skeleton, involving a great many features, can only be compared with early mammals themselves (p. 259). If the tritylodontids are related to the traversodontids, then they must have evolved these and other mammalian features independently of the mammals. Alternatively, it is possible that the postcanine teeth are not as closely comparable to those of traversodontids as supposed, and the comparison is certainly rather

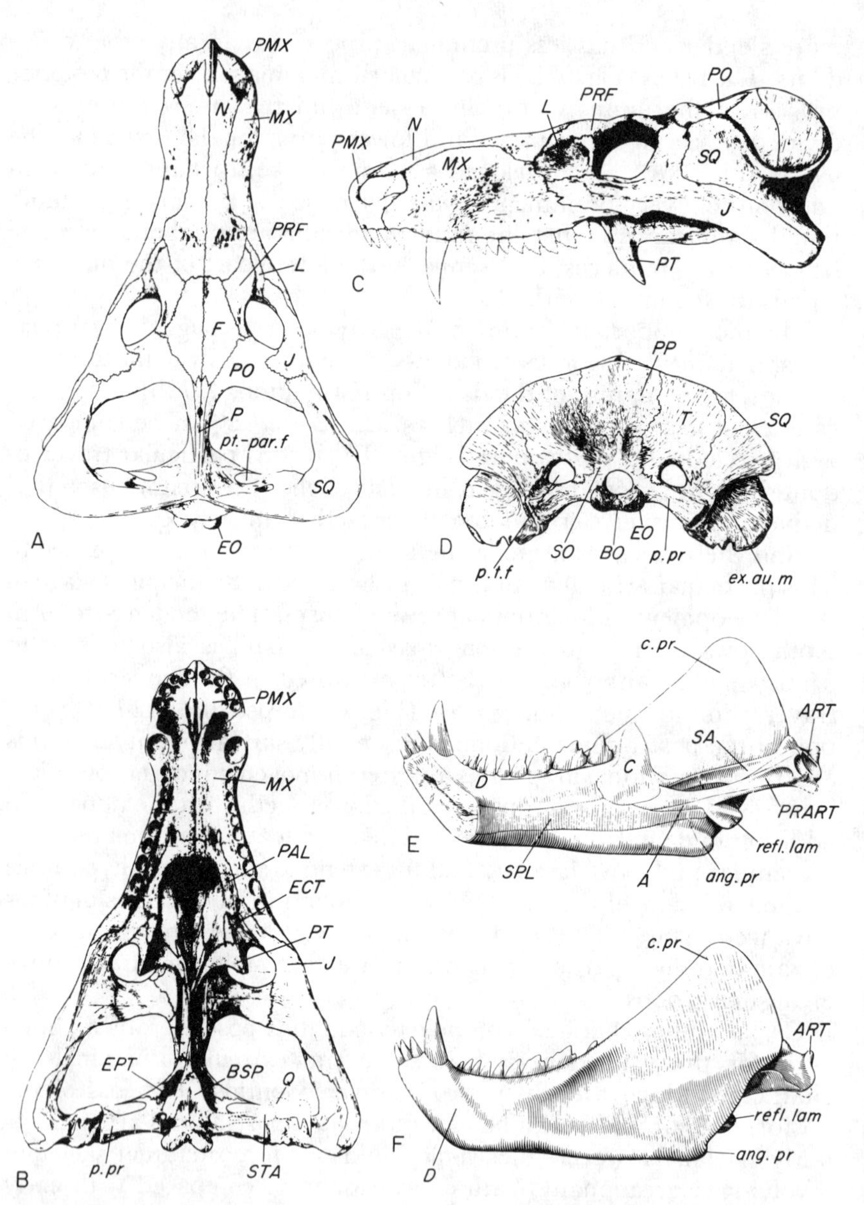


FIG. 82. The skull of *Cynognathus*. A, dorsal view. B, ventral view. C, lateral view. D, posterior view. E, medial view of lower jaw. F, lateral view of lower jaw. (A–D from Broili and Schröder, 1934b; E and F from Kermack *et al.*, 1973, © The Linnean Society of London.)

*A*, angular; *ang.pr*, angular process; *ART*, articular; *BO*, basioccipital; *BSP*, basisphenoid; *C*, coronoid; *c.pr*, coronoid process; *D*, dentary; *ECT*, ectopterygoid; *EO*, vague. In this case, the tritylodontids could be closely related to early mammals (Fig. 87), having evolved herbivorous teeth independently of other cynodonts. This possibility is discussed further in the context of the origin of the mammals (p. 266).

All the remaining advanced cynodonts of the Triassic were carnivorous. The earliest of them are the Cynognathidae, best known of which is *Cynognathus* itself. It occurs in the Lower Triassic *Cynognathus*-zone of South Africa and also from the contemporaneous Puesto Viejo Formation of Argentina. *Cynognathus* (Fig. 82) has a superficial resemblance to *Diademodon*, with its long narrow snout and the various characters typical of all the advanced cynodonts, particularly the enlarged dentary bone. The postcranial skeleton is virtually identical to that of *Diademodon* (Jenkins, 1971a). On the other hand, *Cynognathus* differs from the herbivores in two important respects, the structure of the teeth and of the temporal fenestra. Each of the postcanine teeth, both uppers and lowers, are described as sectorial, having a large main cusp that recurves, and a number of accessory cusps both in front of and behind the main cusp (Fig. 88C). The presence of matching wear-facets indicates that the upper and lower teeth occluded and were used in a shearing action for cutting up meat. The temporal fenestra is substantially smaller than in *Diademodon* for example, and the postorbital bar and zygomatic arch more robust. The occiput (Fig. 82D) forming the hind wall of the fenestra is higher and lacks the characteristic V-shaped notch in its dorsal edge found in the herbivores. This somewhat archaic appearance of the fenestra, coupled with the absence of gomphodont teeth, has led some authors to propose a separate origin of *Cynognathus* from a galesaurid or even procynosuchid level (Crompton and Jenkins, 1973). However, the external auditory meatus of *Cynognathus* is as well developed as in the other advanced herbivores suggesting that the occiput has passed through a diademodontid-like stage. The present cynognathid condition is best seen as a secondary one involving a dorsal growth of the occiput as an adaptation to a carnivorous mode of life, when more extensive posterior pulling jaw muscles were required.

Although a relationship probably exists between cynognathids and

exoccipital; *ex.au.m*, external auditory meatus; *F*, frontal; *J*, jugal; *L*, lachrymal; *MX*, maxilla; *N*, nasal; *P*, parietal; *PAL*, palatine; *PMX*, premaxilla; *PO*, postorbital; *PP*, postparietal; *p.pr*, paroccipital process; *PRART*, prearticular; *PRF*, prefrontal; *PT*, pterygoid; *p.t.f*, post-temporal fenestra; *pt.par.f*, pterygo-paroccipital foramen; *Q*, quadrate; *refl.lam*, reflected lamina of the angular; *SA*, surangular; *SO*, supraoccipital; *SPL*, splenial; *SQ*, squamosal; *STA*, stapes; *T*, tabular; *V*, vomer.

Magnifications: A–D, *c*. × 0.22; E and F, *c*. × 0.17.

the herbivorous cynodonts, no particular group of herbivores appears to have been closer than any other. It is probable therefore that cynognathids separated from the rest before the differentiation of the separate herbivore groups, and before the evolution of the gomphodont postcanine tooth (Fig. 87).

Cynognathids became extinct by the end of the Lower Triassic, and their place was taken in the Middle Triassic by a second advanced carnivorous group, the Chiniquodontidae. They are known almost exclusively from Middle and Upper Triassic deposits in South America, although *Aleodon* (Crompton, 1955b) from the Middle Triassic Manda Formation of Tanzania is possibly a member of the family. Three closely related chiniquodontids occur in South America. *Probelesodon* (Romer, 1969a) from the Middle Triassic Chañares Formation and *Chiniquodon* (Romer, 1969b) from the Chañares, the early Upper Triassic Santa Maria Formation of Brazil, and the Upper Triassic Ischigualasto Formation of Argentina are both fairly small, with skull lengths around 15 cm. *Belesodon* (Romer, 1969b) from the Santa Maria Formation is a fairly large cynodont with a skull about 25 cm long. Chiniquodontids (Fig. 83) differ most markedly from the cynognathids in their very large temporal fenestrae and slender postorbital bars. The secondary palate is exceptionally long, extending back to the level of the end of the tooth rows. The normal number of incisors, four uppers and three lowers are present, followed by powerful canines. The postcanine teeth are all of a sectorial type, with a single dominant main cusp that curves backwards, and a variable number of small accessory cusps both anterior and posterior to the main cusp. All these cusps are in line, and there is no development of a cingulum or internal cusps. Wear facets on the teeth indicate that true occlusion occurred between uppers and lowers. The dentary is extremely well developed, having large coronoid and angular processes, and also a long articular process extending posteriorly above the postdentary bones, superficially like tritylodontids. The postdentary bones are reduced to a compound rod set into the medial surface of the dentary and extending very little posteriorly beyond the dentary. The accessory jaw articulation between the surangular and squamosal is present, alongside the articular-quadrate joint, just as in the other advanced cynodonts (Crompton, 1972a).

The chiniquodontid postcranial skeleton (Fig. 84) is well known only in *Probelesodon* (Romer and Lewis, 1973). Costal plates are entirely absent from all the ribs, and the limbs are relatively long and slender. The shoulder girdle is remarkable for the extreme narrowness of the scapula blade, but otherwise both it and the forelimb bones have a normal cynodont structure. The pelvis is also fairly standard, although the posterior process of the ilium is somewhat reduced. The ischium extends postero-ventrally, rather than horizontally backwards as in traversodontids, but the pubis is unfortunately unknown. The femur

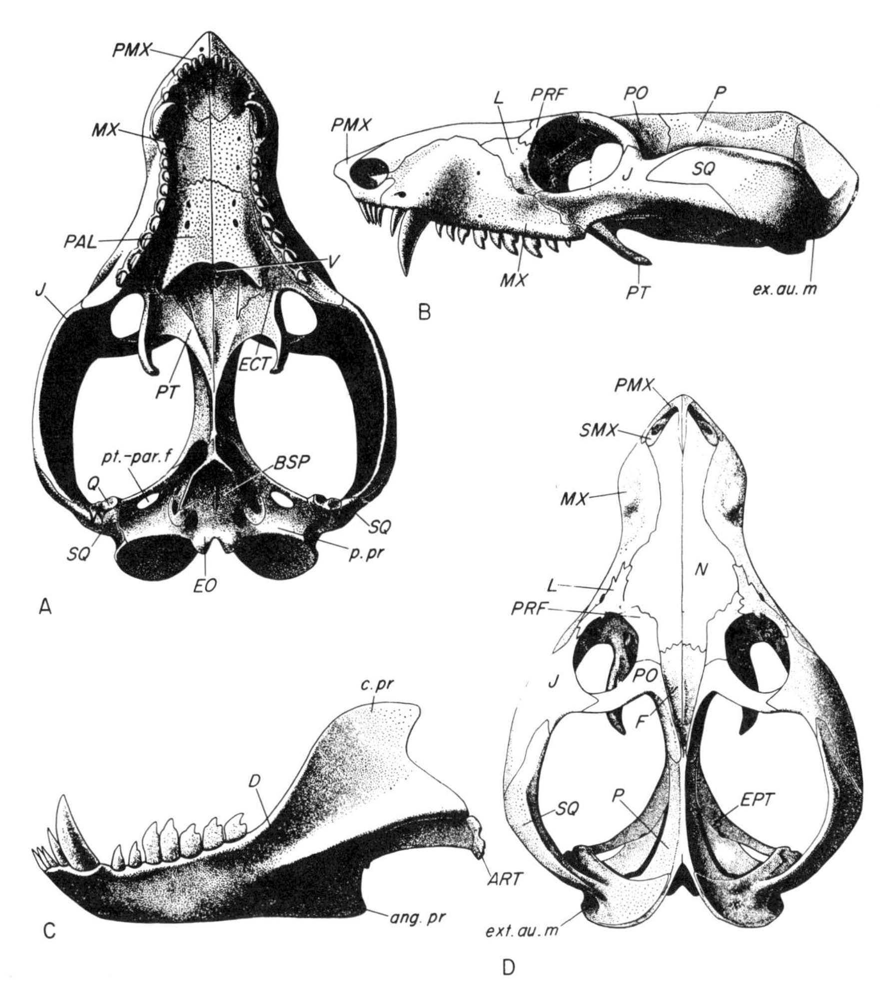


FIG. 83. The skull of the chiniquodontid cynodont *Probelesodon*. A, ventral view. B, lateral view. C, lateral view of the lower jaw. D, dorsal view. (From Romer, 1969a.)

*ang.pr*, angular process; *ART*, articular; *BSP*, basisphenoid; *c.pr*, coronoid process; *D*, dentary; *ECT*, ectopterygoid; *EO*, exoccipital; *EPT*, epipterygoid; *ex.au.m*, external auditory meatus; *F*, frontal; *J*, jugal; *L*, lachrymal; *MX*, maxilla; *N*, nasal; *P*, parietal; *PAL*, palatine; *PMX*, premaxilla; *PO*, postorbital; *p.pr*, paroccipital process; *PRF*, prefrontal; *PT*, pterygoid; *pt.par.f*, pterygo-paroccipital foramen; *Q*, quadrate; *SQ*, squamosal; *V*, vomer.

Magnification *c.* × 0.6.

differs from typical cynodonts only in the slightly more proximal extent of the trochanter major.

The relationship of the chiniquodontids to the other advanced cynodonts, both the herbivorous forms and the cynognathids, is manifested by the enlarged dentary and reduced postdentary bones, the accessory surangular-squamosal jaw articulation, the presence of occluding post-canine teeth, the external auditory meatus, and the characteristic narrowing of the basicranial region of the braincase. Unless it be supposed that all these features evolved in parallel, it must be concluded that chiniquodontids diverged from the same post-galesaurid line that led to these other forms. Furthermore, certain features indicate that the chiniquodontids were phylogenetically closer to the herbivorous groups than to the cynognathids. The large temporal fenestra, and the deep V-shaped emargination between the occiput and the root of the zygomatic arch are characters of the herbivores rather than *Cynognathus*. Admittedly these characters do not demonstrate unequivocally that chiniquodontids were not closer to cynognathids, since the condition of the latter is probably a modification of a more *Diademodon*-like condition. However, other more trivial characters, such as the reduction of the lateral pterygoid flange of the palate, tend to support the idea of a relationship with the herbivores (Fig. 87).

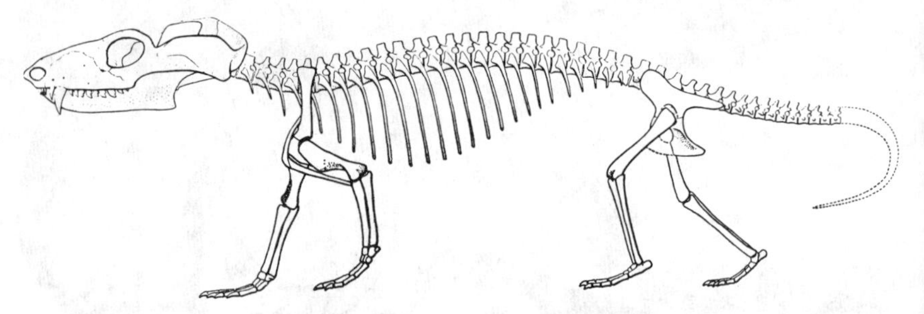

FIG. 84. The skeleton of the chiniquodontid cynodont *Probelesodon*. (From Romer and Lewis, 1973.)
Magnification *c.* × 0.17.

There is one more form, *Probainognathus* (Fig. 85), from the Middle Triassic Chanãres Formation which is often classified as a chiniquodontid (e.g. Bonaparte, 1970). It has, however, certain differences from the other members of the family considered so far, which are of considerable significance in the problem of the origin of mammals, and Romer (1973) has suggested that a separate family, Probainognathidae, should be erected to contain it.

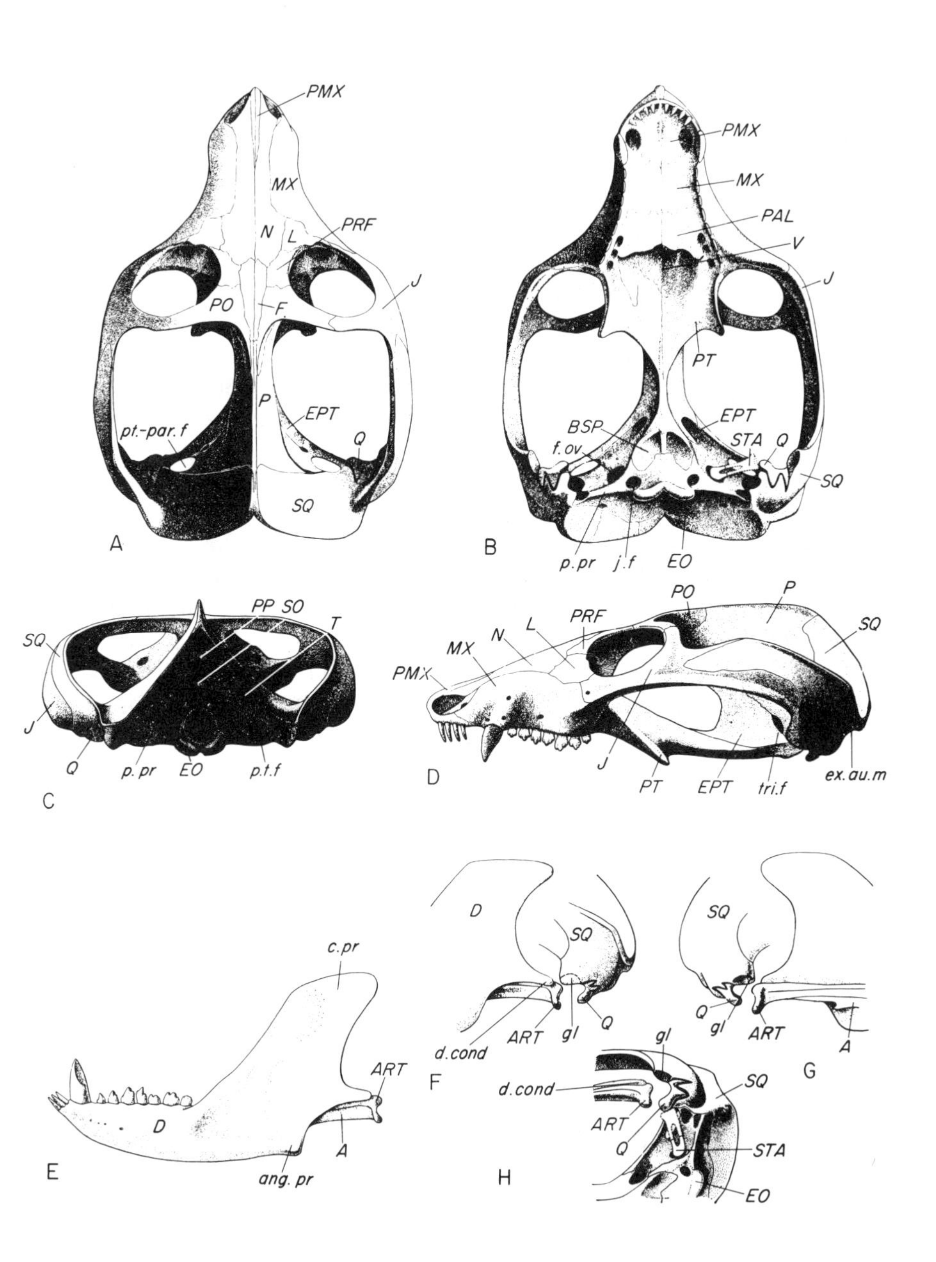

PMX
MX
N
L
PRF
J
PO
F
P
EPT
pt.-par. f
Q
SQ
A
PMX
MX
PAL
V
J
PT
EPT
BSP
f. ov
STA
Q
SQ
p. pr
j. f
EO
B
PP
SO
T
SQ
J
Q
p. pr
EO
p.t.f
C
PO
P
PRF
L
N
MX
SQ
PMX
J
PT
EPT
tri. f
ex. au. m
D
c. pr
ART
D
A
ang. pr
E
D
SQ
d. cond
ART
gl
Q
F
SQ
Q
gl
ART
A
G
gl
d. cond
SQ
ART
Q
STA
EO
H


FIG. 85.

*Probainognathus* (Romer, 1970) is a small, very advanced cynodont with a skull length only about 7 cm. The snout is short and low, the temporal fenestrae relatively enormous, and the zygomatic arches lie almost parallel to the long axis of the skull. The root of the zygomatic arch is more ventral than that of typical chiniquodontids and the arch itself narrower, giving this part of the skull a distinctly mammalian look. The dentition also differs from other chiniquodontids. There are the expected number of incisors, four above and three below, followed by reduced canines. Only seven postcanine teeth are present, which resemble those of the primitive cynodont *Thrinaxodon* (p. 187), except for the important functional distinction that they develop wear facets indicating that they occluded like other chiniquodontid teeth. The single central cusp of each tooth is less dominant compared to the accessory cusps lying both in front of and behind it. Unlike other chiniquodontids as far as is known, there is also a cingulum, bearing cuspules, on the inner side of the tooth. These cuspules are not so well developed as in *Thrinaxodon* and tend to become completely worn off (Crompton, 1972b). The single most dramatic feature of *Probainognathus* is that the dentary may have become involved in the jaw articulation (Fig. 85F–H), for the first time amongst cynodonts (Romer, 1970). The facet of the squamosal, which in other advanced cynodonts receives the surangular, is further forwards in *Probainognathus*, and may be in contact with the posteriormost tip of the articular process of the dentary, as well as the surangular (Crompton, 1972b; Crompton and Jenkins, 1979).

The postcranial skeleton of *Probainognathus* is poorly known, but there are a few bones tentatively identified with it (Romer and Lewis, 1973). The scapula blade has a normal width and well-developed acromion process, and the humerus is relatively longer and more slender than in

FIG. 85. The skull of *Probainognathus*. A, dorsal view. B, ventral view. C, posterior view. D, lateral view. E, lateral view of the lower jaw. F, lateral view of the jaw articulation region. G, medial view of the jaw articulation region. H, ventral view of the jaw articulation region. (From Romer, 1970.)

*A*, angular; *ang.pr*, angular process; *ART*, articular; *BSP*, basisphenoid; *c.pr*, coronoid process; *D*, dentary; *d.cond*, dentary condyle; *EO*, exoccipital; *EPT*, epipterygoid; *ex.au.m*, external auditory meatus; *F*, frontal; *f.ov*, fenestra ovalis; *gl*, glenoid; *J*, jugal; *j.f*, jugular foramen; *L*, lachrymal; *MX*, maxilla; *N*, nasal; *P*, parietal; *PAL*, palatine; *PMX*, premaxilla; *PO*, postorbital; *PP*, postparietal; *p.pr*, paroccipital proces; *PRF*, prefrontal; *PT*, pterygoid; *p.t.f*, post-temporal fenestra; *pt-par.f*, pterygo-paroccipital foramen; *Q*, quadrate; *SO*, supraoccipital; *SQ*, squamosal; *STA*, stapes; *T*, tabular; *tri.f*, trigeminal foramen; *V*, vomer.

Magnification *c.* × 0.64.

*Probelesodon*. Of the pelvic bones only the ilium is known and it is low and not particularly mammalian. The femur is also long and slender.

Despite the *Thrinaxodon*-like teeth of *Probainognathus*, there can be no doubt that it is very closely related to the typical chiniquodontids, as judged from the structure of the skull and jaw, and particularly the very long secondary palate which is characteristic of this family alone amongst cynodonts.

The final group to discuss are the remarkably mammal-like but still poorly known tritheledontids (diarthrognathids). They are all small forms of late Middle and Upper Triassic age, of which the best known is *Diarthrognathus* (Fig. 86A–H) from the Cave Sandstones, at the top of the Karroo sequence in South Africa (Crompton, 1958, 1963). The length of the skull was only about 6 cm and, as in tritylodontids, both the prefrontal and postorbital bones are absent and the orbit and temporal fenestra confluent. The sagittal crest formed by particularly long parietal bones is relatively low and wide compared with other cynodonts, probably in relation to the small size of the animal. The zygomatic arch is very slender and arises from a narrow base low down on the side of the occiput, which is a mammalian feature. Other mammal-like features of the skull are the ventral extension of the frontal to meet a dorsal extension of the palatine in the front wall of the orbit, and the great reduction of the lateral pterygoid processes of the palate. The fenestra ovalis lies entirely within the periotic bone as in mammals and in contrast to all other cynodonts, where it is partly surrounded by the basisphenoid and the basioccipital bones. The articular process of the dentary bone is particularly prominent, extending back above the postdentary bones, which are reduced to the usual compound rod. A small boss on the postero-lateral corner of the articular process actually contacted the inner side of the squamosal, constituting a secondary dentary-squamosal jaw joint, immediately lateral to the quadrate-articular joint, although this alleged contact is doubted by Gow (1980).

The anterior dentition of *Diarthrognathus* is unknown, but seven rather peculiar postcanines are present (Gow, 1980). The upper teeth (Fig. 86L) are slightly widened transversely, and have a single, conical cusp in the centre of the crown. A posterior accessory cusp is present, but no anterior accessory cusp, although there is a tendency to form an anterior cingulum. The lower teeth (Fig. 86K) are rather different. The crown is roughly circular, and behind the single main cusp there are up to three posterior accessory cusps which lie somewhat to the outer side of the tooth, rather than directly behind the main cusp. A well-developed cingulum bounds the internal and posterior edges of the crown.

Two other South African tritheledontids are known, both from the Red

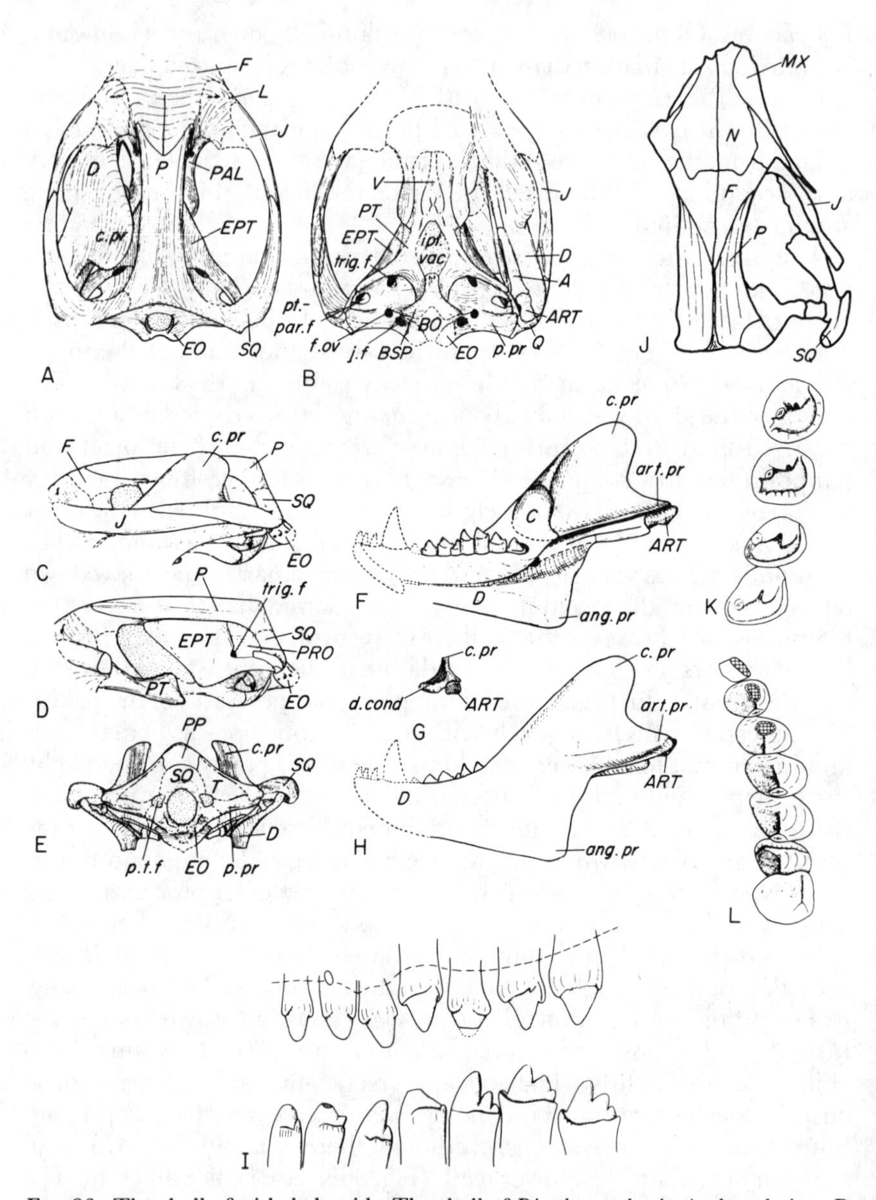


Fig. 86. The skull of tritheledontids. The skull of *Diarthrognathus* in A, dorsal view; B, ventral view; C, lateral view; D, lateral view with the lower jaw and zygomatic arch removed; E, posterior view; F, medial view of lower jaw; G, hind end of lower jaw in posterior view; H, lateral view of lower jaw. I, upper and lower postcanine teeth of *Pachygenelus* in lateral view. J, skull of *Therioherpeton* in lateral view, as preserved. K, left lower postcanines of *Diarthrognathus* in crown view. L, left upper postcanines of

Beds of the Karroo and therefore slightly earlier than *Diarthrognathus*. *Pachygenelus* is now known from several specimens which have not yet been described except for the teeth (Gow, 1980). Only two incisors are present in both the upper and the lower jaws. The canines are unique amongst cynodonts in that the lower canine occludes with the antero-lateral side of the upper canine, a feature of typical mammals. There are seven postcanine teeth in each jaw (Fig. 86I). The upper teeth are circular in crown view, and have a single dominant main cusp. An anterior and posterior accessory cusp are present, but rather than being in line with the main cusp, they are slightly offset to the inner side of the tooth. An external cingulum connects the bases of the two accessory cusps. The lower teeth are larger, and laterally compressed, and the main cusp is followed by a series of up to three accessory cusps behind. In the more posterior teeth there is also an anterior accessory cusp. An internal cingulum is also developed. Wear facets present on the teeth indicate that they occluded and presumably had a shearing function.

The posterior end of the dentary of *Pachygenelus* does not have an articulatory boss, and therefore there was probably no development of a dentary-squamosal joint.

The third South African tritheledontid is *Tritheledon* which is still known only from the pair of maxillae described by Broom (1912). The upper postcanines (Gow, 1980) are transversely widened and have the usual single, large main cusp in the centre of the crown. A series of accessory cusps, decreasing in size, runs inwardly and then forwards around the crown.

One South American form has been described that probably belongs amongst the Tritheledontidae. It is *Therioherpeton* from the Middle-Upper Triassic Santa Maria Formation of Brazil (Bonaparte and Barbarena, 1975), and is known only as a tiny skull about 3 cm long in a poorly preserved, incomplete condition (Fig. 86I). Like *Diarthrognathus*, it lacks the prefrontal and postorbital bones, has a long, low sagittal crest

*Diarthrognathus* in crown view. (A–E from Crompton, 1958; F–H from Crompton, 1963; I, K and L from Gow, 1980; J from Bonaparte and Barbarena, 1975.)

*A*, angular; *ang.pr*, angular process; *ART*, articular; *art.pr*, articular process; *BO*, basioccipital; *BSP*, basisphenoid; *C*, coronoid; *c.pr*, coronoid process; *D*, dentary; *d.cond*, dentary condyle; *EO*, exoccipital; *EPT*, epipterygoid; *F*, frontal; *f.ov*, fenestra ovalis; *ipt.vac*, interpterygoid vacuity; *J*, jugal; *j.f*, jugular foramen; *MX*, maxilla; *N*, nasal; *P*, parietal; *PAL*, palatine; *PP*, postparietal; *p.pr*, paroccipital process; *PRO*, prootic; *PT*, pterygoid; *p.t.f*, post-temporal fenestra; *pt.par.f*, pterygo-paroccipital foramen; *Q*, quadrate; *SO*, supraoccipital; *SQ*, squamosal; *T*, tabular; *trig.f*, trigeminal foramen; *V*, vomer.

Magnifications: A–H, *c*. × 0.9; I,K and L, *c*. × 2.5; J, *c*. × 1.5.

formed mainly from the parietals, and the zygomatic arch is very slender. However, like *Pachygenelus* the dentary probably failed to reach the squamosal. There are eight upper postcanine teeth, each consisting of a linear series of cusps. In the case of the fifth tooth, which is the only one reasonably preserved, the main cusp has one anterior and two posterior accessory cusps, and no cingulum was evidently present. The only lower postcanine preserved is essentially similar. (See note 3 added in proof, p. 351.)

The incompleteness of the various tritheledontid specimens means that there is little enough certainty of their mutual interrelationship, although Gow's (1980) description of the teeth indicates that they are probably a monophyletic family. The phylogenetic relationships of the family to other cynodonts is, however, tentative. That they are indeed related to other cynodonts is now certain, notwithstanding Crompton's (1958) early suggestion that they were related to therocephalian therapsids. A relationship with other advanced cynodonts, rather than direct descent from a galesaurid or procynosuchid level also seems probable (Fig. 87) in view of the advanced structure of the lower jaw and jaw articulation region, large temporal fenestra, reduced lateral pterygoid processes, etc. The only feature against this is the presence of a large interpterygoid vacuity (Fig. 86B) described in *Diarthrognathus* (Crompton, 1958), which is present only in the most primitive cynodonts. However, this part of the skull is extremely badly preserved and the presence of the fenestra is by no means certain.

Among the other advanced cynodont groups, a possible relationship exists with the chiniquodontids, including *Probainognathus*. Both have a postcanine pattern based essentially on a linear series of cusps, and both have a well-developed articular process of the dentary. *Probainognathus* in particular has certain other similarities to tritheledontids, including the slender zygomatic arch arising from a narrow root above the quadrate, the long parietal bone forming a low sagittal crest, and fairly comparable postcanine teeth with cingula. However, until the tritheledontid skull and postcranial skeleton is much better known, the relationships of the group will remain in doubt.

## Functional anatomy

Between the most primitive and most advanced cynodonts many mammalian characters developed, and therefore an understanding of the functional significance of cynodont structure is of the greatest importance in appreciating the nature of the transition to mammals themselves. This applies both to the bony structures preserved and also to whatever inferences can be drawn about soft structures and physiology.

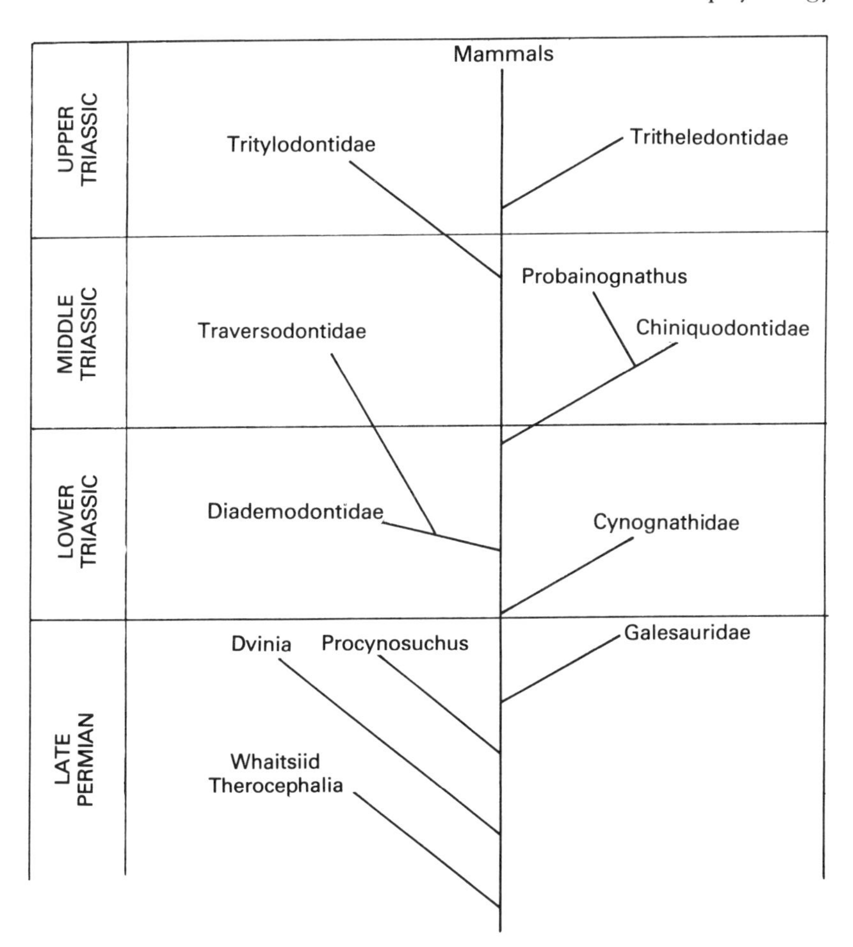


FIG. 87. Phylogeny of the cynodonts.

## *Feeding mechanism*

Differentiation of the dentition into food gathering incisors and canines anteriorly and complex food-processing postcanine teeth posteriorly is as highly developed in cynodonts as in many mammals. Even the postcanine teeth themselves are differentiated in most cases into simpler teeth towards the front and more complex ones behind, which is

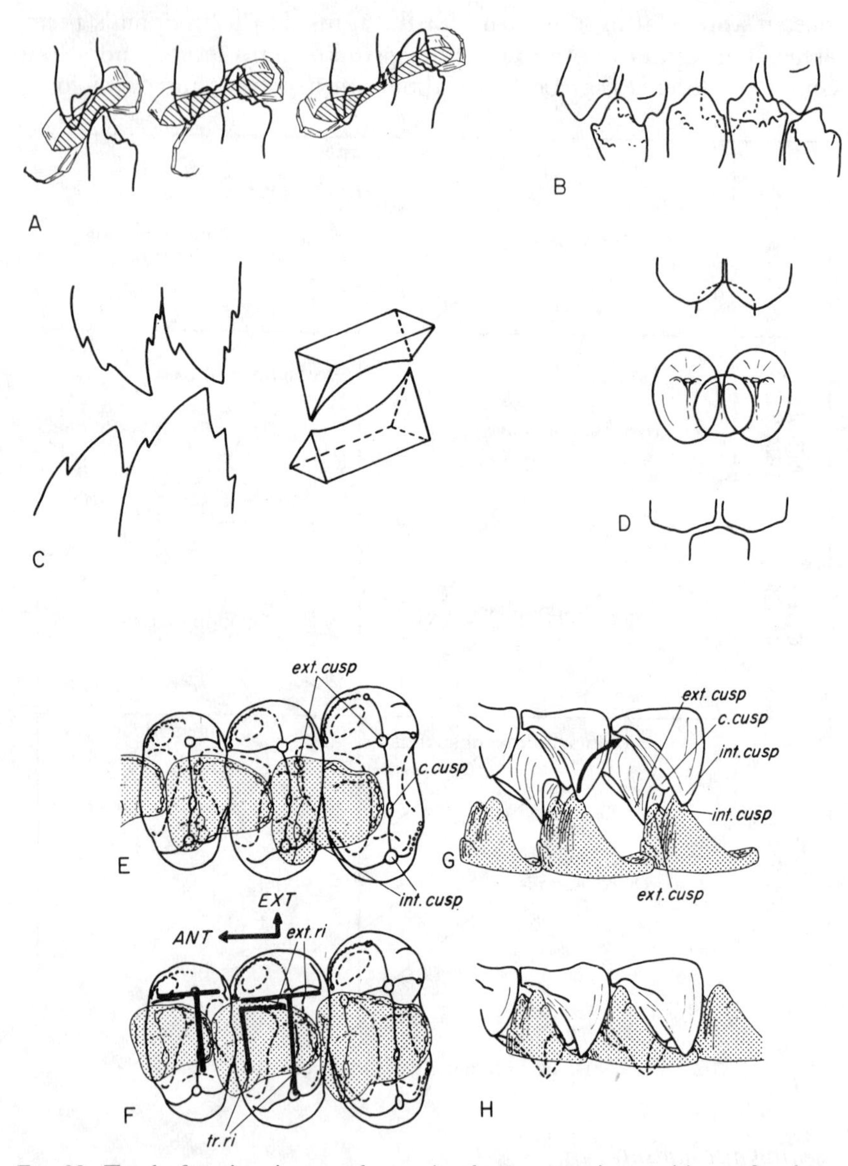


FIG. 88. Tooth function in cynodonts. A, three successive positions of a lower postcanine relative to an upper postcanine when masticating an insect, in *Procynosuchus*. B, left postcanine dentition of *Thrinaxodon*, showing the interdigitation of uppers with lowers. C, opposing upper and lower postcanines of *Cynognathus*, and a model of the opposing shearing edges of sectorial teeth. D, lateral, occlusal and medial views of the opposing postcanines of *Diademodon*. E–H, postcanines of *Scalenodon* (traversodontid): E, occlusal view at the beginning of the power stroke of the bite; F, the same at the end of

related to the greater biting forces available further back along the jaws. The most primitive postcanine teeth were probably those of *Procynosuchus*, where a cingulum complete with cuspules has evolved around the inner side of both the uppers and the lowers. The single large cusp has elongated somewhat along the length of the jaw and has sharp anterior and posterior edges. Opposing upper and lower teeth pierced the food, but then the cingula functioned to anchor the teeth, preventing further penetration. Continued movement of the lower tooth past the upper resulted in a tearing of the food between these teeth (Fig. 88A). This action seems particularly suitable for insects with their tough cuticle (Kemp, 1979). Essentially the same mechanism occurred in *Thrinaxodon* with one further important development. Here, instead of a single upper tooth acting in conjunction with a particular single lower tooth, each upper acted between two adjacent lowers. The tooth rows as a whole interdigitated (Fig. 88B). Instead of a series of separate small tears, the dentition produced a single tear along the length of several teeth. However, even at this stage the teeth did not truly occlude, for the lower teeth passed upwards internal to the upper teeth.

In all the later cynodonts, specifically designed tooth-to-tooth contacts did occur. In principal, such occluding teeth can be used in three different ways, all of which are to be found amongst the various kinds of cynodont teeth. First, opposing pairs of sharp, usually concave edges can cause a cutting, or shearing action, analogous to scissors. Second, opposing flat or basined surfaces can cause crushing of food in a mortar-and-pestle manner. Third, a file-like action is possible between opposing surfaces that are covered with many small cusps. The obvious potential of *Thrinaxodon*-like teeth for shearing was developed in the sectorial teeth of *Cynognathus*, where the sharp edges of the upper main cusp sheared against the edges of the main cusps of two adjacent lower teeth (Fig. 88C). Specific wear facets develop on these edges, which were probably important in maintaining the sharpness. The chiniquodontid teeth represent a similar expression of this simple mode of action.

The evolution of the crushing action is seen in the gomphodont cynodonts, where the postcanine teeth are expanded transversely

the power stroke; G, antero-medial view at the start of the power stroke; H, the same at the end of the power stroke. (A redrawn from Kemp, 1979; B and D redrawn after Crompton, 1972b; C redrawn after Broili and Schröder, 1934b; E–H after Crompton, 1972b.)

*c.cusp*, central cusp; *ext.cusp*, external cusp; *ext.ri*, external ridge; *int.cusp*, internal cusp; *tr.ri*, transverse ridge.

across the jaw, particularly the upper teeth. In the case of *Diademodon* (Fig. 88D), the rather ill-defined ridges which run across each upper tooth bound a series of basins shared between adjacent teeth, into which the respective lower teeth bit. These upper basins are bounded laterally by the main external cusp and its ridges, which served to prevent food from being squeezed out laterally from between the teeth. Instead therefore, food crushed between the teeth was forced inwards into the buccal cavity, where a well-developed tongue was no doubt present to collect it. There appears to have been little sophistication of the crushing system in *Diademodon* and rapid wear of the teeth soon reduced them to flat-topped dentine pegs. This was, however, an important feature, since each upper tooth became in effect "worn-in" to its specific lower teeth, whereupon they matched one another more closely.

The gomphodont postcanine teeth of the traversodontids (Fig. 88E–H) realised certain potentials of *Diademodon*-like teeth, from which they evolved. The transverse ridges of the teeth became more prominent. Those of the upper teeth shifted backwards and those of the lower teeth forwards, and together they formed opposing pairs of shearing edges. A second set of shearing edges evolved at the lateral edges of the teeth. The upper tooth developed a longitudinal ridge running forwards along the outer edge of the crown. A similar longitudinal ridge running backwards along the outer edge of the lower tooth also appears, and together these two longitudinal crests formed a pair of shearing edges. In action, the lower tooth moved upwards until the front face of its transverse crest contacted the back face of the upper transverse crest (Fig. 88E,G). Further upward movement of the lower tooth caused the cutting of food caught between the two crests. Simultaneously, the outer face of the longitudinal crest met the inner face of the upper longitudinal crest, again causing a cutting action. Traversodontid teeth were also capable of crushing, for after completion of the shearing by the crests the broad posterior part of the lower tooth faced the similarly broad anterior part of the upper crown (Fig. 88F,H). Both were slightly concave, at least after a degree of wear had occurred, and food trapped between them was crushed by the further upward movement of the lower tooth. It will be seen that the action of the lower tooth involved not only dorsal but also a certain amount of posterior movement.

Although traversodontid-like teeth were not ancestral to any mammalian teeth, they show remarkable analogies. In many mammal groups, the teeth combine the shearing action of sharp crests with a crushing action between broader areas exactly like these cynodonts. There are, in fact, minor differences in the structure and functioning of the teeth in the different traversodontid genera, although they all appear to have operated on the same principles (Crompton, 1972b). Some, such as *Luangwa* (Kemp, 1980c), wore down the transverse crests very quickly indeed, and the crushing function was the more important. In contrast, the short but very widened teeth of *Exaeretodon* (Fig. 77D) relied more on the shearing function and presumably reduced the crushing function somewhat. These differences must reflect different herbivorous diets, although they are yet to be analysed in detail.

The teeth of the tritylodontids show the third mode of operation of complex teeth. The essential movement of the teeth was the same as in traversodontids, with the lower tooth occluding between two adjacent upper teeth and moving posteriorly (Fig. 77E). The action of the teeth is comparable to filing. As the two lower rows of cusps were dragged backwards between the two respective longitudinal valleys formed between the three upper rows of cusps, opposing pairs of cusps each had a small shearing action. Taking the postcanine dentition as a whole, the large number of such cusp-pairs would cause a fine shredding of the food.

The evolution of complex, occluding postcanine teeth in cynodonts was accompanied by a modification of the pattern of tooth replacement. Most reptiles, with an undifferentiated dentition, replace each tooth at frequent intervals, in a pattern known as alternate replacement. An apparent wave of replacement commences at the anterior end of the jaw, causing the shedding and replacement of alternate teeth along the tooth row. This is followed by the next wave, which affects the other set of alternate teeth, and so on. Each successive wave causes a set of slightly larger teeth to develop, and so the tooth-size keeps pace with the growth of the animal. And by staggering the replacement in this way, long gaps in the tooth row are avoided. However, this pattern of replacement is unsuitable for an animal which possesses complex, matching pairs of upper and lower teeth, particularly when a certain amount of "wearing-in" is necessary to produce completely accurate matching. Also, when the postcanine row of teeth is itself differentiated into more than one morphological type of tooth, simple addition of new teeth at the back would alter the relative proportion of each type present.

In *Thrinaxodon*, which has been studied extensively from this point of view (Parrington, 1936; Osborn and Crompton, 1973), the basic alternate replacement pattern was present, including the addition of post-canine teeth at the back of the tooth row. There were two modifications. Firstly, the anteriormost postcanine tooth was shed but not replaced, at about the same rate as addition of a new posterior postcanine. The number of teeth in the jaw remained approximately constant at about seven therefore. Secondly, each type of tooth was not necessarily replaced by the same type. Normally there are two simple anterior teeth, three intermediate types, and two posterior, complex types. This ratio is maintained because a complex type of tooth could be replaced by an intermediate type, and an intermediate type by a simple type. Thus, in one full replacement cycle, the anteriormost simple type of tooth was lost, the first intermediate type was replaced by a simple type, the first complex type was replaced by an intermediate type, and finally a complex tooth was added at the back of the tooth-row (Fig. 89A).

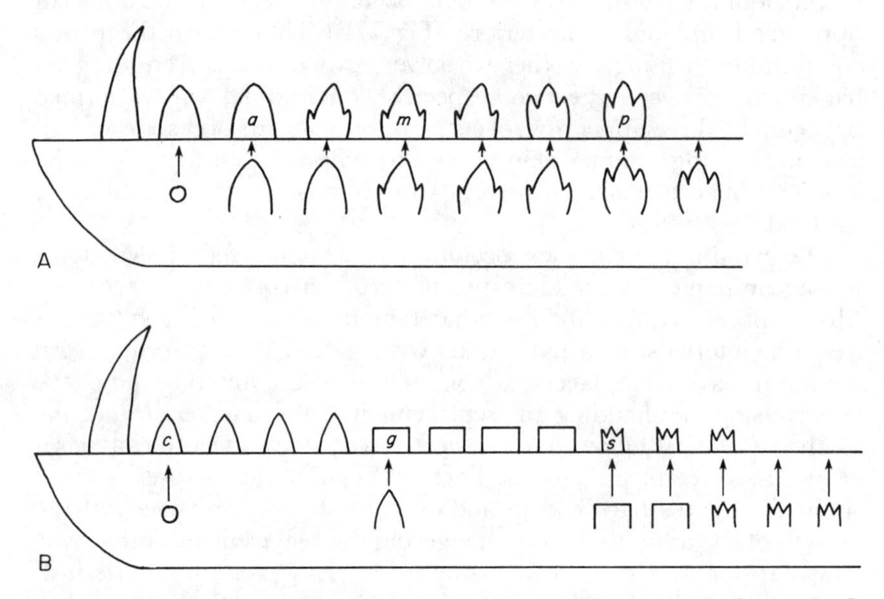


Fig. 89. Pattern of tooth replacement in cynodonts. A, *Thrinaxodon*. B, *Diademodon*. In each case, a single cycle of replacement consists of the dentition shown being replaced by the dentition figured within the jaw.

*a*, anterior tooth-type; *c*, conical tooth-type; *g*, gomphodont tooth-type; *m*, middle tooth-type; *p*, posterior tooth-type; *s*, sectorial tooth-type.

While this method of maintaining both the number of teeth and the overall morphology of the tooth-row constant was probably common to all primitive cynodonts, including *Procynosuchus* (Kemp, 1979), the herbivorous, gomphodont cynodonts evolved further. *Diademodon* has a series of simple, conical teeth at the front of the postcanine tooth-row, followed by the gomphodont teeth, and finally sectorial teeth at the back (Fig. 89B). As in *Thrinaxodon*, the anterior teeth were shed and not replaced, the anterior gomphodont teeth were replaced by conical teeth, and the anterior sectorial teeth were replaced by gomphodont teeth. New sectorial teeth were added at the back of the tooth-row (Hopson, 1971). However, the rate of addition of teeth at the back exceeded the rate at which the teeth were lost at the front, and therefore the total number of teeth rose as the animal grew. The number of both conical and sectorial teeth remained more or less constant. The main effect of the uncoupling of the stimulus for replacement at the front part of the row from that at the rear part was that the rate of replacement of the gomphodont teeth was reduced, although new ones were being added relatively rapidly. They were therefore present in the jaw for long enough to become fully "worn-in". The appearance of the gomphodont part of the tooth row is of teeth increasing in size posteriorly, because newer teeth are larger than older ones. Also the observed degree of wear decreases from the anterior to the posterior teeth, because anterior teeth are older than more posterior teeth. The pattern of replacement in the more advanced gomphodont cynodonts, the traversodontids and tritylodontids, has not been so thoroughly investigated. However members of both these groups do show the same gradient of tooth-size and wear as *Diademodon*, and probably had the same pattern of replacement. The only obvious difference is that at least in well-grown individuals both the anterior conical type of tooth and the posterior sectorial type have been suppressed.

The adductor jaw musculature of cynodonts evolved in conjunction with the development of more elaborate, occluding teeth, which required both greater and more accurate biting forces for their operation. The primitive cynodont arrangement of the jaw muscles (Fig. 90A) is that found in the Permian forms *Dvinia* and *Procynosuchus* (Kemp, 1979), and resembles quite closely the therocephalian condition. The external adductor, which constituted much the greater part of the musculator of the jaws, was in two parts, a deeper one homologous to the mammalian temporalis muscle and a superficial one which was the earliest manifestation of the mammalian masseter. The temporalis muscle arose from the broad, lateral-facing surface of the sagitttal crest and from the posterior wall of the temporal fenestra formed by the squamosal. There was probably a temporal aponeurotic sheet present, attached to the medial and posterior edges of the fenestra, which covered much of the fenestra and gave origin to more temporalis fibres. The muscle fibres converged antero-ventrally and laterally to insert on the upper and inner surfaces of the postdentary bones, particularly the surangular, and into a presumed bodenaponeurotic sheet arising dorsally from the coronoid process (Barghusen, 1968). Most significantly, some of the

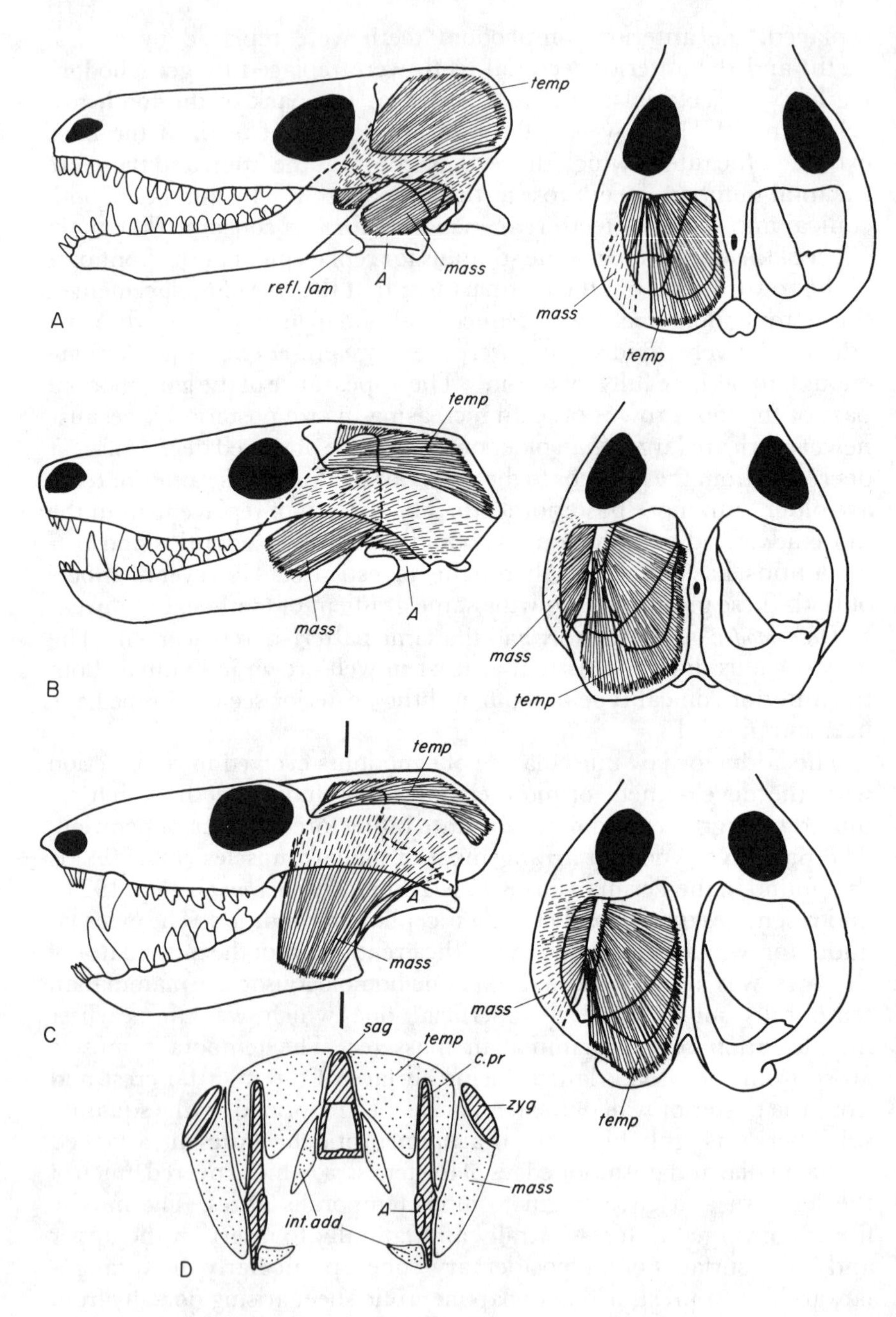
temp
refl.lam
A
mass
mass
temp
A
temp
A
mass
mass
temp
B
temp
A
mass
mass
temp
C
sag
temp
c.pr
zyg
mass
int.add
A
D

FIG. 90.

muscle had invaded the lateral surface of the coronoid process itself, inserting into the small adductor fossa. No such development as this occurs in therocephalians, or any other therapsid group and it is a significant step towards mammalian structure. The masseter muscle took its origin from the inner surface of the zygomatic arch, particularly the more posterior part as indicated by the dorso-lateral bowing of the arch, and a slight hollowing of this part of the squamosal bone. The most probable region of the lower jaw to which the masseter attached is the smooth, broad fossa that occupies the whole lateral surface of the angular bone, behind the attachment of that bone to the dentary. The reduction of the reflected lamina of the angular made this area available.

The internal adductor musculature, which was much less developed than the external adductor, was represented mainly by the pterygoideus complex originating from the posterior face of the lateral pterygoid process of the palate and the length of the pterygoid bone alongside the braincase. The epipterygoid bone of even these primitive cynodonts is broadly expanded alongside the sidewall of the braincase (Fig. 71A–C) and part of the internal adductor may very well have attached to it. The insertion of this complex of muscles was generally the ventral and inner surfaces of the postdentary bones. The reflected lamina of the angular of *Procynosuchus* and other primitive cynodonts is already very reduced compared to other therapsids and its function as part of the jaw muscle system must have been diminished.

Functionally the primitive cynodont jaw muscles behaved much as those of the therocephalians with two quantitative modifications. There was a relatively greater mass of adductor musculature, so the biting force available to the teeth was larger. Equally important was the relatively large size of the masseter muscle acting on the lateral side of the lower jaw, and also the appearance of more laterally placed fibres of the temporalis muscle, inserting in the adductor fossa. There must have been a tendency for the lateral component of the force produced by these parts of the musculature to balance the medial component produced by the rest of the temporalis and the internal adductor complex. Therefore the overall tendency of the muscles to pull the hind ends of

FIG. 90. Cynodont jaw musculature. A, *Procynosuchus*. B, *Thrinaxodon*. C, *Probelesodon*. D, transverse section through the temporal region of an advanced cynodont such as *Probelesodon*. (Outline of A redrawn after Kemp, 1979; of B after Parrington, 1946; of C after Romer, 1969a.)

*A*, angular; *a.pr*, angular process; *c.pr*, coronoid process; *int.add*, internal adductor muscle; *mass*, masseter muscle; *refl.lam*, reflected lamina of the angular; *sag*, sagittal crest; *temp*, temporalis muscle; *zyg*, zygomatic arch.

the lower jaws inwards was reduced, and the possibility of finer control of the movements of the jaws in a transverse direction began to arise. It can be no coincidence that this property of the muscles appeared incipiently in *Procynosuchus* and other primitive forms at the same time as the postcanine dentition was beginning to be used in a more accurate manner than in all previous therapsids.

The galesaurid cynodonts such as *Thrinaxodon* show further advances towards a mammalian arrangement of the jaw musculature (Fig. 90B). The temporal fenestra is larger, indicating a further incremental increase in the volume of adductor musculature. The temporalis muscle originated from the medial and posterior walls of the temporal fenestra and presumably a temporal aponeurosis as in *Procynosuchus*. One difference is that the posterior wall of the fenestra is quite deeply incised between the occiput and the root of the zygomatic arch. This part of the skull lies directly behind the greatly expanded coronoid process of the dentary which now rises high into the fenestra. Muscle fibres from the coronoid process could not attach to the back of the temporal fenestra immediately behind the process as they would be too short, and would therefore limit the gape of the jaws. The very large coronoid process has in effect grown upwards to replace the bodenaponeurosis of more primitive forms (Barghusen, 1968), and the temporalis muscle inserted onto both the inner surface and also the upper half of the external surface of the process.

The masseter muscle of *Thrinaxodon* was relatively even more enlarged than the temporalis. The zygomatic arch from which it originated is deeper and more strongly bowed upwards and outwards, indicating that the masseter attachment extended for its full length, almost to the level of the orbit. Some of these muscle fibres probably still attached to the lateral face of the angular bone, but the great majority had shifted onto the dentary. The dentary has expanded posteroventrally compared to the primitive condition, and the adductor fossa has also expanded to occupy the whole of the posterior ramus of the dentary, right down to its ventral edge. The masseter muscle must have inserted into the lower half of this fossa, just as in mammals. The internal adductor musculature of *Thrinaxodon*, like that of *Procynosuchus*, cannot be described accurately. To judge from the reduction of the lateral pterygoid processes of the palate, this muscle complex was of reduced importance, and in volume it must certainly have been but a small fraction of that of the external adductor musculature.

*Thrinaxodon* and the other galesaurids represent an intermediate stage in the evolution of the jaw musculature, between forms like *Procynosuchus* on the one hand and the various advanced cynodont groups on the other. In these latter, the external adductor muscle was somewhat larger again (Fig. 90C), but the major advance was another increase in the size of the dentary and concomitant reduction of the postdentary bones. The coronoid process retained the height that it had in *Thrinaxodon*, but extended posteriorly further towards the jaw articulation. In the Middle Triassic forms, the traversodontids and chiniquodontids, the base of the coronoid process expands backwards above the postdentary bones as a process termed the articular process, although it was only in a few terminal members that it actually became part of the jaw articulation. As in *Thrinaxodon*, the whole of the coronoid process, inner and outer surfaces, received temporalis muscle fibres. Enlargement of the masseter is indicated by the further development of the angular process of the dentary, which now dwarfs the angular bone, and appears to have taken over the insertion of virtually the whole of this muscle. In some forms, particularly the herbivores, a special process for the origin of the anteriormost part of the masseter has developed on the jugal bone, just below and behind the orbit (Fig. 76C).

The functional significance of the trend in jaw evolution of cynodonts has been debated for many years. There are in fact two phenomena to account for, first the rearrangement of the muscles and change in overall shape of the lower jaw, and second the enlargement of the dentary and reduction of the postdentary bones within the lower jaw. On the face of it, either of these anatomical changes could have occurred without being accompanied by the other. The changes in the muscles and shape of the lower jaw served the purpose of increasing the degree of balance of the muscle forces in the horizontal plane, while at the same time concentrating the forces at the teeth and relieving the jaw articulation of stress. Transverse balance of the forces was achieved, probably completely in advanced cynodonts, by the development of the masseter muscle to the extent that the laterally directed component of its force was comparable in magnitude to the medially directed component of the temporalis muscle. The internal adductor musculature was probably important in this connection too. The temporalis muscle attached to the jaw at a higher level than did the masseter (Fig. 90D), so that even if the transverse components of the muscle forces of these two were exactly equal, the lower jaw would still tend to be rotated about its longitudinal axis, the lower edge passing laterally and the upper edge medially. The internal adductor muscle attached along the ventral edge of the jaw and its force would have been directed largely medially. As it attached below the level of most of the masseter muscle, it would counteract the rotation of the jaw (Fig. 91A). The virtue of balancing the jaw muscles transversely in this way is not only that it permitted reduction of the bony structures such as the lateral pterygoid processes and the jaw articulation which in primitive forms were necessary to prevent unwanted movements, but also it made possible very accurate side-to-side movements of the jaws.

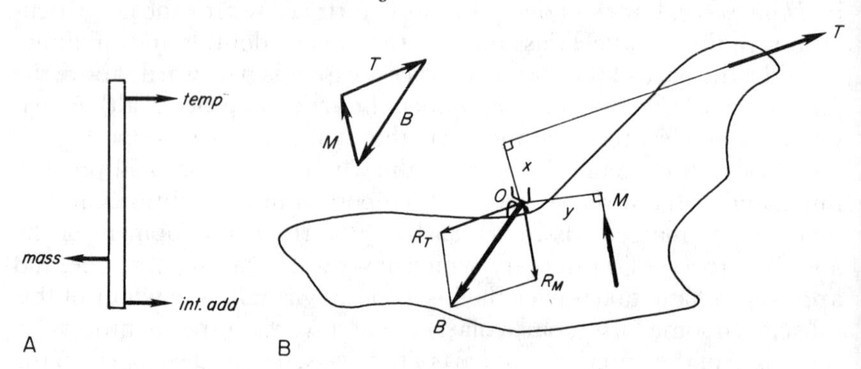


FIG. 91. Jaw mechanics of advanced cynodonts. A, simplified transverse section through the lower jaw of an advanced cynodont, showing how the internal adductor musculature (*int.add*) tends to balance the couple produced from the medial component of the temporalis muscle (*temp*) and the lateral component of the masseter muscle (*mass*). B, the vector of the temporalis muscle ($T$) acts about a moment arm ($x$), producing a clockwise couple ($T.x$) about the point of bite ($O$). The vector of the masseter muscle ($M$) acts about a moment arm ($y$), producing an anticlockwise couple ($M.y$) about $O$. If $T.x = M.y$, then the couples cancel, the lower jaw does not move, and there is no reaction force at the hinge. The muscle forces produce reaction forces($R_T$ and $R_M$) at $O$, with a total magnitude and direction $B.B$ is reconstructed from the triangle of forces (inset).

The concentration of the adductor jaw muscle forces at the teeth depends on the development of both the high coronoid process for the temporalis muscle, and the large angular process for the masseter muscle. Of the several methods of analysing the forces involved, that of Bramble (1978) is perhaps the most elegant (Fig. 91B). The point of occlusion of a particular pair of upper and lower teeth is considered as a fulcrum about which the whole of the lower jaw may potentially rotate ($O$). The temporalis muscle was close to horizontal because of its attachment to the high coronoid process. The force it produced was therefore almost posteriorly directed at a level well above the teeth. This force ($T$) therefore tended to rotate the left jaw clockwise about the point of contact of the teeth, thereby depressing the hind end of the jaw. The force produced by the masseter muscle ($M$) was directed almost vertically from the angular process behind the teeth. Its contraction tended to rotate the jaw anti-clockwise about the tooth-contact and thus to raise the hind end of the lower jaw. Acting together, it is theoretically possible for these two antagonistic moments to equal one another, and therefore to cancel. In such a case, the back end of the jaw will be subjected to a zero force, the lower jaw will be in equilibrium and not move, and all the force of the muscles will in effect concentrate at the point of tooth contact. The magnitude and orientation of the force at the teeth can be found most simply by completing the triangle of forces composed of the two respective muscle forces (*T* and *M*) and the bite force (*B*). (For any object in equilibrium, the vectors of all the forces acting on the object always form a closed polygon, in this case with only three forces involved the polygon is three-sided.) It is seen that this theoretical bite force is large, and is oriented forwards and downwards, rather than simply downwards, implying that the most effective bite occurs when the lower tooth presses postero-dorsally against the upper tooth. It is no coincidence therefore that in the traversodontids the teeth are designed so that the lower tooth, with its transverse crest at the front of the tooth, faces partly backwards. It meets an upper tooth whose crest is at the back, and which therefore faces partly forwards (Fig. 88H). The occlusal plane between the two teeth is not horizontal, but tilted antero-dorsally (Kemp, 1980c). In the case of the advanced carnivorous cynodonts, the shearing action of the postcanine teeth requires that the lower tooth moves slightly backwards rather than directly upwards against its respective upper tooth, again showing that the bite force is directed forwards and downwards.

This model of the jaw action of advanced cynodonts is of course idealised, and no doubt complete balance of the muscle forces in the horizontal plane and complete concentration of the net force at the teeth did not always occur.

The second question about cynodont jaw evolution concerns the enlargement of the dentary bone and reduction of the postdentary bones. As described so far, the mechanical system could have worked equally well with a lower jaw that had a more normal ratio of dentary to postdentary bones, as long as the outline shape of the whole jaw was the same. Two current theories, not necesarily mutually incompatible, offer functional explanations for dentary enlargement.

One theory (e.g. Kemp, 1972a) supposes that the sutures between separate bones are potential lines of weakness in a structure subjected to large forces produced by muscle attached at different points. In the case of the primitive cynodont jaw, the temporalis muscle attached to both the dentary and to the surangular bone, while the masseter attached to the angular. Internal adductor muscles attached to the prearticular, among other bones. Since these muscles all pulled on the jaw in different directions, stresses would inevitably be generated at the sutures between these various bones, tending to pull them apart from one another. As the muscles enlarged, these stresses would increase and perhaps reach a critical level. However, any relative increase in the size of the dentary at the expense of the postdentary bones would cause some of the muscle fibres attaching to the postdentary bones to be "captured" by the dentary. The size of the stress across the sutures between the bones would therefore be slightly diminished. In this way, a continuous process of dentary enlargement and capture of more and more muscle fibres by that bone would progressively decrease the tendency for the jaw to disarticulate at the sutures. As has been seen, this process went hand in hand with the reorientation of the muscles in such a way as to reduce the forces acting at the jaw articulation. The need for strong postdentary bones connecting the jaw articulation to the dentary was also therefore reduced. The final result is where almost all the jaw muscles attach to the dentary, through which virtually all their force was transmitted directly to the dentition. The second theory accounting for the changes in the jaw bones is that of Allin (1975). To look ahead a little, it is known that the reduced postdentary bones became incorporated in the middle ear of mammals as the accessory ear ossicles, which conduct sound vibrations from the eardrum to the fenestra ovalis in the side wall of the braincase. Allin's suggestion is that in cynodonts the postdentary bones were already involved in hearing, and that reduction in the size of these bones resulted from selection for increasingly acute hearing. Enlargement of the dentary was then simply a necessary correlate of the change. This radical view rests entirely on the belief that the postdentary bones were already involved in sound conduction, and is considered in more detail later (p. 236).

Evidence about the jaw opening musculature is much more elusive than that for the adductor muscles. The usual view is that cynodonts possessed a depressor mandibuli muscle running between the occiput and the downturned retroarticular process of the articular bone, similar to the condition in modern reptiles. There is, however, a difficulty associated with this interpretation (Kemp, 1979, 1980c). The reptilian depressor mandibuli muscle is innervated by the facial nerve (VII) but in the monotreme mammals the rather similarly placed jaw opening muscle is innervated by cranial nerve V, the trigeminal. Innervation is apparently a good indication of homology among muscles, and therefore the monotreme muscle cannot be homologous to the depressor mandibuli muscle, and is accordingly termed the detrahens. In the other modern mammals, the therians, the jaw opening musculature is different again, for there is neither depressor mandibuli nor detrahens, but instead a digastric muscle composed of two parts. The anterior part is innervated by nerve V, while the posterior part is innervated by nerve VII, and anatomically it is quite different for it does not attach to the occiput, but acts via the hyoid apparatus in the throat. If the cynodonts did possess a depressor mandibuli muscle, then in the line leading to the mammals it was lost, to be replaced by the superficially similar neomorphic detrahens in monotremes. It seems more reasonable to assume that the cynodonts already had a trigeminally innervated detrahens which the monotremes retained. The line leading to the therians lost this muscle, did not replace it by a similarly oriented muscle, but instead elaborated the pre-existing complex of throat muscles as a digastric.

The nature of the cynodont jaw articulation is intimately associated with the organisation of the jaw muscles. The primitive cynodont condition, again represented by *Procynosuchus* (Kemp, 1979), was similar to that of the therocephalians. The articulating surface of the articular bone faces postero-dorsally and somewhat inwards, and the axis of the hinge was not transverse but antero-medially oriented. Thus it is clear that the adductor musculature still pulled medially and the hinge joint was still designed to prevent the hind ends of the lower jaws from being forced inwards. As in therocephalians (p. 173), the quadrate-quadratojugal complex attached to the squamosal in a way that allowed it to rotate about the longitudinal axis, thereby compensating for the oblique hinge axis. The quadrate simply lay in a shallow recess in the front face of the squamosal, while the quadratojugal lay partially in a transverse slit.

A number of important changes had occurred to the jaw articulation in *Thrinaxodon* (Fig. 92A–C), including a relative reduction in size of both the quadrate and the articular bones. The axis of the jaw hinge had become more or less transversely oriented, reflecting the approach to a balance between the medially directed component of the temporalis muscle and the lateral component of the masseter. The method of attachment of the quadrate complex differed, for the quadrate has evolved a posterior flange which lies within a slit in the ventral margin of the squamosal. The quadratojugal is also held in a deep slit in the squamosal, which is sagittal rather than transverse (Parrington, 1946). It appears therefore that the kind of mobility of the quadrate had changed, so that it was now capable of moving backwards and forwards and therefore imparting a propalinic or antero-posterior shift to the lower jaw. The degree of such movement was small in *Thrinaxodon*, but sufficient to allow the lower teeth to move postero-dorsally past the upper teeth, rather than simply dorsally, which probably increased the cutting action of the teeth. A related development in *Thrinaxodon* was the development of a secondary jaw articulation, albeit in an incipient form at this stage (Crompton, 1972a). A flange of the squamosal bone, the articular flange, has grown down lateral to the quadrate complex at the hind end of the zygomatic arch. Meanwhile, the surangular bone,

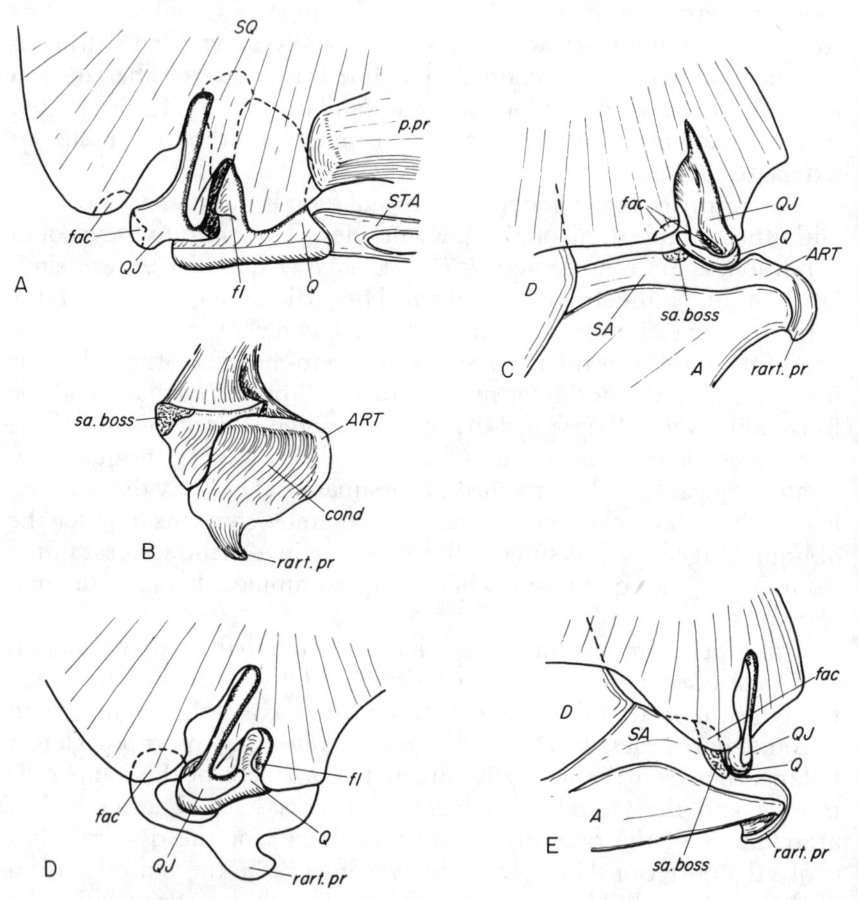


FIG. 92. The jaw articulation of cynodonts. A, posterior view of the quadrate complex and associated structures of *Thrinaxodon*. B, posterior view of the hind end of the left lower jaw of *Thrinaxodon*. C, lateral view of the jaw articulation region of *Thrinaxodon*. D, posterior view of the jaw articulation region of *Trirachodon*. E, lateral view of the jaw articulation region of *Trirachodon*. (Redrawn after Crompton, 1972a.)

*A*, angular; *ART*; articular; *cond*, articular condyle; *D*, dentary; *fac*, facet for reception of surangular boss; *fl*, flange of quadrate; *p.pr*, paroccipital process; *Q*, quadrate; *QJ*, quadratojugal; *rart.pr*, retroarticular process; *SA*, surangular; *sa. boss*, boss on surangular; *SQ*, squamosal; *STA*, stapes.

which is the postdentary bone immediately next to the articular, has developed a small boss lateral to the articular (Fig. 92B). The articular flange of the squamosal and the boss of the surangular lie very close to one another, and were probably connected by ligaments and cartilage. They thus form a new jaw articulation, immediately alongside the original quadrate-articular joint. The advantage of the new surangular-squamosal connection was that it prevented the quadrate from being pulled too far out of its socket when the lower jaw shifted forwards.

In the advanced cynodonts (Fig. 92D,E), the articular flange of the squamosal is substantially larger and its articulation facet extends further forwards. It was now in direct contact with the similarly enlarged boss of the surangular, the two forming a strong joint. Here, particularly in the traversodontids, the two bones forming this new joint remained in contact even when the quadrate and the lower jaw together moved backwards and fowards, and therefore the secondary joint actually controlled the propalinic movements. The final stage occurs in certain very advanced forms. In both the traversodontids and the chiniquodontids, an articular process of the dentary evolves, growing backwards from the base of the coronoid process and supporting the reduced postdentary bones. Inevitably, the posteriormost tip of the articular process approached close to the secondary, squamosal-surangular articulation. In the chiniquodontid *Probainognathus* (Fig. 85F–H), the tritheledont *Diarthrognathus* (Fig. 86F–H), and possibly the tritylodontid *Tritylodontoides* (Fourie, 1968), the articular process may have invaded the secondary articulation and may make direct contact with the squamosal. Technically, this is the moment at which the mammalian dentary-squamosal jaw articulation was achieved.

## *Middle ear*

A lively controversy exists concerning the mechanism by which cynodonts detected sound. The essence of the problem lies in the very different means used by most modern reptiles (and birds) and by mammals, when the cynodonts possess unique features of the middle ear region that do not correspond obviously to either. In typical modern reptiles, the stapes inserts medially into the fenestra ovalis in the otic region of the braincase and extends laterally towards a tympanic membrane behind and partly supported by the large quadrate bone. The actual connection to the tympanum is via a cartilaginous extension to the stapes, the extrastapes. The stapes functions by matching the impedance of sound waves in air to the impedance of waves in the fluid filling the cochlea canal of the inner ear. In practice, this means

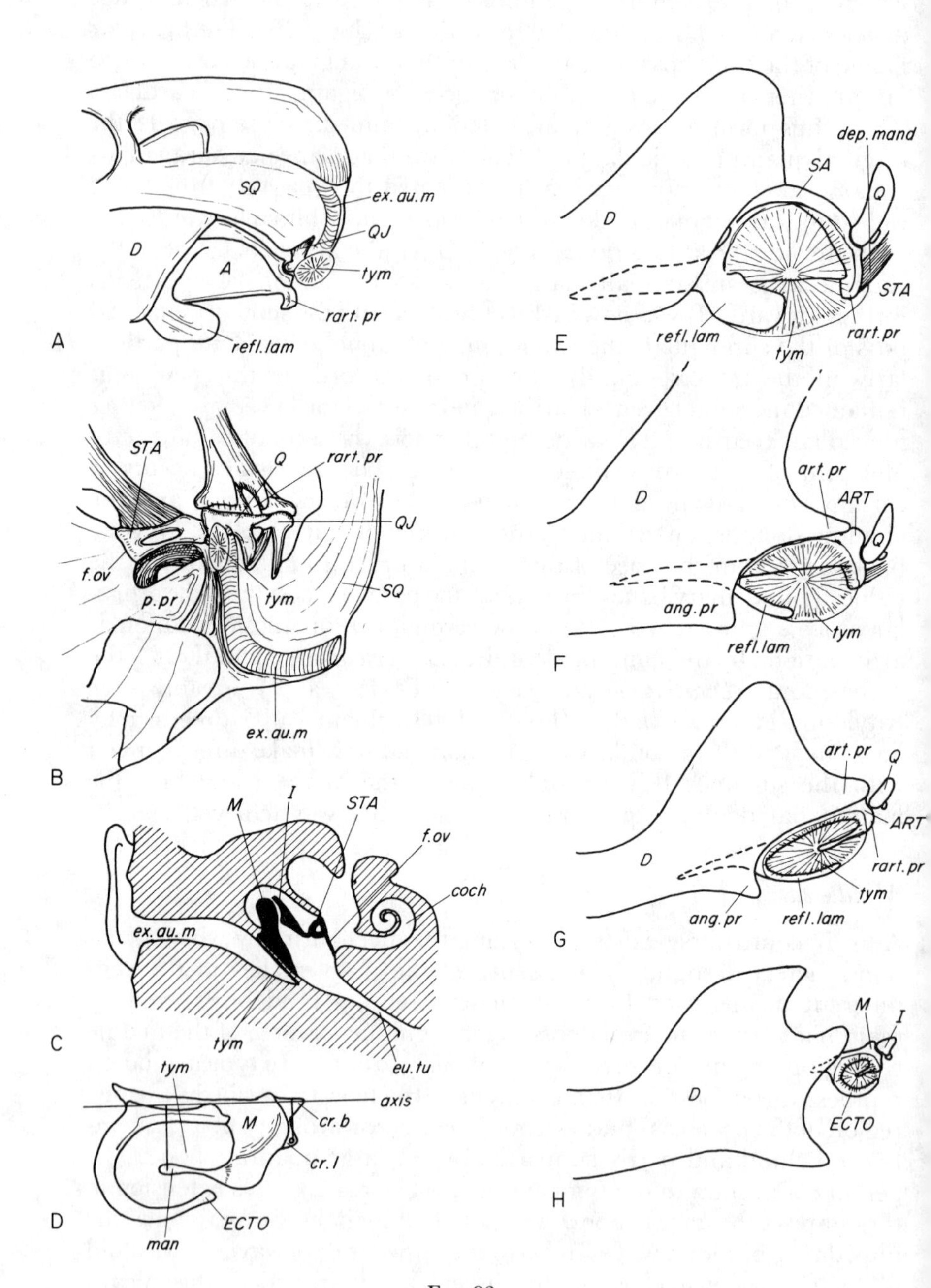

SQ
ex. au. m
D
QJ
A
tym
rart. pr
A
refl. lam
D
SA
dep. mand
Q
STA
refl. lam
tym
rart. pr
E
STA
Q
rart. pr
QJ
f. ov
tym
SQ
p. pr
ex. au. m
B
D
art. pr
ART
Q
ang. pr
tym
refl. lam
F
M
I
STA
f. ov
coch
ex. au. m
tym
eu. tu
C
art. pr
Q
ART
D
rart. pr
tym
ang. pr
refl. lam
G
tym
axis
M
cr. b
cr. l
ECTO
man
D
M
I
D
ECTO
H


FIG. 93.

converting the low pressure airborne waves into high-pressure waves suitable for transmission in fluid, which is achieved by having the tympanic membrane of a much greater area than the fenestra ovalis (a "stiletto heel" effect). There is also a lever arrangement between the extrastapes and the tympanum, further increasing the pressure magnification. In the case of the modern mammals (Fig. 93C), the stapes is not connected directly to the tympanum, but via two ear ossicles, the incus and the malleus. Only the malleus actually contacts the tympanum. Again this system acts by impedance matching, with a much larger tympanum than fenestra ovalis. There is also a lever arrangement. The malleus connects to the tympanum by means of a long arm, the manubrium, while the conjoined incus attaches to the distal end of the stapes by a shorter arm (Fig. 93D). Therefore there is a lever ratio of about 2:1, doubling the pressure between tympanum and stapes. The origin of the two extra ossicles in mammals has been demonstrated from embryological studies. The malleus is an ossification of the back end of the primary lower jaw and is therefore the homologue of the cynodont articular. The incus is formed at the back of the primary upper jaw and corresponds to the quadrate of the cynodont. It has also been demonstrated that the mammalian tympanic bone, which holds the tympanic membrane, is the homologue of the angular and its reduced reflected lamina. It follows that the evolution of the extra ossicles in the mammals occurred as a result of the enlargement of the dentary to form the complete functional lower jaw, with the resulting reduction of the post-dentary bones and their loss of a function associated with the feeding apparatus.

One view of the functioning of the cynodont middle ear is that an essentially reptilian system was present. The stapes ran laterally from

FIG. 93. Sound reception in cynodonts and mammals. A, lateral view of the hind part of the skull of *Thrinaxodon*, showing standard reconstruction of a post-quadrate tympanic membrane. B, the same in ventral view. C, schematic transverse section through the ear region of a modern mammal. D, the ear ossicles of a modern mammal. E–H, Allin's (1975) conception of the evolution of hearing in cynodonts and mammals: lateral views of the lower jaws of E, primitive cynodont; F, advanced cynodont; G, early mammal; H, modern therian mammal. (A–D redrawn after Hopson, 1966; E–H redrawn after Allin, 1975.)

*A*, angular; *ang.pr*, angular process; *ART*, articular; *art.pr*, articular process; *axis*, functional axis of the ear ossicles; *coch*, cochlea; *cr.br*, crus brevis of the incus; *cr.l*, crus longus of the incus; *D*, dentary; *dep.mand*, depressor mandibuli; *ECTO*, ectotympanic; *eu.tu*, eustachian tube; *ex.au.m*, external auditory meatus; *f.ov*, fenestra ovalis; *I*, incus; *M*, malleus; *man*, manubrium of the malleus; *p.pr*, paroccipital process; *Q*, quadrate; *QJ*, quadratojugal; *refl.lam*, reflected lamina of the angular, *rart.pr*, retroarticular process; *SA*, surangular; *SQ*, squamosal; *STA*, stapes; *T*, tabular; *tym*, tympanic membrane.

the fenestra ovalis and connected via an unossified extrastapes to a tympanum situated behind the region of the quadrate bone. The main piece of evidence (Parrington, 1979) is the apparent external auditory meatus, which is the deep groove in the back of the squamosal. An air-filled tube supposedly ran from an external opening high up the side of the head and ended immediately lateral to the point where the distal end of the stapes lies, and where a reptile-like tympanum would be expected (Fig. 93A,B). In a form such as *Diademodon* the lower end of the meatus expands in a trumpet-like fashion which is difficult to explain as anything except a sound-conducting tube, terminating at a tympanum. Under this interpretation, the transition to the mammalian condition occurred very late in evolution, when the postdentary bones were practically vestigeal and the new dentary-squamosal jaw articulation had been established. Hopson (1966) has shown how the anatomy of the postdentary bones and quadrate of tritylodontids for example resembles quite closely the arrangement of the mammalian ear ossicles. All that has to happen is for the stapes to lose its contact with the tympanum, and the retroarticular process of the articular gain an attachment, becoming the manubrium of the malleus.

The second theory of hearing in cynodonts is that of Allin (1975). Here it is supposed that even in cynodonts the angular, articular and quadrate bones were already involved in sound conduction, and that the mammal condition is merely a culmination of the continuous refinement of a pre-existing functional arrangement. A tympanic membrane (Fig. 93E–H) is supposed to exist in the gap between the reflected lamina of the angular and the retroarticular process of the articular, overlying an air-filled recess (and incidentally, the function of the reflected lamina in more primitive synapsids is held to be maintenance of the patency of such a recess). Vibration of this tympanic membrane was transmitted to the articular bone, the quadrate bone and finally the stapes.The evidence for this view includes the observation that the stapes of cynodonts normally abuts strongly against the quadrate, and the absence of a clear-cut site for a post-quadrate tympanum. Allin argues that this theory explains the structure of the angular bone with its reflected lamina, and also makes more sense of the transition to the mammalian arrangement of the middle ear. The lower jaw was subjected to selection for improved hearing acuity, which took the form of a reduction in the size and mass of the postdentary bones and quadrate. The external auditory meatus, so essential for the first theory, is regarded as no more than the site of origin of a depressor mandibuli muscle.

There are drawbacks to both ideas. In the case of the first, the tympanum as usually restored looks suspiciously small compared to the size of the fenestra ovalis to allow an adequate impedance-matching ability, and the stapes is much more massive than apparently desirable to detect higher frequency sound. And why could not the system be improved in the modern reptilian fashion by lightening the stapes and developing a simple lever arrangement? The sound reception of many modern reptiles and birds is as acute as of mammals. On the other hand, Allin's theory requires a most unsatisfactory explanation of the putative external auditory meatus, since it would represent an excessively tortuous, overlarge origin for a jaw-opening muscle; and at least in the more primitive cynodonts the size of the angular and postdentary bones indicates that the vibrating system would have had very high inertia and be heavily damped in consequence. The system would hardly be capable of detecting anything but very low frequency airborne sound.

## *Locomotion*

In the course of cynodont evolution, a number of mammalian features of the postcranial skeleton evolved, culminating in the extremely mammal-like tritylodontids and the mammals themselves. The most primitive form is *Procynosuchus* (Fig. 72) which resembles the other therapsids, particularly the gorgonopsids and therocephalians in the structure of the limb girdles and limbs (p. 115). The pattern of locomotion was presumably similar to these other groups, with a permanently sprawling forelimb gait, and a hindlimb capable of either a sprawling or a more erect gait. Only the vertebral column of *Procynosuchus* shows the unmistakable development of definitive cynodont characters. The zygapophyses of the vertebrae in the thoracic region restricted lateral undulation of the column even more than occurred in other therapsids (Fig. 94A–C). The postzygapophyses are almost peg-like and fit tightly into the trough-like prezygapophyses of the next vertebra. The ribs have become more strongly attached to their respective vertebrae too. The two heads of each rib have run together to form a single, long facet called a synapophysis, which articulates with the similarly elongated facet of the vertebra. This probably indicates that the ribs were used as levers to prevent adjacent vertebrae from moving relative to one another. In contrast to this tendency to stiffen the vertebral column, the neck region became more flexible, and the first two vertebrae, the atlas and the axis, specialised to increase the dorso-ventral and rotational movements of the head (Kemp, 1969b).

However, *Procynosuchus* is difficult to interpret in detail as a represen-

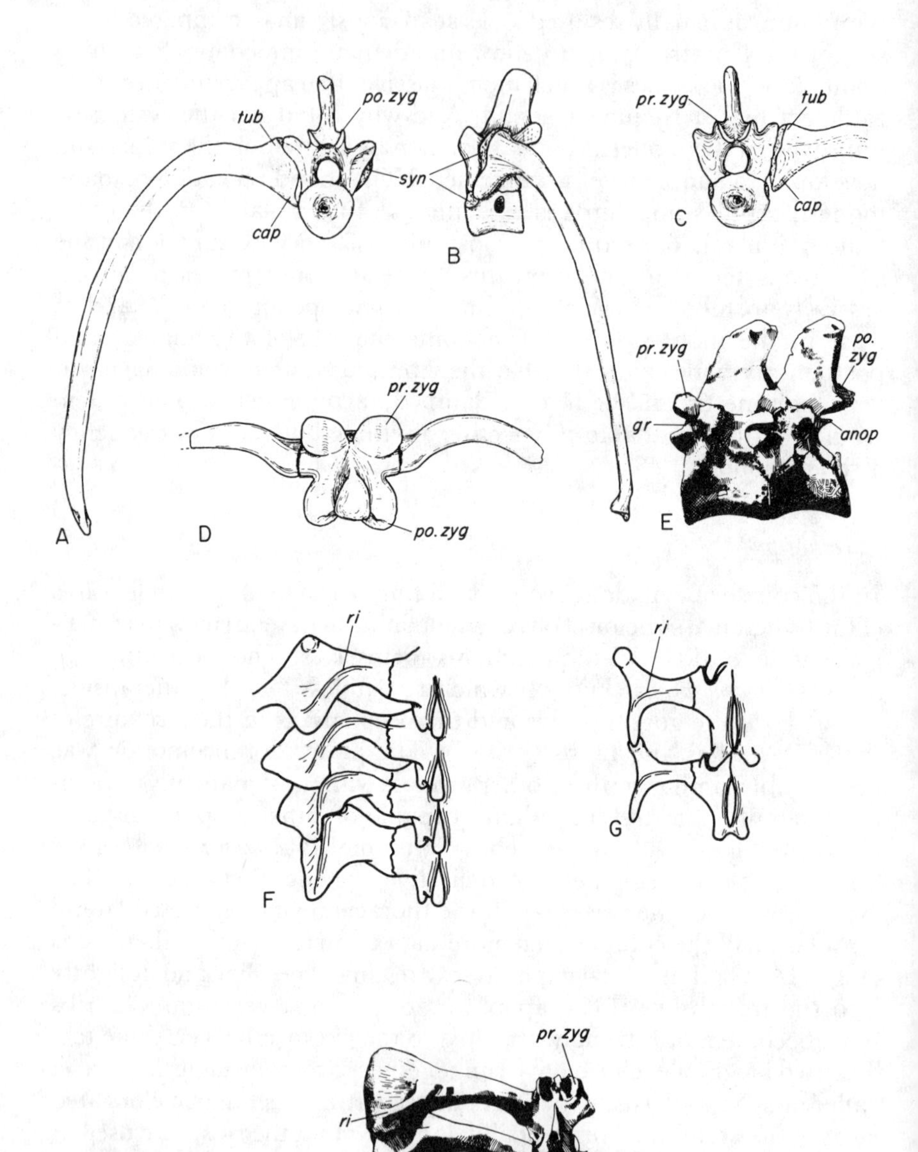


FIG. 94. The axial skeleton of cynodonts. Thoracic vertebra and left rib of *Procynosuchus* in A, posterior view; B, lateral view; C, anterior view. D, dorsal view of lumbar vertebra and ribs of *Procynosuchus*. E, lateral view of two lumbar vertebrae of the traversodontid *Luangwa*. F, dorsal view of the last thoracic and first three lumbar vertebrae and left ribs

tative of the ancestral cynodont because it possesses certain specialisations not found in any other cynodonts, or indeed any therapsids at all, which were adaptations for swimming. It appears to have been a fully amphibious animal. Thus the zygapophyses of the lumbar vertebrae are broad and horizontal (Fig. 94D) and allowed the posterior part of the column and the tail to be thrown into sinusoidal waves. Rather elongated haemal arches between the tail vertebrae increased the surface area of the tail as a swimming organ, while the limb bones and feet were flattened, showing that a paddling action was also possible (Kemp, 1980b).

The next stage of cynodont evolution is represented by the galesaurids such as *Thrinaxodon*. The most dramatic development (Fig. 75) is an expansion of the inner part of the rib shaft to form a series of closely interlocking costal plates. Several functional interpretations of cynodont expanded ribs have been suggested, including that of Brink (1956) that they increased the resistance of the rib cage to a newly evolved diaphragm. Jenkins (1971a) believed that they both strengthened the vertebral column, and made more effective the action of muscles causing lateral bending of the vertebrae. A more likely possibility (Kemp, 1980c) is that they helped to maintain the rigidity of the vertebral column and actually prevented lateral bending. This would be correlated with a more powerful locomotory force produced by the hindlimbs, while the forelimbs were still more or less passive as in other therapsids. In such circumstances there would be greater stresses acting along the column and tending to force it to bend. As in *Procynosuchus*, the nature of the zygapophyses inhibit bending anyway, and there are now accessory zygapophyses, the anapophyses (Fig. 94E), developed below the pre- and post-zygapophyses, which also inhibit bending movements between adjacent vertebrae. The obvious advantages of such a rigid vertebral column are that the forward momentum of the animal would be maintained, and possibly the creature was more manoeuvrable compared to reptiles whose vertebral column oscillates from side to side during locomotion. The limbs and limb girdles of *Thrinaxodon* (Jenkins, 1971a) do not differ substantially from those of

of *Cynognathus*. G, dorsal view of two lumbar vertebrae and left ribs of the traversodontid *Massetognathus*. H, dorsal view of two lumbar vertebrae and left ribs of the traversodontid *Luangwa*. (A–D from Kemp, 1980b; E and H from Kemp, 1980c; F and G from Jenkins, 1970.)

*anap*, anapophysis; *cap*, capitulum; *gr*, groove for articulation with anapophysis; *po.zyg*, postzygapophysis; *pr.zyg*, prezygapophysis; *ri*, ridge; *syn*, synapophysis; *tub*, tuberculum.

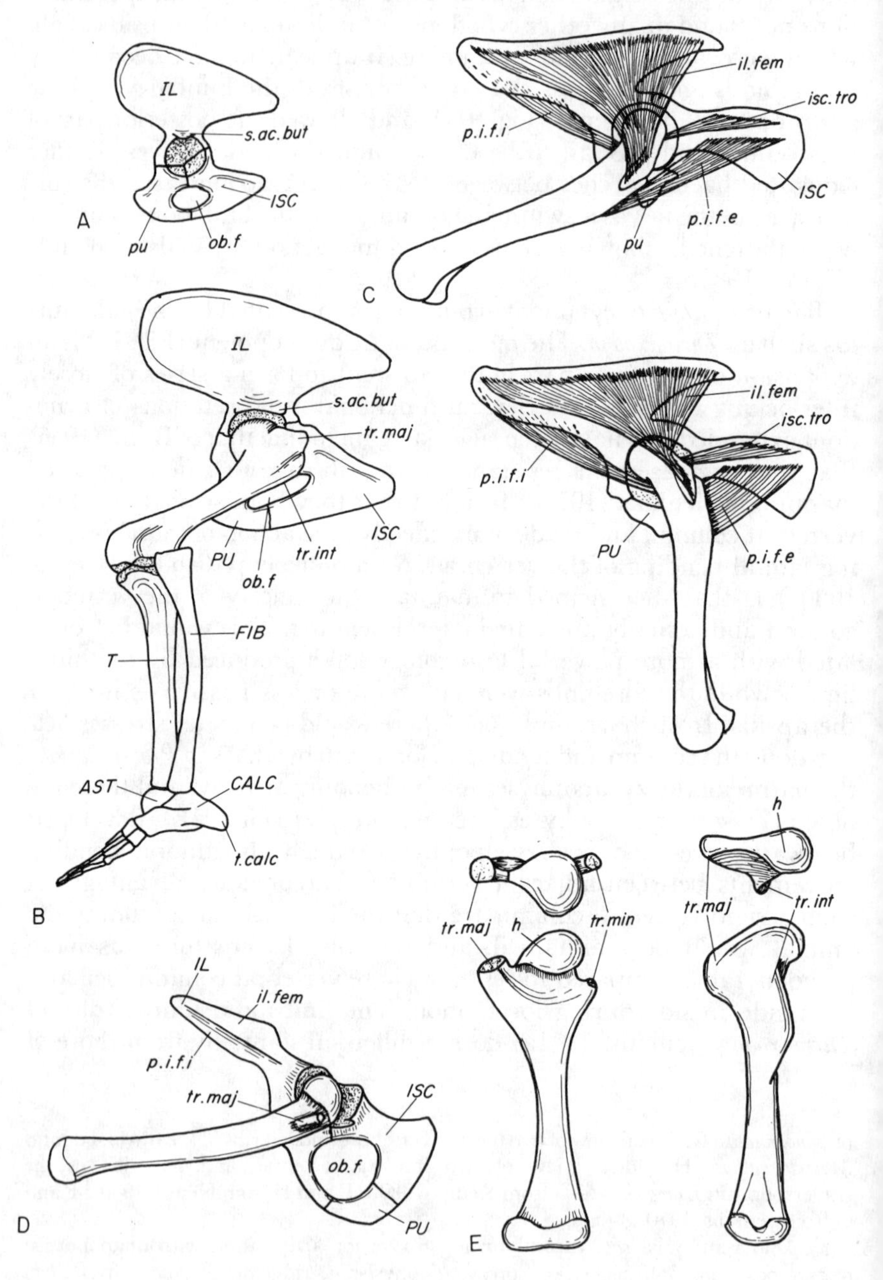
IL
s.ac.but
ISC
A
pu
ob.f
il.fem
isc.tro
p.i.f.i
ISC
p.i.f.e
pu
C
IL
s.ac.but
tr.maj
ISC
PU
tr.int
ob.f
FIB
T
AST
CALC
t.calc
B
il.fem
isc.tro
p.i.f.i
PU
p.i.f.e
h
tr.maj
tr.int
tr.maj
h
tr.min
IL
il.fem
p.i.f.i
tr.maj
ISC
ob.f
PU
D
E

FIG. 95.

*Procynosuchus* although they do show incipient developme
features fully expressed in the advanced cynodonts.

In the advanced cynodonts of the Lower Triassic, *D...*
*Cynognathus* alike (Jenkins, 1971a), a very important modification to the pattern of locomotion had occurred. The structure of the hindlimb and pelvis (Fig. 95B) shows that the ability of the hindlimb to operate in a primitive, sprawling fashion had been lost. The acetabulum is deeper and smaller than in primitive cynodonts, while the articulating head of the femur is a large sphere set off at an angle to the shaft of the bone. The femur could only fit into the acetabulum if the knee joint lay fairly close to the body throughout the stride, and therefore the hindlimb must have moved in a plane approaching a parasagittal plane. The stride commenced with the femur extending antero-laterally, and then the bone swung backwards and downwards until it extended ventro-laterally from the hip joint.

The hip musculature was modified, for those parts which were most important for the sprawling gait of earlier therapsids were reduced and the whole arrangement became more mammal-like. The anterior process of the ilium is expanded and the trochanter major of the femur enlarged, indicating development of the more anterior part of the ilio-femoralis muscle, which became the main retractor of the hindlimb (cf. Fig. 95C). The most posterior part of the ilio-femoralis muscle, which was used mainly for rotation of the femur about its long axis during the sprawling gait, was reduced as shown by the reduction of the posterior process of the ilium. The ischium remained large and the main muscle originating from it, the pubo-ischio-femoralis externus, was a second important retractor of the femur at this stage, and is associated with the retention of a prominent internal trochanter on the underside of the femur (cf. Fig. 95E). These two main retractor muscles, ilio-

FIG. 95. The pelvis and hindlimb of cynodonts. A, lateral view of the left pelvis of *Thrinaxodon*. B, lateral view of the left pelvis and hindlimb of *Cynognathus*. C, left pelvis and femur of the traversodontid *Luangwa* in the protracted position (above) and the retracted position below). D, lateral view of the left pelvis and femur of the tritylodontid *Oligokyphus*. E, proximal and ventral views of the femora of *Oligokyphus* (left) and *Luangwa* (right). (A and B redrawn after Jenkins, 1971a; C and E redrawn after Kemp, 1980c; D and E redrawn after Kühne, 1956.)

*AST*, astragalus; *CALC*, calcaneum; *FIB*, fibula; *IL*, ilium; *il.fem*, ilio-femoralis muscle; *h*, head of the humerus; *ISC*, ischium; *is.tr*, ischio-trochantericus muscle; *ob.f*, obturator fenestra; *p.i.f.e*, pubo-ischio-femoralis externus muscle; *p.i.f.i*, pubo-ischio-femoralis internus muscle; *PU*, pubis; *s.ac.but*, supra-acetabular buttress; *t.calc*, tuber calcis; *tr.int*, trochanter internus; *tr.maj*, trochanter major; *tr.min*, trochanter minor; *T*, tibia.

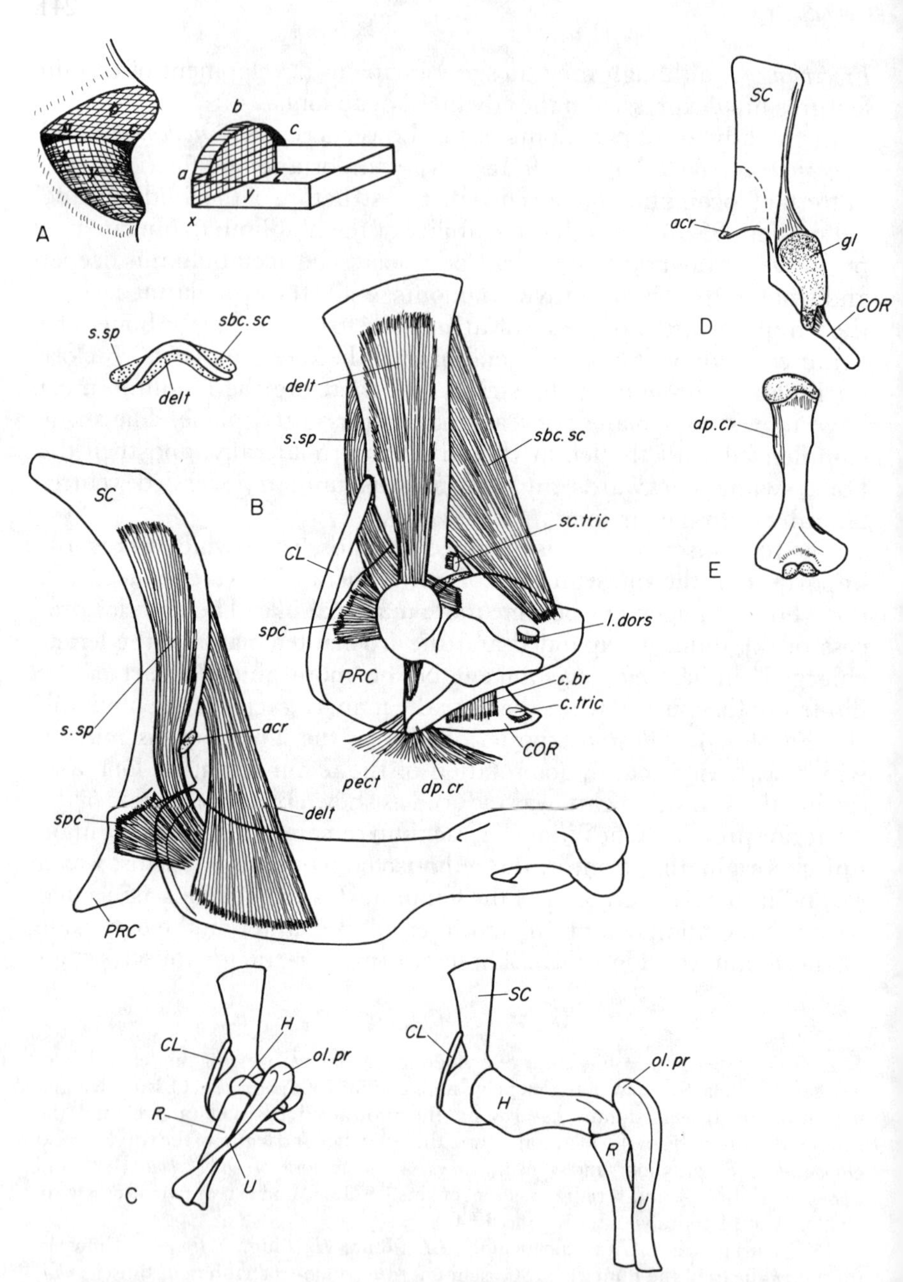


FIG. 96. Shoulder girdle and forelimb of cynodonts. A, diagram of the left glenoid and head of the humerus. At the start of the stride, the humerus extends laterally from the glenoid and points *a–a* and *x–x* are in contact. As the humerus retracts it also rotates about its long axis. The upper part of the humerus head rolls over the upper part of the

femoralis and pubo-ischio-femoralis externus appear to have almost completely replaced the old reptilian caudi-femoralis muscle running from the tail to the femur. The pubis of these Lower Triassic cynodonts is reduced, showing that the protractor muscle, the pubo-ischio femoralis internus had almost completely migrated dorsally, to the body facia, ribs and vertebrae of the lumbar region, and possibly the inner surface of the ilium as well. In the primitive therapsids, it was the more dorsal part of this muscle which functioned most effectively as a retractor during the more erect gait.

In contrast to this marked change towards mammalian organisation of the hindlimb, the forelimb of the Lower Triassic cynodonts remained persistently primitive and sprawling. There is no possibility that the humerus could have adopted anything other than a horizontal orientation. Indeed, the glenoid restricted the movements of the humerus even more than in primitive cynodonts and therapsids generally. The upper half of the glendoid (Fig. 96A), formed by the scapula, is an almost horizontal shelf against which the bulbous dorsal surface of the humerus head rested. As the humerus retracted, the contact controlled the rotation of the bone about its long axis, the humerus head rolling like a wheel across the glenoid. The ventral part of the glenoid formed by the coracoid is similar to that of other therapsids, and controlled the protraction–retraction movments of the humerus as in those forms (p. 117) (Kemp, 1980c). Despite its basically primitive action, there was an important innovation of the shoulder musculature, started in *Thrinaxodon* but taken to greater lengths in the later forms. The front edge of the scapula blade is turned outwards, and this includes the acromion process at the base of the blade, to which the clavicle attached. There is therefore a gap between the clavicle and acromion laterally and the coracoid plate and lower part of the scapula medially. The area of

glenoid, while the lower part of the head rolls over the lower part of the glenoid. At mid-stroke, points *b–b* and *y–y* are in contact. At the end of the stroke points *c–c* and *z–z* are in contact. B, lateral and anterior views of the shoulder girdle and humerus in the retracted position (a section through the scapula blade is also shown), of an advanced cynodont. C, lateral view of the forelimb in the protracted (left) and the retracted (right) positions. D, lateral view of the shoulder girdle of the tritylodontid *Oligokyphus*. E, dorsal view of the left humerus of *Oligokyphus*. (A and B redrawn after Kemp, 1980c; C redrawn after Jenkins, 1971a; D and E redrawn after Kühne, 1956.)

*acr*, acromion; *c.br*, coraco-brachialis muscle; *CL*, clavicle; *COR*, coracoid; *c.tric*, coracoid head of the triceps muscle; *delt*, deltoideus muscle; *dp.cr*, delto-pectoral crest; *gl*, glenoid; *H*, humerus; *l.dors*, latissimus dorsi muscle; *ol.pr*, olecranon process; *pect*, pectoralis muscle; *PRC*, procoracoid; *R*, radius; *sbc.sc*, subcoraco-scapularis muscle; *SC*, scapula; *sc.tric*, scapula head of the triceps; *spc*, supracoracoideus; *s.sp*, supraspinatus muscle; *U*, ulna.

origin of the supracoracoideus muscle expanded from its primitive limits on the coracoid and the lower part of the scapula, through the gap and onto the front part of the inner surface of the scapula (cf. Fig. 96B). This development was the first appearance of what was to become the supraspinatus muscle of the mammals. In the therian mammals, the new area of origin of the muscle became everted as a lateral-facing supraspinatus fossa. In the case of the advanced cynodonts, the origin still lay topologically on the medial surface of the scapula. Functionally, this change indicates that a more powerful protraction of the humerus occurred, although still in a horizontal plane. A rather similar change occured to the posterior part of the scapula, as the hind edge of the blade is similarly turned outwards. This is associated with the enlargement of the subcoraco-scapularis complex, which was the main retractor muscle and originated from the inner face of the scapula, emerged posteriorly, and inserted on the humerus. Again this is essentially a move towards the mammalian organisation of the shoulder musculature.

The lateral reflection of both anterior and posterior edges of the scapula blade led to a deeply concave central part. This has been taken as evidence that the supracoracoideus had invaded the outer face of the scapula blade, forming the homologue of the mammalian infraspinatus muscle (Fig. 108C), (Romer, 1922; Jenkins, 1971a). There is no apparent functional advantage in such a development, however, so long as the humerus still moved in a horizontal plane. It seems more probable that the concavity is a simple consequence of the reflection of the edges of the scapula, which was associated only with making the inner surface of the scapula available for enlarged protractor, and retractor muscles (Kemp, 1980c). The outer scapula surface was probably still occupied by the deltoideus muscle and its derivative the teres minor, as in more primitive therapsids. This is still the case in living monotreme mammals. The changes in shoulder musculature seem to have affected equally the protractor and the retractor muscles, and therefore selection was probably not for production of larger locomotory forces, but for increasing the speed of the stride. In general, the advanced cynodont forelimb was still not involved in the production of locomotory forces so much as maintaining the front of the animal off the ground in a fairly passive manner.

Overall, the advanced cynodonts developed the more erect hindlimb gait as the sole mode of locomotion, and this presumably increased the locomotory power. Also the closer spacing of the hindfeet to one another must have greatly improved the manoeuvrability needed for activities such as the capture of prey, evasion of predators, and traversing uneven ground.

In the Middle Triassic traversodontid cynodonts, the trends in hindlimb evolution continued, and a form such as *Luangwa* (Kemp, 1980c) had a virtually mammalian arrangement. The anterior part of the ilium (Fig. 95C) has expanded even further forwards and is a long, low process, while the posterior part is further reduced. The ilio-femoralis muscle occupied the whole lateral surface of the ilium and the greater anterior development of the muscle indicates that it retracted the femur further than in earlier cynodonts. The ischium is still a very large bone and has become aligned almost horizontally, which again suggests that the hindlimb could retract further back. The pubis is also more mammal-like, for it is further reduced in size and now extends postero-ventrally. No part of this bone extends anterior to the acetabulum and therefore no retractor muscle fibres could have attached to it. A significant new feature of the pelvis is the strong out-turning of the lower part of the ilium. The internal surface of this lower part faces partly downwards and forwards, and was probably an area of origin of the pubo-ischio-femoralis internus muscle. This muscle had already commenced an antero-dorsal migration in the earlier cynodonts, and here it seems to have achieved a position equivalent to that in mammals, from the ilium blade itself. In fact, this muscle in mammals, the iliacus (Fig. 108E), originates from the lower half of the outer ilium surface. The internal origin in traversodontids suggests that the mammalian condition was achieved by a later eversion of the lower part of the ilium so that the muscle-bearing surface became topologically lateral. *Luangwa* also evolved a larger supraspinatus muscle in the shoulder region. The gap between the acromion process and the lower part of the scapulo-coracoid is larger, and the fossa on the inner face of the front of the scapula is much more extensive (Fig. 96B).

There was a trend within the cynodonts of reduction of the costal plates of the ribs. In *Cynognathus* and *Diademodon* (Fig. 94F), these were present and well developed on all the dorsal vertebrae except the anterior thoracics, and also the accessory zygapophyses noted in *Thrinaxodon* were present. The same is true of the traversodontid *Luangwa*, but in *Massetognathus* (Jenkins, 1970), which is closely related to *Luangwa* as far as skull and tooth structure is concerned, all the thoracic ribs lack costal expansions, and those of the lumbar region are reduced to a series of very small, delicate overlapping processes (Fig. 94G). The accessory zygapophyses are also absent. In *Exaeretodon* the process has gone further (Fig. 79B), for none of the ribs show any sign of costal plates (Bonaparte, 1963b). The significance of this trend is not clear. Jenkins (1971a) relates the loss of costal expansions to the development of mammalian features of the muscles and ligaments which strengthened the vertebral column, and functionally replaced the osteological devices of the earlier cynodonts.

The culmination of evolution of the postcranial skeleton of cynodonts is found in the tritylodontids, of which the skeleton of *Oligokyphus* (Fig. 81) is well known (Kühne, 1956). Here the posterior process of the ilium is virtually absent, and the long anterior process is divided by an external longitudinal ridge (Fig. 95D). This separates the area of origin of the ilio-femoralis (mammalian gluteal) muscle dorsally from the pubo-ischio-femoralis internus (mammalian iliacus) muscle ventrally, as in modern mammals. The trochanter major of the femur (Fig. 95E) is prominent and very close to the proximal end of the bone. Another mammalian feature has also evolved, the trochanter minor, which is for the insertion of the pubo-ischio-femoralis internus muscle, and lies in front of the head of the femur. The tritylodontid ischium was reduced compared to other cynodonts, and the internal trochanter is absent, and therefore the posterior part of the pubo-ischio-femoralis externus (mammalian obturator externus) muscle was reduced in size and importance. The greatest part of the retraction of the femur must have now been due to the ilio-femoralis muscle alone, as in modern mammals. The hindlimb must have operated in a plane close to parasagittal very like small mammals of today. The forelimb and girdle is also advanced over traversodontids, in particular by the extreme eversion of the acromion process of the scapula (Fig. 96D). The passage forwards for the supraspinatus muscle is very large, and the origin of this muscle from the medial surface of the scapula is indicated by a deep fossa. The glenoid fossa is more widely open and the humerus (Fig. 96E) relatively a great deal more slender. *Oligokyphus* had a much more mobile and unrestricting shoulder joint, and it seems likely that for the first time the humerus operated in a plane approaching parasagittal, and was not restricted to the horizontal, sprawling gait. The likely advantage to the animal of this revolutionary development was the extra manoeuvrability due to the approach of the two forefeet towards the midline, as had already occurred in the hindlimb. In general therefore, the tritylodontids had achieved a fully mammalian type of locomotion.

The final group of advanced cynodonts to consider are the chiniquodontids. The postcranial skeleton of *Probelesodon* (Fig. 84) is known from more or less complete but indifferently preserved specimens (Romer and Lewis, 1973). Structurally it was at about the same level as the traversodontid *Luangwa*, except that it lacked all traces of expanded ribs, and presumably therefore its locomotion was similar with an advanced hindlimb gait associated with a sprawling forelimb gait. One peculiarity which has not been accounted for is an extraordinary slenderness of the scapula. There are also a few specimens of incomplete postcranial skeletons tentatively assigned to *Probainognathus* (Romer and Lewis, 1973). Here the shoulder girdle seems to be similar to other groups of advanced cynodonts. The most noticeable feature, however, is the relative length and slenderness of both the humerus and the femur. Nevertheless, neither of these forms has yet been studied from a functional point of view.

## *General biology*

Between the most primitive and the latest cynodonts many features of the skeleton characteristic of mammals evolved at least in an incipient form, which can be interpreted as a sign of the appearance of functional characteristics of mammals. It is likely that several or even many features of the soft anatomy and physiology of mammals had also evolved in cynodonts, although in most cases there is little or no direct evidence. Nevertheless it is important to consider what few pointers there are.

A question on which there is possible evidence is whether a mammalian type of diaphragm was present, bounding the thoracic rib-cage posteriorly and assisting the breathing by its muscular movements. To do this the abdomen must be unconstricted by a rib-cage so that changes in the thoracic volume can be compensated for by appropriate changes in the abdomen volume (Brink, 1956). In *Procynosuchus* (Fig. 72), the pair of ribs on the twentieth vertebra (approximately the thirteenth dorsal vertebra) are long and curved exactly as the preceding series of ribs. The twenty-first pair of ribs are about the same length, but are much straighter and thereafter the ribs rapidly attenuate in length and come to lie more and more horizontally. It seems likely that the relatively sudden change between the twentieth and the twenty-first ribs marks the position of a diaphragm, the edges of which were supported by the twentieth ribs (Kemp, 1980b). In later cynodonts (Fig. 75), this reduction of the ribs in the lumbar region is more marked still, while the posterior boundary of the thoracic region is indicated by the abrupt termination of a rib-cage of very mammalian appearance (Brink, 1956; Jenkins, 1971a), and therefore a diaphragm was almost certainly present. One curious exception is the late, very large traversodontid *Exaeretodon* (Bonaparte, 1963b), where the lumbar ribs only shorten gradually from front to back, and remain more or less vertically oriented (Fig. 79B). It can only be supposed that a strong lumbar rib-cage developed because of the increase in size in this form. Other advanced cynodonts which lost the costal plates, such as *Probelesodon*

(Romer and Lewis, 1973) and the tritylodontids (Kühne, 1956), retained short lumbar ribs.

Concerning the important question of whether the cynodonts had evolved hair and therefore an insulative outer layer, there is no satisfactory direct evidence. Hair being proteinaceous, it is seldom preserved in fossils, and no report exists of hair-like impressions associated with any mammal-like reptile. The well-developed foramina that open onto the snout and front part of the dentary indicate that there was a rich blood and nerve supply, to the external surface of the head. Furthermore, the absence of grooving of the bone surfaces adjacent to the foramina shows that a thick layer of fleshy tissue covered the lips. (Tatarinov, 1967). This kind of evidence has been used to argue that sensory vibrissae of a mammalian kind were present, and therefore hair had evolved since vibrissae are assumed to be modified body hairs (Watson, 1931; Brink, 1956).

Whether or not vibrissae were present, another character of the mammalian snout was undoubtedly developed. Much of the external surfaces of the bones of the snout and dentary are finely sculptured with minute foramina often associated with tiny grooves. This sort of surface is typical of modern reptiles where the skin is tightly applied to the bone surface. However, the bone surface of the premaxilla, maxilla and dentary adjacent to the teeth is prefectly smooth and the minute foramina are sparse. This corresponds to the presence of soft, muscular lips on both the upper and the lower jaws, a feature not unexpected in an animal which kept its food in its mouth for chewing, rather than swallowing it immediately. Equally certain, although not directly observable, must have been the presence of a well-developed tongue to assist the movements of food in the mouth.

The mammalian heart is characterised by a complete division of the ventricles, and corresponding double circulation, whereby the pulmonary circuit to the lungs is separated from the systemic circuit to the body. There is, of course, no direct evidence bearing on heart structure in cynodonts, but in two respects the circulation of the head had developed incipient mammalian characteristics. This probably reflects an increase in blood pressure which double circulation gives. The first feature is the infraorbital arterial system (Fig. 97A). A large foramen lies in the floor of the orbit and opens into a canal which runs forwards through the bones, the infraorbital canal. Smaller canals radiate from the main canal and open externally as the external foramina of the maxilla (Fourie, 1974). This arrangement is similar to that seen in mammals, except that there the infraorbital canal opens externally as a single foramen in the maxilla, the infraorbital foramen. The second possible development of the head circulation concerns the venous drainage of the braincase (Fig. 97B) and therefore of the brain. In typical reptiles, the main vessel for head drainage is the vena capitis lateralis, which lies alongside the braincase and connects to a series of veins emerging from the main foramina of the braincase. Although still present in cynodonts as indicated by a channel, the vena capitis lateralis was fairly modest. At the same time, the size of the foramina of the

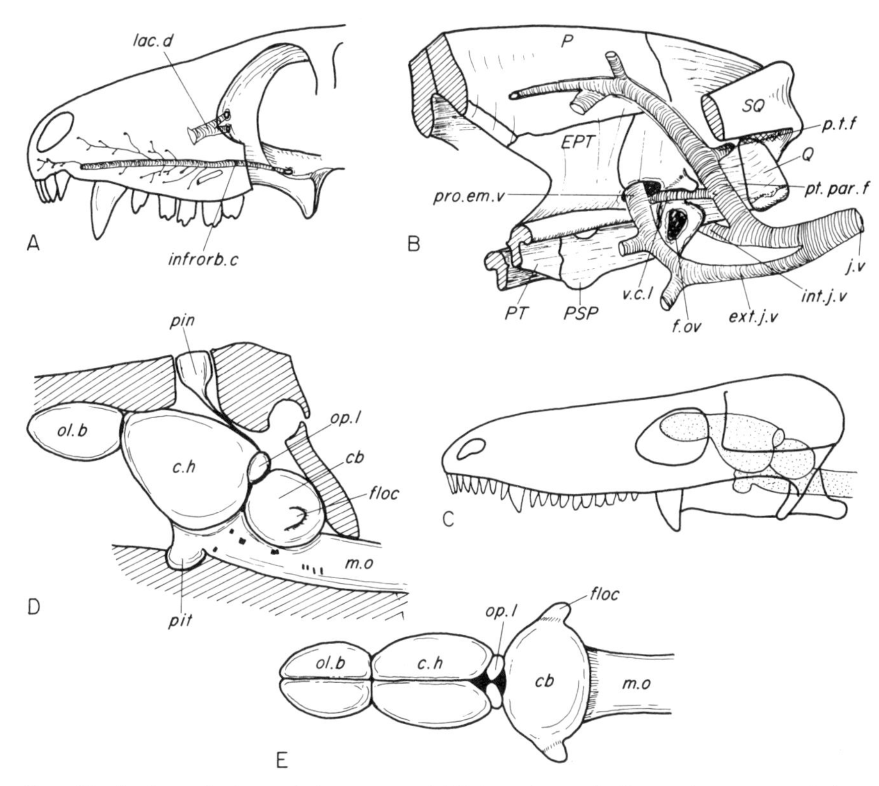


FIG. 97. A, lateral view of the snout of *Thrinaxodon* to indicate the course of the infraorbital canal. B, posterior part of the skull of *Procynosuchus* viewed obliquely from the side, front and below, to show the main veins. C, lateral view of the skull of *Procynosuchus* to show the possible proportions of the brain. D, reconstruction of the brain of *Procynosuchus* in lateral view. E, the same in dorsal view. (A modified after Fourie, 1974; B–E redrawn after Kemp, 1979.)

*cb*, cerebellum; *c.h*, cerebral hemisphere; *ext.j.v*, external jugular vein; *floc*, flocculus; *f.ov*, fenestra ovalis; *infrorb.c*, infraorbital canal; *int.j.v*, internal jugular vein; *j.v*, common jugular vein; *lac.d*, lachrymal duct; *mo*, medulla oblongata; *ol.b*, olfactory bulb; *op.l*, optic lobe; *P*, parietal; *pin*, pineal body; *pit*, pituitary; *pro.em.v*, prootic emissary vein; *PSP*, parasphenoid; *PT*, pterygoid; *p.t.f*, post-temporal fenestra; *pt.par.f*, pterygo-paroccipital foramen; *Q*, quadrate; *SQ*, squamosal; *v.c.l*, vena capitis lateralis.

braincase for the trigeminal, facial and vagus nerves are relatively very large, indicating that large veins passed through in addition to the nerves. It is possible that these emerging veins joined up directly to form a new, external jugular vein, rather than connecting to the reduced vena capitis lateralis. If this is so, then the drainage is much more like that of mammals (Kemp, 1979).

One of the most important physiological characters of mammals is the ability to produce hypertonic urine and thus reduce water loss in a relatively arid climate. The only comment on whether cynodonts had developed this ability is by Grine *et al.* (1979), who argued that certain Lower Triassic forms such as *Diademodon* possessed a salt-secreting gland in a hollowing in front of the orbit. This supposedly secreted a salt solution into the external nostril, as in a number of modern reptiles and birds. The possession of the gland implies that the kidney was not capable of producing hypertonic solution and hence the need for an extrarenal excretory system. However, this interpretation must be regarded as highly speculative.

As in earlier therapsids, the nasal cavity of cynodonts was probably a well-developed olfactory organ. Ridges for turbinal bones are still present, and a maxillary sinus in the internal surface of the maxilla bone appears to be a lateral extension of the nasal capsule, increasing the area of olfactory epithelium (Kemp, 1980c).

The size and structure of the cynodont brain (Fig. 97C–E) has proved very difficult to elucidate. The problem lies in the absence of a bony wall over the sides and floor of the brain in front of the level of the pituitary fossa. Therefore, although the shape of the dorsal surface of the brain is impressed on the undersurface of the skull roofing bones, the depth of the anterior parts of the brain, including particularly the cerebral hemispheres, is unknown. At one extreme, the cynodont brain has been reconstructed as tubular, with no more depth than width (Jerison, 1973; Hopson, 1979). This gives it a volume which, when allowance is made for body size, is no greater than in typical reptiles. On the other hand, it is possible to reconstruct a brain relatively very much deeper, which could have been accommodated in the space available. In *Procynosuchus*, a brain which approaches primitive mammalian size was possible (Kemp, 1979). Although the cerebral hemispheres could not have been quite as wide as in a mammal, they could have been just as deep. Certainly in this form the cerebellum must have been close to mammalian size, since the form of this part of the brain is almost completely impressed on the roof, sides and floor of the braincase.

Of the reproductive habits and physiology of cynodonts even less can be said. There is certainly no evidence that points positively towards the evolution of such mammalian features as lactation and altricial young. The presence in even the smallest discovered juveniles such as *Diademodon* (Hopson, 1973) of a functional dentition, and continuous tooth replacement from then onwards, suggests that the young fed independently right from birth, much as in modern reptiles.

The evidence that the cynodonts had an endothermic temperature physiology is very strong. Most of the previous therapsids were probably inertial homiotherms, which is to say they depended simply on large body size to reduce the rate of heat loss, and therefore could maintain an approximately constant body temperature with a low metabolic rate. The small forms may have hibernated during the cold season, or may possibly have had a higher metabolic rate, although the evidence is indecisive. In the case of the cynodonts, however, the conclusion that the metabolic rate was high is inescapable. The whole design of the feeding apparatus, in all the different adaptive types, was clearly to increase the rate of food assimilation. The rate of oxygen uptake was also probably increased. A secondary palate separated the respiratory air passage from the buccal cavity, so that the animal could continue breathing whilst masticating its food, and the diaphragm would have allowed a larger lung capacity and probably a higher rate of breathing as well. The more speculative arguments that the blood pressure increased, hair evolved, and a larger brain developed all point towards the same conclusion that a more advanced, mammal-like temperature physiology occurred.

It is not certain whether the changes in the locomotory system of cynodonts can be correlated directly with the temperature physiology. Even if the changes were designed to increase the speed of running, it is possible that fast locomotion only occurred in short spurts, and therefore it does not follow that a mammalian type of physiology had evolved. Many modern lizards can produce high speeds for short periods. However, there are two possible aspects of the cynodont locomotion that indirectly support the hypothesis that they were endothermic. The first is that the permanently more erect gait implies a greater degree of neuromuscular control of locomotion. This in turn means that the central nervous system was more complex and therefore likely to be more affected by temperature fluctuations; accurate temperature control was of greater importance and the most likely way this was achieved was the mammalian method of becoming endothermic. The second argument is that at least one of the purposes of a progressive locomotion system must have been to increase the efficiency and rate of food gathering, again a feature necessary to an endotherm.

The successful radiation of the cynodonts during the Triassic was associated with a number of ecological changes compared to the Permian, although how all these factors interrelate is not yet known. The Period is noted for a general rise in temperatures, and the gradual development of increasingly arid conditions (Robinson, 1971). There was a gradual change in the characteristic flora from the hitherto dominant *Glossopteris* flora of the Late Palaeozoic to the so-called *Dicroidium* flora. There was no sudden change, but the large lycopod and horsetail elements declined, while a great variety of gymnosperms including conifers, cycads, gingkos, etc., began to dominate the habitats. The most important change in the fauna other than the mammal-like reptiles was the gradual rise of the archosaur reptiles. In the basal Triassic *Lystrosaurus*-zone there were the semi-aquatic proterosuchians, to to be replaced in the Lower and Middle Triassic by a variety of more advanced groups of thecodonts. By the Upper Triassic both groups of dinosaurs, and the crocodiles were well established.

# 11 Mammals

THE FIRST FOSSILS universally regarded as mammals occur in deposits of Upper Triassic (Rhaetic) age. Throughout the rest of the Mesozoic Era, the Jurassic and the Cretaceous, the mammals radiated quite widely and were probably much more common than formerly supposed. However, possibly because of competition with the dinosaurs, the mammals remained small, probably nocturnal forms. It was not until the extinction of the dinosaurs at the end of the Cretaceous that the great mammalian radiation commenced, culminating in the diverse variety of Tertiary forms.

## Triassic mammals

At least three distinct groups of supposed mammals occur in deposits of late Triassic or Liassic age. The most abundant and best known forms belong to the family Morganucodontidae, where the whole skeleton has been described. A second form *Kuehneotherium* is known only from its teeth and lower jaw. The only remains of haramiyids found are peculiar isolated teeth, but because of the multi-cusped and multi-rooted nature of these haramiyids are generally regarded as mammals. A possible fourth type of mammal has so far been described from a single multi-rooted tooth from France (Russell *et al.*, 1976). Most of the material of Upper Triassic mammals occurs in fissure deposits of South Wales, which represent the infilling of underground water courses that were present in the Carboniferous Limestone hills of the Upper Triassic period. The specimens are exceedingly abundant but always very fragmentary, although the bone itself is very finely preserved.

### *Morganucodontids*

By far the commonest of the British fissure mammals is *Eozostrodon*,

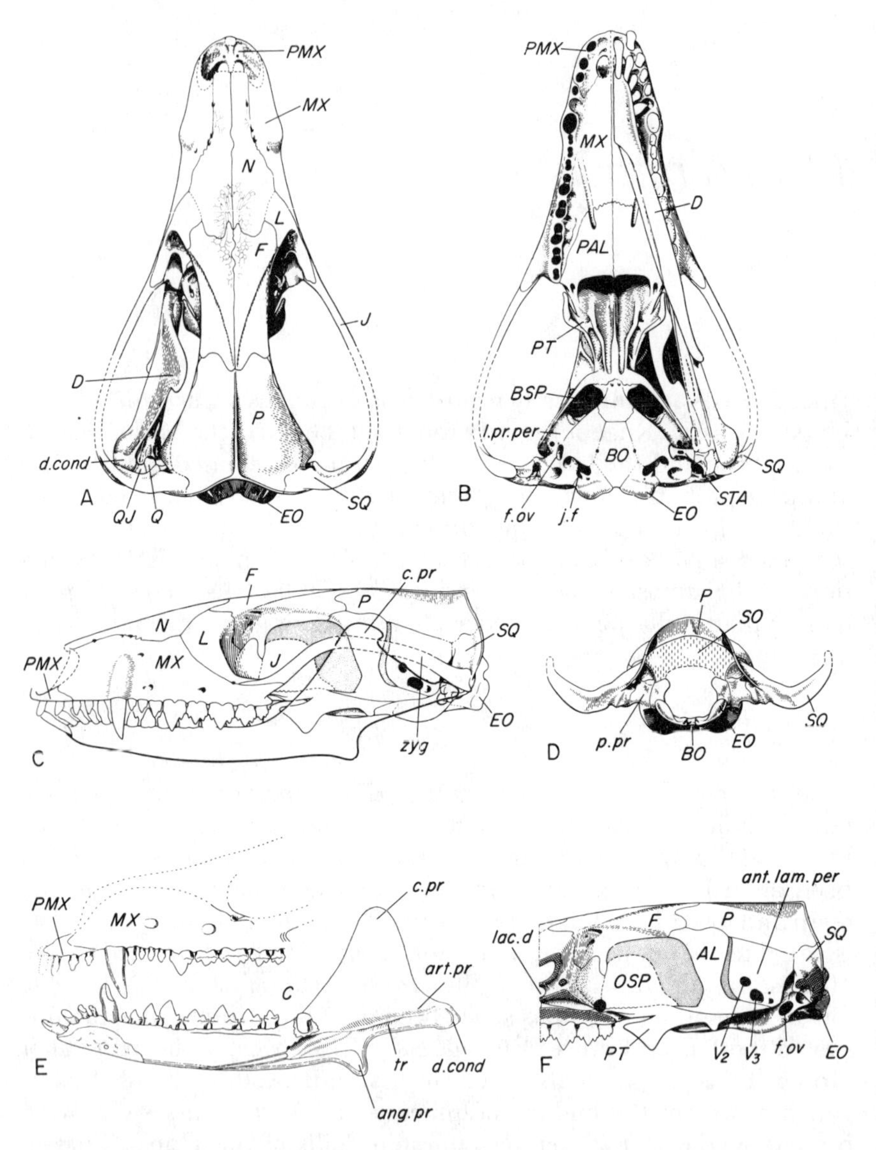


FIG. 98. The skull of *Eozostrodon* (=*Morganucodon*). A, dorsal view. B, ventral view. C, lateral view. D, posterior view. E, lateral view of snout and medial view of lower jaw. F, lateral view of hind part of skull without the zygomatic arch. (A–D and F from Kermack, Mussett and Rigney, 1981; E from Parrington, 1971.)

*AL:*, alisphenoid (epipterygoid). *ang.pr*, angular process; *ant.lam.per*, anterior lamina of the periotic; *art.pr*, articular process; *BO*, basioccipital; *BSP*, basisphenoid; *C*, coronoid; *c.pr*, coronoid process; *D*, dentary; *d.cond*, condyle of dentary; *EO*, exoccipital;

otherwise referred to as *Morganucodon*. There is an unfortunate nomenclatural disagreement about which name is valid and different authors have used different ones (Parrington, 1971, 1978; Clemens, 1979a). However, the most recent review of the group is by Jenkins and Crompton (1979) who accept the validity of the generic name *Eozostrodon*, and this usage has been followed here.

The animal itself is known from thousands of teeth, jaws, and fragments of skull and postcranial skeleton from Wales (Parrington, 1971). A complete skull and jaws of another specimen comes from the Late Triassic of the Lufeng Red Beds of China (Kermack *et al.*, 1973, 1981), and a possible second morganucodontid from China is *Sinoconodon* (Patterson and Olson, 1961). Two more related genera have been found in the Red Beds of the South African Karroo, *Erythrotherium* (Crompton, 1964) and *Megazostrodon* (Crompton and Jenkins, 1968). These latter two are of particular importance because they are both associated with a postcranial skeleton, which, although not very well preserved, aided Jenkins and Parrington (1976) in their important interpretation of the fragmentary postcranial bones of the fissure-deposit *Eozostrodon*.

The morganucodontids, like all Triassic mammals, are very small. Skull length is around 2–3 cm and the presacral length of the animals of the order of 10 cm. In general appearance they must have resembled the smaller shrews in life. The skull (Fig. 98) is typically mammalian in many features, such as the loss of the prefrontal and postorbital bones, the extensive parietal replacing the squamosal in the postero-dorsal region, and the slender, relatively low zygomatic arch. The lateral pterygoid processes of the palate are lost and the ectopterygoid bone reduced. The brain was enclosed laterally by bone, the largest part consisting of a broad anterior lamina of the periotic, which was pierced by foramina presumably for the maxillary and the mandibulary branches of the trigeminal nerve respectively (Kermack, 1963). The epipterygoid, or alisphenoid to give it its mammalian name, is high and fairly expanded antero-posteriorly but it has been displaced forwards by the periotic. The cranial process of the squamosal plays a very minor

*F*, frontal; *f.ov*, fenestra ovalis; *J*, jugal; *j.f*, jugular foramen; *L*, lachrymal; *lac.d*, lachrymal duct; *l.pr.per*, lateral process of the periotic (prootic); *MX*, maxilla; *N*, nasal; *OSP*, orbitosphenoid; *P*, parietal; *PAL*, palatine; *PMX*, premaxilla; *p.pr*, paroccipital process; *PT*, pterygoid; *Q*, quadrate; *QJ*, quadratojugal; *SO*, supraoccipital; *SQ*, squamosal; *STA*, stapes; *tr*, trough for post-dentary bones; $V_2$, foramen for maxillary branch of the trigeminal nerve; $V_3$, foramen for the mandibulary branch of the trigeminal nerve; *zyg*, zygomatic arch.

Magnification *c.* × 2.7.

role in the side wall of the braincase, behind the prootic (Kermack *et al.*, 1981).

The lower jaw of *Eozostrodon* (Kermack *et al.*, 1973) is very slender as befits such a small animal (Fig. 98C,E). The shape of the dentary is characteristically mammalian, with large coronoid and angular processes. The adductor jaw musculature appears to have resembled that of the advanced cynodonts (p. 227), with complete differentiation of a laterally placed masseter muscle associated with the angular process and a medial temporalis attaching to most of the coronoid process. The articular process of the dentary is more extensive than in cynodonts, however, and its posterior end is in the form of a conspicuous, bulbous condyle. It articulates with a deep glenoid fossa in the anterior face of the squamosal (Parrington, 1978). The postdentary bones of *Eozostrodon* are still very cynodont-like. They form a compound rod lying in a trough on the medial side of the dentary, supported above by the articular process of the dentary. The articular bone at the hind end of the postdentary rod still forms a jaw hinge with the small quadrate bone lying in a pocket in the squamosal, immediately internal to the glenoid. The axes of the two jaw hinges, dentary-squamosal and articular-quadrate, coincide along a lateral-medial line, and therefore the double jaw articulation of the most advanced cynodonts is still present.

Thus the postdentary bones and quadrate had not formed a set of ear ossicles independent of the lower jaw as occurs in modern mammals. However, if Allin's (1975) theory that the advanced cynodonts already utilised these bones for sound conduction to the stapes is accepted (p. 236), then the system had improved in morganucodontids. The secondary, dentary-squamosal jaw hinge had enlarged and took a greater proportion if not all of the stresses at the jaw articulation. The articular-quadrate hinge was free to function solely in sound conduction. The reflected lamina of the angular is unknown, but could have been a support for the tympanic membrane (Crompton and Parker, 1978). The cochlea housing is relatively large in morganucodontids, which supports the interpretation that the middle ear structure had increased auditory sensitivity (Crompton and Jenkins, 1979).

It is the nature of the dentition that most clearly distinguishes morganucodontids from such advanced cynodonts as the tritylodontids and tritheledontids. The normal number of incisors is four in both upper and lower jaws, although this may vary in different forms. The upper incisors are perpendicular, but the lower ones tend to be procumbent, and extensive wear of the teeth occurs (Parrington, 1971), indicating their importance in food-gathering. The canines are modest and, unlike many later mammals, the lower canine still bites inside the upper, as shown by wear facets. A small depression in the palate received the lower canine when the jaws closed, a primitive feature. The maximum number of postcanine teeth in *Eozostrodon* and probably in other forms is nine, although because of the pattern of tooth replacement and development, less are frequently observed. The premolars (defined in terms of tooth replacement, as well as morphology) number five in the upper jaw and four in the lower jaw (Parrington, 1971), and are of a simpler construction than the molars. They consist of a large main cusp with a varying development of a posterior accessory cusp, but no anterior accessory cusp and virtually no cingulum. Both the upper and the lower molars (Fig. 99A,B) have a similar, laterally compressed but sharp main cusp. Two posterior accessory cusps and a single anterior accessory cusp are also present; they lie in a line parallel to the jaw. The upper molar possesses a cingulum completely surrounding the crown and bearing fine cuspules. The lower molars differ for they have a cingulum restricted to the internal edge of the crown. Its cuspules are larger but less numerous than in the case of the upper cingulum. Accurate and extensive occlusion occurred between the upper and lower molars (Crompton and Jenkins, 1968; Mills, 1971). The main cusp of the lower molar wore against the depression between the main cusp and the anterior accessory cusp of the upper tooth. Simultaneously, the main upper cusp occluded between the main cusp and the posterior accessory cusp of the lower tooth. Although respective pairs of unworn teeth do not match very well, after a period wear facets developed on the crowns, and the match improved. The free edges of the flat wear facets are sharp, and approached one another like the blades of a pair of scissors. Food caught between the edges would be cut, or sheared. It is generally accepted that this need for the teeth to spend a period "wearing in" before they are fully efficient is a reflection of the primitive level of such teeth, compared to those of other mammals, where the teeth are a better match to start with. As has been seen (p. 219), certain advanced cynodonts achieved equally effective shearing teeth. However, a fundamental difference between them and the morganucodontids is that in the cynodonts the lower tooth moved in a posterior direction as it worked against the upper tooth. Here the lower tooth passed in a slightly medial direction as it sheared against the upper. There was no antero-posterior movement of the lower jaw at all.

The pattern of tooth replacement of morganucodontids was almost certainly typically mammalian diphyodonty (Parrington, 1971, 1978). This is characterised by a single replacement of the incisors, canines and premolars, and no replacement at all of the molars. It is very difficult to demonstrate in a positive way such a limited amount of replacement in actual specimens. However, unerupted or incompletely erupted teeth have been found in jaw fragments of small individuals of *Eozostrodon* (Fig. 99C). These include incisors, canines and anterior postcanines, but never the posterior postcanines. Assuming that diphyodonty did indeed occur in morganucodontids, it can be accounted for as a further reduction in the extent of tooth replacement seen in advanced cynodonts (p. 221). One feature very reminiscent of the cynodonts is the addition of the last molar to the tooth row relatively late in life. Many apparently adult specimens are found in which this tooth is absent, and others in which it is in various stages of eruption. The functional significance of reduced tooth replacement is that the specific pairs of matching upper and lower molars were retained together throughout life, which is necessary if they must "wear in" before they are efficient. A consequence, however, is that the molars must last for the entire life of the individual, and therefore the animal's lifespan was probably very short (Parrington, 1971). This would be expected in any case in such

FIG. 99. Molar teeth of *Eozostrodon* (=*Morganucodon*). A, internal view. B, crown view of opposing molars. C, internal view of a premolar showing a developing replacement tooth, as preserved. (A and B from Crompton and Jenkins, 1968, © Cambridge University Press; C from Parrington, 1971.)

small animals. Hopson (1973) has also suggested that diphyodonty is evidence of lactation. Animals in which the main teeth are not replaced must achieve more or less full size before those teeth erupt. There has to be the characteristic mammalian pattern of growth, with very rapid growth during a relatively brief juvenile phase, and no further increase in size thereafter. Such rapid juvenile growth implies a source of extremely rich nourishment, such as maternal milk.

The morganucodontid postcranial skeleton (Fig. 100) combines a basically mammalian structure with retention of several primitive, cynodont-like features (Jenkins and Parrington, 1976). There are no traces of expanded costal plates on any of the ribs, and the thoracic rib-cage is fully mammalian in appearance. There is a fairly abrupt transition between the vertebrae of the thoracic region and those of the lumbar region, indicated by a change in orientation of the neural spines from posteriorly directed to vertical. This may indicate that the characteristic mammalian feature of dorso-ventral flexibility of the vertebral column occurred at the point of transition. The first two vertebrae, the atlas and the axis, are still cynodont-like in so far as the elements of the

FIG. 100. Skeleton of the morganucodontid *Megazostrodon*. (From Jenkins and Parrington, 1976.)
Magnification *c.* × 0.9.

atlas have not fused to form a ring-shaped bone. However, they were probably held together tightly by ligaments, because an odontoid process attached to the front of the centrum of the axis is present. In modern mammals the atlas ring rotates around this process to produce rotation of the head on the vertebral column. Whilst the movement is incipiently present in the cynodonts (Kemp, 1969b), it appears to have been much more extensive in morganucodontids. Only three sacral vertebrae are present, but this is probably related to the small size of the animal compared to cynodonts, where there are five sacrals. The pectoral girdle and forelimb are comparable to cynodonts, but have evolved certain new features indicating that freer movements of the limb occurred (Fig. 101D–F). The coracoid plate is reduced and the acromion process at the base of the scapula blade is large. Therefore a well-developed supraspinatus muscle was undoubtedly present, originating from the medial face of the scapula as in cynodonts, although there is no development of the lateral facing supraspinatus fossa of modern therian mammals. The glenoid is much more open than in cynodonts, restricting the movements of the humerus to a far lesser extent, while the humerus itself is a slender version of the cynodont humerus, with essentially the same processes for muscle attachment and a similarly bulbous head. The locomotory movements of the forelimb probably involved dorsal and ventral movements of the distal end of the humerus, while the bone remained in a retracted position. Most of the stride would have occurred as a result of flexion–extension movements of the elbow, much as in small modern mammals (Jenkins, 1971b). The advantage of this pattern of stride is that the forefoot would have remained more or less underneath the body, increasing the manoeuvrability of the animal.

The morganucodontid pelvis and hindlimb have a typically mammalian structure (Fig. 101A–C). The anterior ilium blade is separated into dorsal and ventral parts by a longitudinal ridge on the lateral surface. This separated the area of origin of the more dorsal gluteal muscle complex from the ventral iliacus muscle. In certain advanced cynodonts (p. 245) such as the traversodontid *Luangwa*, the iliacus (reptilian pubo-ischio-femoralis internus) originated from the medial surface of the ilium. Here, as in tritylodontid cynodonts, the lower medial part of the ilium appears to have been reflected outwards to form a lateral-facing surface. The posterior process of the ilium has been finally supressed in the morganucodontids. The femur has developed a completely mammalian form. The trochanter major stands well clear of the bulbous, inturned head of the bone, while a trochanter minor has appeared on the anterior side of the head for the iliacus muscles. Loss of

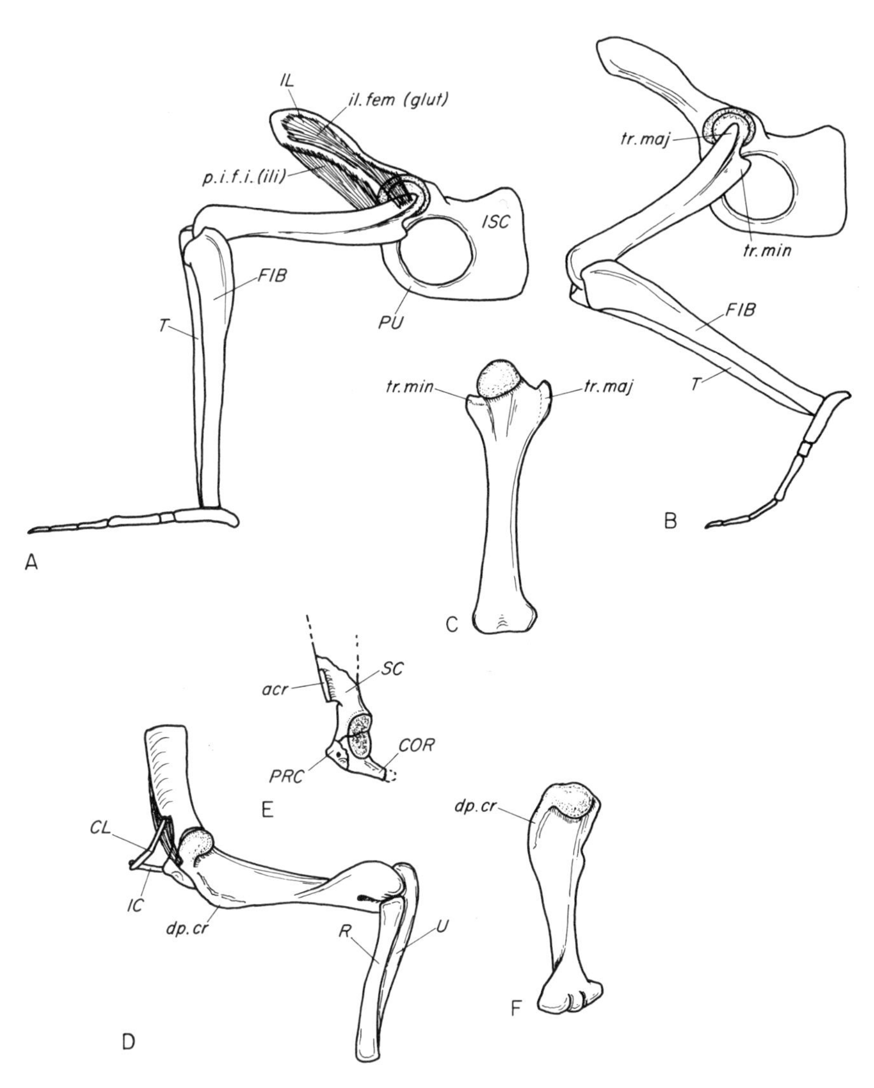


FIG. 101. The limbs and girdles of *Eozostrodon* (*Morganucodon*). A, left pelvis and hindlimb in the protracted position. B, the same in the retracted position. C, dorsal view of the left femur. D, left shoulder girdle and forelimb in lateral view. E, lateral view of the left scapulo-coracoid (as preserved). F, dorsal view of the left humerus. (Based on Jenkins and Parrington, 1976.)

*acr*, acromion; *CL*, clavicle; *COR*, coracoid; *dp.cr*, delto-pectoral crest; *FIB*, fibula; *IC*, interclavicle; *IL*, ilium; *il.fem(glut)*, ilio-femoralis (gluteal) muscle; *ISC*, ischium; *obt.f*, obturator fenestra; *p.i.f.i.(ili)*, pubo-ischio-femoralis internus (iliacus) muscle; *PRC*, procoracoid; *PU*, pubis; *R*, radius; *SC*, scapula; *T*, tibia; *tr.maj*, trochanter major, *tr.min*, trochanter minor; *U*, ulna.

the internal trochanter, so prominent in cynodonts, indicates that the pubo-ischio-femoralis externus muscle (corresponding to the mammalian obturator externus muscle) was of reduced importance in locomotion. Movements of the hindlimb (Fig. 101A,B) seem to have been fully mammalian, with the femur directed more or less anteriorly. Small dorsal and ventral movements of the distal end of the femur probably occurred in correlation with flexion and extension of the lower leg, much as was the case for the forelimb.

Locomotion in morganucodontids (Jenkins and Parrington, 1976) appears therefore to have been fully mammalian. Both front and hind feet were placed close to the midline and the humerus and femur underwent relatively small dorso-ventral movements, coupled with extension of the lower leg during the power phase of the stride and flexion during the recovery phase. This gave the animal a highly agile, variable gait. Sustained speed was probably of less significance than agility in catching insects, and the typical habitat of these forms was probably irregular, forested terrain and possibly trees.

Many of the features of both the skull and the postcranial skeleton indicate that the morganucodontids were essentially like modern mammals in their general biology. The relatively well-developed sectorial dentition compares functionally with the dentition of modern insectivore mammals, and this coupled with the very small size of the animals indicates that their main diet was probably insects. As with modern forms, other food such as soft invertebrates and various parts of plants were no doubt also eaten. The locomotion, with its stress on agility as much as speed and efficiency, also suggests that they actively hunted. Certainly the morganucodontids were endothermic, in view of their apparently high food intake. It is widely believed that these small early mammals were adapted for a nocturnal existence, and indeed this may have been a reason for the refinement of the temperature regulation (Crompton *et al.*, 1978). One of the reasons for thinking they were nocturnal is that the brain was evidently enlarged, not as much as in modern mammals of a similar size, but distinctly more than is usually supposed to be the brain-size in cynodonts. The enlargement has affected particularly the cerebral hemispheres, which are related primarily to the sense of smell (Jerison, 1973). This, of course, is one of the most important senses for typical nocturnal animals. The other important sense is hearing, and the morganucodontids may very well have improved upon the hearing system of cynodonts, as has been noted.

If these animals were indeed endotherms, then they must have been insulated by fur in order to reduce the otherwise high rate of heat loss that would have occurred, particularly at night. One of the most characteristic features of all modern mammals is lactation and complex maternal care of the offspring. This is necessary for a small endotherm, since the rate of heat loss of the juveniles would be intolerably high. The evidence that morganucodontids had the mammalian pattern of tooth replacement, diphyodonty, and the likely relationship of this to lactation and high early growth rates has been noted.

The wide geographical distribution of the group, including both North and South Hemispheres, and their apparent abundance at least in Europe, indicate that whatever their particular adaptation they were a very successful group.

### *Kuenhneotheriids*

The sole form of this important group is *Kuehneotherium* and possibly a second as yet undescribed genus (Kermack *et al.*, 1968) which occurs as a rare element in the South Wales fissure deposits. Only isolated teeth and jaw fragments have so far been identified as kuehneotheriid. It is possible that some of the fragments of skull and postcranial skeleton which are attributed to *Eozostrodon* actually belong to *Kuehneotherium*, assuming that the structure of the two is practically identical.

The lower jaw of *Kuehneotherium* (Fig. 102A–C) is relatively more slender than *Eozostrodon*, and it lacks the angular process of dentary of the latter form. No postdentary bones have ever been found in association with the dentary, but a broad trough in the medial side of the dentary exactly like that in *Eozostrodon* is presumed to have held a compound postdentary rod.

The number of postcanine teeth is not known for certain, but lies between nine and eleven. Also it is not known whether the molariform teeth were replaced or not and therefore whether the mammalian pattern occurred. (See note 4 added in proof, p. 351.) The essential feature of the molars, both uppers and lowers, is that the cusps form a triangle rather than the straight line characteristic of *Eozostrodon* teeth. The upper molar (Fig. 102D) has a large main cusp towards the inner side of the tooth. A somewhat smaller anterior accessory cusp lies well towards the outer side, while the posterior accessory cusp lies rather less towards the outer edge. There is also a relatively well-developed second posterior accessory cusp lying at the extreme postero-lateral corner of the crown. A cingulum, devoid of cuspules, surrounds the crown completely, and it is in the form of a distinct shelf between the main cusp and the posterior accessory cusp on the inner side. The lower molar (Fig. 102E) is taller than the upper, and again consists of a large main cusp and anterior and posterior accessory cusps. The latter two lie towards the inner side of the tooth. A cingulum is present on the

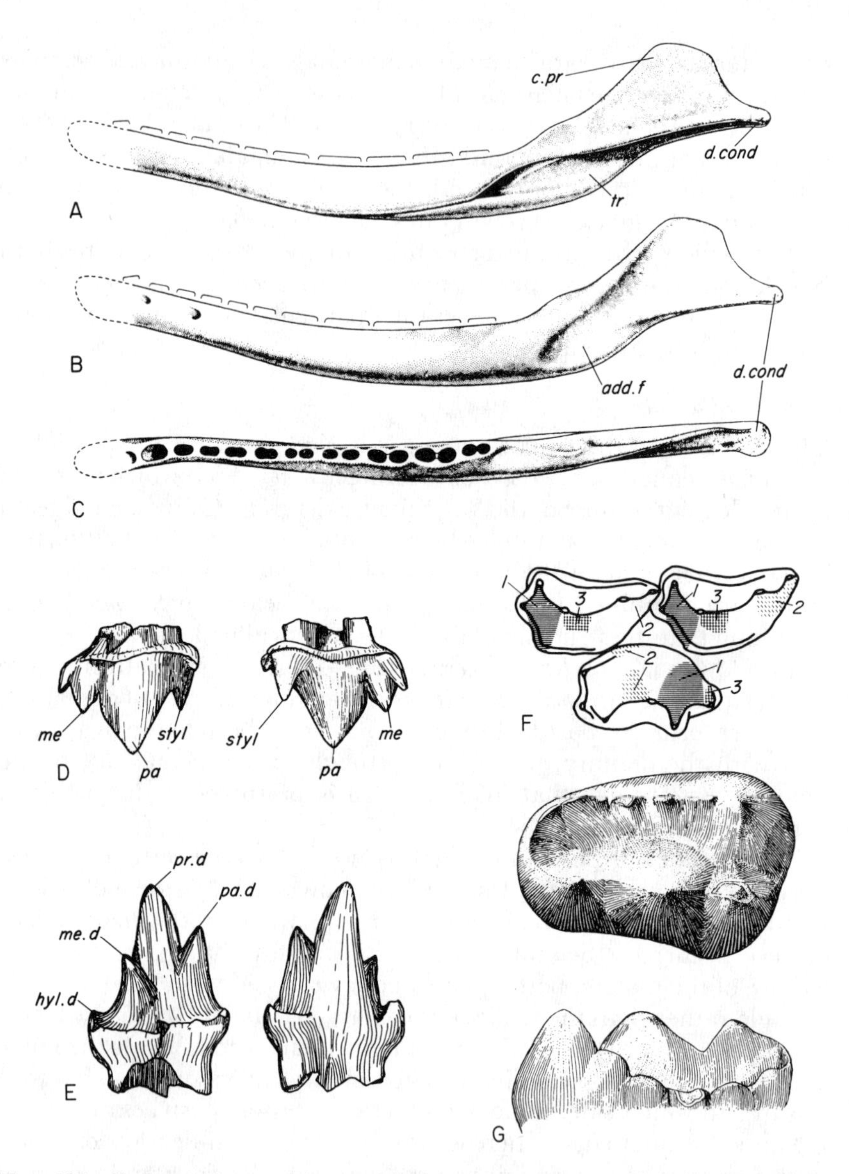


FIG. 102. *Kuehneotherium* and haramiyids. Lower jaw of *Kuehneotherium*; A, medial view; B, lateral view; C, dorsal view. D, medial and lateral views of a left upper molar of *Kuehneotherium*. E, medial and lateral views of a left lower molar of *Kuehneotherium*. F, crown views of occluding upper (above) and lower molars of *Kuehneotherium*, showing corresponding wear-facets. G, crown and side views of a tooth of the haramiyid *Microcleptes*. (A–E from Kermack *et al.*, 1968, with permission from *J.Linn.Soc.*(*Zool.*) ©The Linnean Society of London; F from Bown and Kraus, 1979; G from Simpson, 1928.)

posterior, outer and anterior sides of the crown, and extends around the inner side as far as the base of the main cusp. Small cusps are present, two on the anterior section of the cingulum which lock against the next tooth in front, and one on the posterior cingulum region. Although all the molars have the basic structure just described, they do vary quite considerably, particularly in the degree of angulation between the three main cusps. Presumably this reflects the position of the teeth in the jaw, although the exact way in which the dentition varies from front to back is not yet clear.

The way in which *Kuehneotherium* molar teeth worked (Fig. 102F) can be deduced from the wear facets which developed on the individual teeth (Kermack *et al.*, 1968; Crompton, 1971, 1974). During the occlusion of the teeth which, as in morganucodontids, occurred on one side of the jaw at a time, the molar teeth formed an interlocking series of alternately reversed triangles. The main cusp of the upper molar occludes outside the posterior accessory cusp of the lower molar. Simultaneously, the main cusp of the lower tooth lies inside the anterior accessory cusp of the upper tooth. The main shearing action occurred between the crest connecting the upper main and anterior accessory cusps, and the similar crest between the main and posterior accessory cusps of the lower tooth. These two opposing crests are aligned almost transversely across the line of the jaw because of the angulation of the cusps. A similar but somewhat less extensive shearing action also occurred between the upper crest connecting the main to the posterior accessory cusp on the one hand, and the crest between the main and anterior accessory cusps of the lower tooth on the other. These crests are aligned obliquely across the jaw line. In addition to these two main pairs of shearing edges, there are a number of other pairs on different parts of the teeth. These were probably of relatively minor functional importance to *Kuehneotherium*, but of considerable significance when this form is compared with certain more advanced mammalian teeth of the Jurassic and Cretaceous.

The molar teeth of *Kuehneotherium* have two apparent advantages over those of morganucodontids in terms of the effectiveness of the shearing function. The first is that by triangulating the cusps the upper teeth match the corresponding lower teeth more closely, even before any tooth-wear has commenced. Therefore a lesser degree of wear is needed

*add.f*, adductor fossa; *c.pr*, coronoid process; *d.cond*, dentary condyle; *hyl.d*, hypoconulid; *me*, metacone; *me.d*, metaconid; *pa*, paracone; *pa.d*, paraconid; *pr.d*, protoconid; *styl*, stylacone; *tr*, trough of dentary for the postdentary bones.

Magnifications: A–C, *c.* × 7; D and E, *c.* × 16; G, *c.* × 17.

before the opposing crests are fully "worn in" to form exactly matching pairs of occluding faces. The second difference is that by arranging for the shearing crests to lie transversely or obliquely relative to the line of the lower jaw it is easier to prevent food from forcing the opposing crests apart and thereby preventing shearing.

The premolar teeth are much simpler than the molars, and resemble closely those of morganucodontids. The main cusp is recurved, an anterior accessory cusp is barely present, and the one or two posterior accessory cusps are very small.

### *Haramiyids*

Even less is known of these apparently mammalian forms than of *Kuehneotherium*, since only isolated teeth have been found. The majority of specimens are from fissure deposits in Somerset of Rhaetic age, although a few teeth are known from Germany (Clemens and Kielan-Jaworowska, 1979). All the teeth (Fig. 102G) are similar in structure (Parrington, 1947), and there is no way of distinguishing uppers from lowers. Each tooth is multiple-rooted, and the crown consists of an occlusal basin surrounded on one side by three fairly large cusps, and on the other by about five smaller cusps. The ridge connecting all these cusps continues around one end of the crown.

Occlusion between the upper and lower teeth of a generally mammalian kind occurred, indicated by the wear facets developed. The upper teeth must have been oriented in reverse to the lower teeth. The row of five cusps of the upper tooth lay in the valley of the lower tooth, and, simultaneously, the row of five cusps of the lower tooth met the valley of the upper tooth (Parrington, 1947). Judging from the exact nature of the wear-facets, the lower jaw moved slightly backwards during the occlusion of the teeth, which contrasts with the slightly medial movement of the jaw in both morganucodontids and kuehneotheriids (Clemens and Kielan-Jaworowska, 1979). If this is true then it is a feature of cynodonts rather than of typical mammals.

### *Interrelationships and origin*

Having established the nature of the earliest mammals, two questions arise regarding their phylogenetic interrelationships. The first is whether they form a monophyletic group and the second is: Which of the mammal-like reptiles is most closely related to them? Thirdly there is the semantic question of exactly which forms should be included in the taxon Mammalia and how therefore should it be diagnosed (p. 293).

Two quite distinct views have been expressed in recent years about the relationship between morganucodontids and kuehneotheriids. One is that the differences between them are so considerable that their latest common ancestor must have been at best a primitive cynodont, and possibly even a therapsid below the cynodont level. The triangular arrangement of the main cusps and the absence of cuspules on the cingula of *Kuehneotherium* teeth are contrasted with those of morganucodontids. Also the exact regions of the teeth where upper molars contact the respective lower molars differ (Mills, 1971). A further argument concerns the structure of the sidewall of the braincase (Kermack, 1967; Kermack *et al.*, 1981). In the case of the morganucodontids, the sidewall is dominated by an anterior expansion of the periotic bone which contains the foramina for the trigeminal nerve (Fig. 103B). The alisphenoid (epipterygoid) is only slightly expanded and it lies in front of the periotic, from which it is separated by a slight cleft. This contrasts with the braincase of the modern therian mammals (Fig. 103C), to which *Kuehneotherium* is related, where the sidewall is constructed from an expanded alisphenoid perforated by the trigeminal foramina. There is no anterior lamina of the periotic, while the squamosal has come to form most of the sidewall behind the alisphenoid. Kermack argues that the morganucodontid condition could have neither evolved from the cynodont type of braincase (Fig. 103D), nor given rise to the therian type. Morganucodontids and the therians must have evolved independently from a very primitive stage. While it is accepted that the therians, including *Kuehneotherium*, could have been derived from cynodonts on the evidence of the expanded alisphenoid in both, morganucodontids could not.

These arguments are countered by those who believe that morganucodontids and *Kuehneotherium* form a monophyletic group, their latest common ancestor having achieved mammalian status (e.g. Parrington, 1973; Crompton and Jenkins, 1979). The similarities between the two respective groups include molar teeth with an apparently homologous set of three main cusps and a basically similar arrangement of the cingulum. Among the known morganucodontids, the South African form *Megazostrodon* has a molar tooth structure particularly comparable to that of *Kuehneotherium*, for the main cusps are actually slightly triangulated and the lower teeth meet the upper teeth in a very similar fashion (Crompton, 1974). To these direct anatomical comparisons between the two Triassic types must be added an important functional similarity. In both forms, the lower molar moves slightly medially as well as upwards as it occludes with the upper molar. During the recovery stroke the jaw of the lower tooth must therefore move slightly laterally

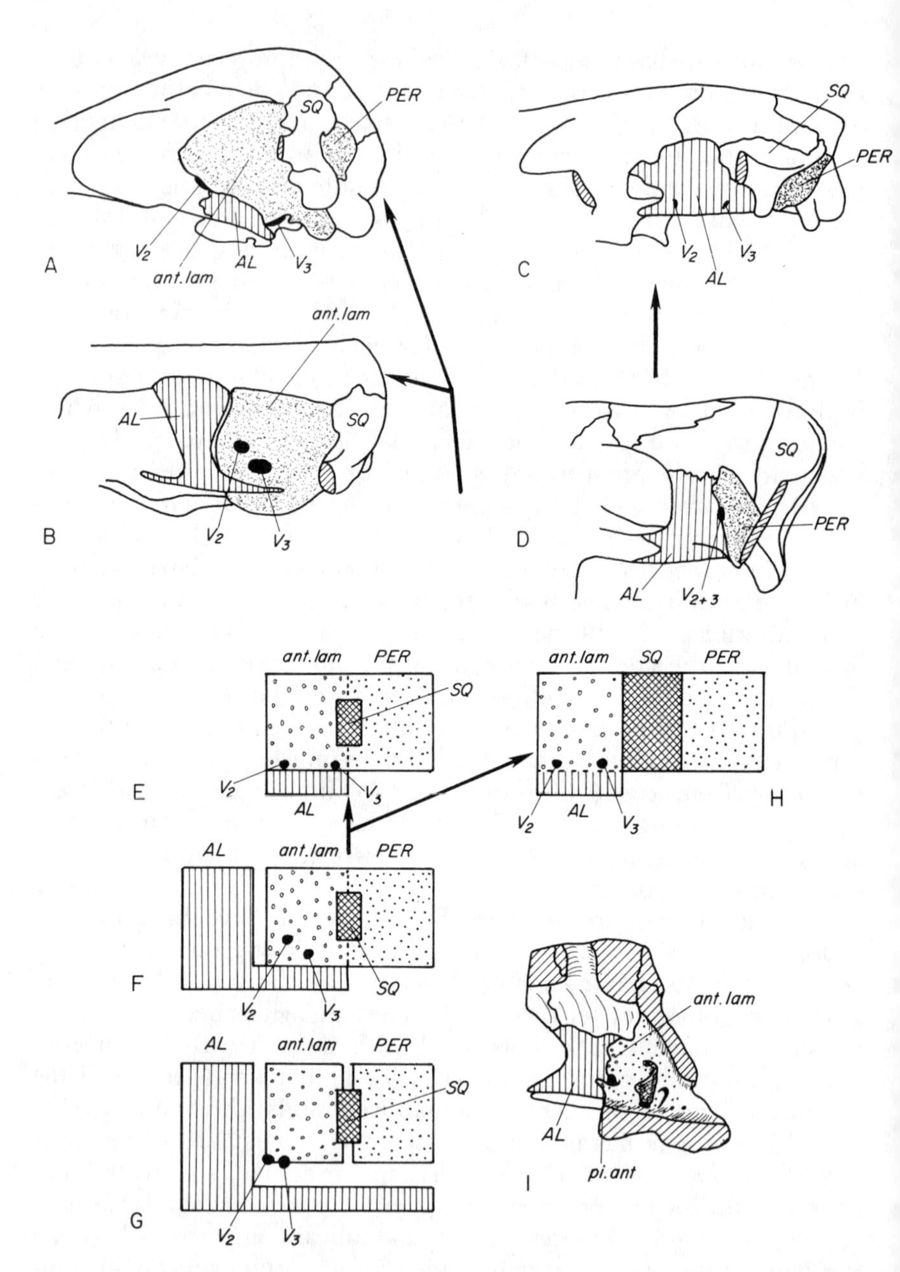


FIG. 103. Structure of the sidewall of the mammalian braincase. A–D, standard interpretation: A, monotreme (*Ornithorhynchus*); B, primitive non-therian mammal (*Eozostrodon*); C, modern therian mammal; D, cynodont. E–H, alternative view: E, monotreme; F, primitive mammal, therian or non-therian; G, basic components; H,

ready for the next bite. As viewed from behind, the motion of the lower jaw is not simply up and down, but is a triangular orbit. This feature contrasts with jaw movement in cynodonts, where the lower jaw has a slight anterior–posterior movement instead. Other characters shared by morganucodontids and *Kuehneotherium* are the virtual identity of their premolar teeth, and the form of the trough in the medial face of the dentary that houses the reduced postendary bones.

On the question of braincase structure, it must be stressed that the braincase or indeed any part of the skull of *Kuehneotherium* is unknown, and therefore the possibility that it actually had a morganucodontid-like braincase cannot be ignored. Cynodonts (Fig. 103I) possess both an expanded alisphenoid and a small anterior lamina of the periotic (Hopson, 1964), and the trigeminal nerve emerged between the two. Since both these structures appear to be ossifications within the same membrane lateral to the brain, there is little difficulty in supposing that the relative size of the two bones could change fairly readily. Either of the two mammal conditions could be derived from the cynodonts, or one from the other. In fact, as discussed later (p. 291), there are reasons for believing that *Kuehneotherium* actually did possess a braincase exactly like that of morganucodontids (Fig. 103F).

In the absence of very much information about *Kuehneotherium* at present, the balance of the evidence strongly favours the view that morganucodontids and kuehneotheriids are closely related.

Very little can be said about the possible relationships of the haramiyids. A vague similarity exists between the row of three cusps of the haramiyid tooth and the main row of cusps of morganucodontid teeth; the lesser cusp row of the haramiyids might be considered as an enlarged cingulum. Thus this group could perhaps be traced to the same common ancestor as the other two Triassic mammal groups. Against this, however, is the possibly cynodont-like motion of the lower jaw, with an antero-posterior movement rather than a lateral-medial one.

Assuming a monophyletic origin of the Upper Triassic mammals, their relationship to the mammal-like reptiles can be considered. In fact, the mammals share such a range of derived characters with the advanced cynodonts that a relationship between the two seems beyond

modern therian mammal. I, internal view of the sidewall of the braincase of the primitive cynodont *Procynosuchus*. (A–D redrawn after Kermack and Kielan-Jaworowska, 1971; I redrawn after Kemp, 1979.)

*AL*, alisphenoid (epipterygoid); *ant.lam*, anterior lamina; *PER*, periotic (petrosal); *pi.ant*, pila antotica; *SQ*, squamosal; $V_2$, foramen for the maxillary branch of the trigeminal nerve; $V_3$, foramen for the mandibulary branch of the trigeminal nerve.

question. Other views about the origin of mammals have been expressed, but these mainly pre-date the present level of knowledge about both advanced cynodonts and the earliest mammals. They have mostly been in the form of suggestions that mammals evolved from certain of the small, advanced therocephalians (e.g. Broom, 1932; Brink, 1956). The only seriously expressed challenge to the close relationship between cynodonts and mammals in recent years has been that of Kermack and his colleagues already referred to. If true, then it implies an incredible degree of parallel evolution between the two respective groups.

The most strikingly similar character complex of advanced cynodonts and mammals is the feeding apparatus, with enlargement of the dentary and its coronoid, angular and articular processes. In at least two cynodont lineages, the dentary forms a new articulation with the squamosal bone, a feature unique to the mammals otherwise. Reduction of the postdentary bones of cynodonts does not go as far as in the modern mammals, where these bones are incorporated into the middle ear, but they are closely comparable to the postdentary bones of the earlier mammals. If the theory of Allin (1975) that the postdentary bones of cynodonts functioned in hearing is accepted, then the comparison between cynodonts and mammals is even closer. Associated with the jaw structure, the differentiation of the teeth into incisors, canines, relatively simple anterior postcanines and complex posterior postcanines, and the development of an extensive laterally placed masseter muscle occurs only in these two groups. Behind these osteological and myological similarities there must also have been a profound similarity in the neurological basis of jaw action. Other features of the cynodont skull which indicate mammalian affinities include the secondary palate, a reduction of the lateral pterygoid processes of the palate and ectopterygoid, and a reduction in the size of the prefrontal and postorbital bones, which are completely absent in the mammals. The expansion of the alisphenoid (epipterygoid) bone to form an integral part of the side wall of the functional braincase occurs only in cynodonts and mammals, as does the development of an anterior lamina of the periotic.

Alone among all the therapsids, only the cynodonts evolved a significant suite of mammalian postcranial characters. For example, the structure of the hindlimb and pelvis, is virtually mammalian in some of the advanced cynodonts. The ilium extends forwards, the pubis has turned backwards, and the ischium has become almost horizontal. The cynodont shoulder girdle and forelimb is less like that of modern therian mammals, although it compares well with both the earliest mammals and with the living monotremes, and includes the development of the supraspinatus muscle for example. The differentiation of the vertebral column into a thoracic region with a well-developed ribcage and a lumbar region in which the ribs are reduced is a feature of both cynodonts and mammals.

Attempts to determine which of the known cynodonts is the most closely related to mammals have been confused with a search for "the mammal ancestor". Thus certain cynodonts with obvious specialisations not expected in the mammal ancestor have been dismissed, without considering whether they might nevertheless have a close phylogenetic relationship to mammals. There are actually at least three kinds of cynodont which are possible candidates for sister-group status of mammals.

The most widely held view at present is that the small, advanced carnivore *Probainognathus* (p. 210) is the cynodont closest to the ancestry of the mammals (Romer, 1970; Hopson and Kitching, 1972; Crompton and Jenkins, 1979). The main argument concerns the postcanine teeth, which have a linear arrangement of three cusps, a cingulum, and occlusion between upper and lower teeth. They are therefore comparable to the molars of morganucodontids. However, they are almost as comparable to the postcanine teeth of the primitive cynodont *Thrinaxodon* and probably represent the primitive cynodont tooth structure. A second much quoted feature of *Probainognathus* that relates it to mammals is the secondary contact between the dentary and the squamosal. In fact there is some doubt about whether there is an actual contact between these two bones (Crompton and Jenkins, 1979), although certainly the articular process of the dentary comes very close to invading the pre-existing (and typical advanced cynodont) secondary jaw hinge between the surangular and the squamosal. Apart from these two comparisons, and its small size, *Probainognathus* has no more similarity to mammals than any other advanced cynodont.

The tritheledontids ("ictidosaurs") are another group of small, carnivorous forms (p. 213). At least one member, *Diarthrognathus*, has a double jaw articulation involving contact between the dentary and the squamosal. Certain mammalian features are present in the skull which had not evolved in *Probainognathus*. The postorbital and prefrontal bones are absent, and the zygomatic arch is very slender. The South American Upper Triassic *Therioherpeton*, which is probably a tritheledontid, has a particularly mammalian arrangement of the skull bones. Tritheledontid postcanine teeth are not closely similar to those of any of the early mammals, although the variation in tooth structure now known to exist within the group (Gow, 1980) makes it perfectly conceivable that there was an unknown member with mammal-like teeth. What is more, alone amongst the mammal-like reptiles the tritheledontids have evolved prismatic enamel on their teeth, which is otherwise present only in mammals, including morganucodontids (Grine and Vrba, 1980). Until the skull and postcranial skeleton of tritheledontids becomes better known, it remains possible, if not likely, that they have a closer phylogenetic relationship to the mammals than does *Probainognathus*.

The third group of advanced cynodonts to consider are the tritylodontids (p. 203), which possess the greatest number and range of mammalian characters of any cynodonts. The skull is as mammal-like as that of the tritheledontids, with loss of prefrontal and postorbital bones, reduction of the lateral pterygoid processes of the palate and low zygomatic arch. The sidewall of the braincase has a larger anterior lamina of the periotic than is known in any other cynodont, and the general structure of this part of the skull is very similar indeed to that of *Eozostrodon* (Hopson, 1964). It is not certain whether the dentary contacted the squamosal, although this has been claimed in one form (Fourie, 1964) and could have occurred in others (Crompton and Jenkins, 1979). The postcanine teeth are multi-rooted. In the case of the postcranial skeleton, there is an extraordinarily detailed similarity between trityodontids such as *Oligokyphus* (Fig. 81) and *Eozostrodon* (Fig. 100), involving both the limbs and girdles, and the axial skeleton.

Thus a strong case can be made that the tritylodontids are more closely related to the mammals than is *Probainognathus*, although structurally they had diverged into a herbivorous habit. Set against this view is the evidence that tritylodontids are related to the traversodontid herbivores (p. 205). There are suggestive, if not compelling similarities between the postcanine teeth of the two groups (Crompton, 1972b), and a few minor points of similarity in the braincases, which may actually prove to be normal advanced cynodont features. If the tritylodontids are indeed related to the traversodontids, then all the mammalian characters of the tritylodontids must have evolved in parallel with the line leading to mammals. This seems to be an unnecessarily complex hypothesis, and it is simpler to conclude that the general resemblance between the herbivorous teeth of the two cynodont groups is a result of convergence, and that the tritylodontids are closely related to mammals.

It is unfortunate that at present the tritheledontids are insufficiently well known to decide with confidence whether they or the tritylodontids are closest to mammals. It is also possible that these two groups together constitute the sister-group of mammals. There is one derived character of mammals possessed by tritheledontids but not tritylodontids and that is prismatic enamel of the teeth. Tritylodontids have a non-prismatic enamel identical to that of other cynodonts (Grine and

Vrba, 1980). On the strength of this fact, it may be concluded tentatively that the tritheledontids are phylogenetically most closely related to the mammals (Fig. 104). The structure of the tritheledontid postcranial skeleton, as yet undescribed, would be of great importance in testing this hypothesis.

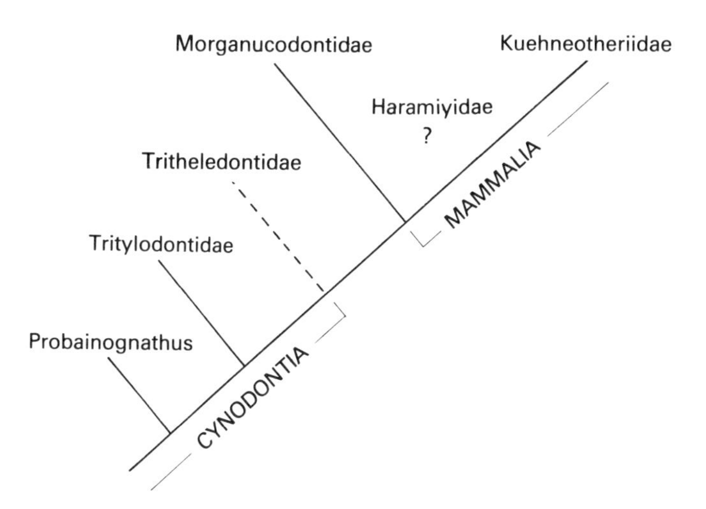


FIG. 104. Tentative phylogenetic relationships of advanced cynodonts and early mammals.

## Therian mammals

During the Jurassic and Cretaceous Periods, a variety of types of mammals evolved. They are known mainly from jaw and tooth fragments, although several more complete specimens have been reported but await description. The whole field of Mesozoic mammals has been reviewed recently (Lillegraven *et al.*, 1979) and only the more general aspects need be discussed here. All these mammals fall into two general groups based mainly on tooth structure. One, the Theria, includes *Kuehneotherium* as a primitive member, several intermediate kinds, and the modern marsupials and placentals. The other, variously referred to as the non-therians, the Atheria and the Prototheria includes morganucodontids, several more advanced groups, and the living monotremes, but is probably not a monophyletic group (Fig. 112).

At the present time, only characters of the teeth can be used to define the therian mammals, since no significant skull or postcranial remains have been described prior to the Late Cretaceous. The important

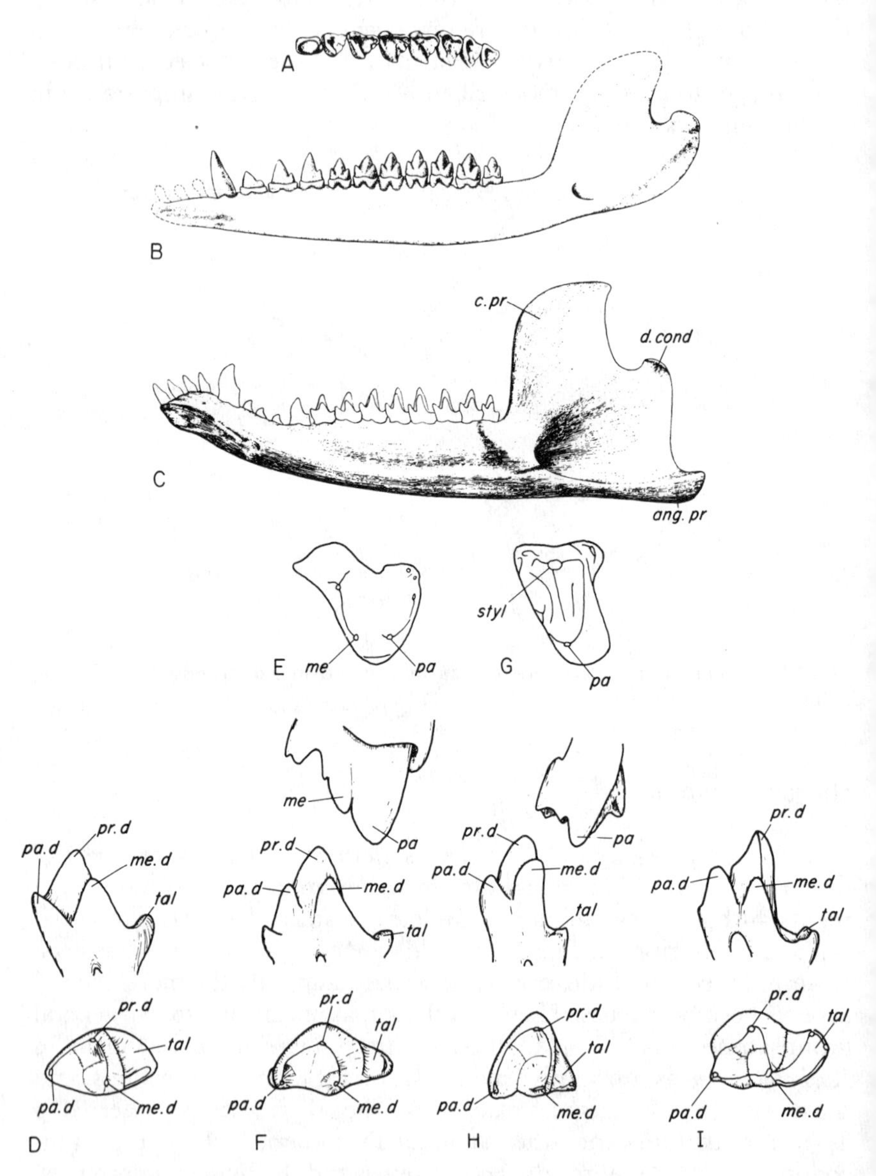


FIG. 105. Primitive therian mammals. A, occlusal view of the lower last premolar and molars of the symmetrodont *Peralestes*. B, medial view of the lower jaw of the symmetrodont *Spalacotherium*. C, medial view of the lower jaw of the drylestid eupantothere *Crusafontia*. D, right lower molar of the primitive eupantothere *Amphitherium* in internal and crown views. E, upper molar of *Peramus* in crown and

feature of the dentition is that the molar teeth are based upon a triangulated arrangement of three principal cusps, the most primitive expression of which is that found in *Kuehneotherium* of the Upper Triassic. During the Jurassic period an adaptive radiation of therian mammals occurred as manifested by several rather different variations on this basic tooth type. One of the lines culminated in the essentially modern types of therian mammals, the marsupials and the placentals, which first appear in the Early Cretaceous.

### *Symmetrodonts*

Symmetrodont teeth and lower jaws (Fig. 105A,B) are known from the Middle Jurassic to the Early Cretaceous, when the group became extinct. So far as comparison is possible, they resemble *Kuehneotherium* closely and several authors feel that *Kuehneotherium* should be included in the Symmetrodonta (Parrington, 1971; Cassiliano and Clemens, 1979). The structure and the mode of wear of the molars is similar, and both lack an angular process of the dentary on the lower jaw. The only difference is that the molars of symmetrodonts tend to be rather simpler; the cingula and their associated cusps are not so well developed.

### *Eupantotheres*

The earliest known member of this group also appears in the Middle Jurassic, as *Amphitherium* of the classic Oxfordshire locality of Stonesfield. During the rest of the Jurassic and the Early Cretaceous, a modest adaptive radiation of eupantotheres occurred and one of the lines produced gave rise to the modern therians (Kraus, 1979). The molar teeth of eupantotheres evolved from a *Kuehneotherium*-like ancestor by accentuating the more transverse crest of the upper tooth, and enlarg-

internal views. F, lower right molar of *Peramus* in internal and crown views. G, upper molar of the dryolestid eupantothere *Melanodon* in crown and external views. H, lower right molar of the dryolestid eupantothere *Phascolestes* in internal and crown views. I, lower right molar of the aegialodontid *Kielantherium* in internal and crown views. (A and B from Casiliano and Clemens, 1979; C from Krebs, 1971; D–I redrawn after Clemens and Mills, 1971; Crompton, 1971; Kielen-Jaworowska *et al.*, 1979b; Simpson, 1928; and Simpson, 1929.)

*ang.pr*, angular process; *c.pr*, coronoid process; *d.cond*, dentary condyle; *me*, metacone; *me.d*, metaconid; *pa*, paracone; *pa.d*, paraconid; *pr.d*, protoconid; *styl*, stylar cusp; *tal*, talonid.

Magnifications: A and B, *c.* × 2.3; C, *c.* × 3.9.

ing the posterior cingulum of the lower tooth to form the talonid, or posterior basin (Fig. 105D).

The Upper Jurassic pantothere *Peramus* is fairly typical (Clemens and Mills, 1970), and its teeth have been studied from a functional point of view. The upper molar (Fig. 105E) has a large principal cusp called the paracone. A long, concave crest runs from the tip of the paracone to the antero-lateral corner of the tooth or parastyle. It is almost transversely oriented relative to the line of the jaw and is much larger than the equivalent crest of *Kuehneotherium*. A second crest runs posteriorly from the paracone to the posterior accessory cusp, or metacone, and on to the postero-lateral corner of the tooth, the metastyle, via a number of small cusps. Again, this crest is almost transverse. The lower molar tooth of *Peramus* (Fig. 105F) is dominated by a trigon of three cusps, comparable to *Kuehneotherium* but more triangulated still. In the case of *Kuehneotherium*, there is a single, small cusp at the back of the tooth, which assists in interlocking adjacent lower molars, and also occludes with the posterior part of the large paracone of the upper molar. In *Peramus*, this incipient talonid is enlarged as a significant part of the crown structure, and bears two cusps. Functionally, the dentition of *Peramus* shows an elaboration of the basic mechanism of *Kuehneotherium* teeth (Crompton, 1971). Again the basis of the action is a number of opposing pairs of sharp, concave edges bounding flat, matching planes (Fig. 106C). The largest pair of edges is the anterior crest of the upper molar (paracone to parastyle) and the posterior crest of the trigon of the lower molar (protoconid to metaconid). The next largest pair is the posterior crest of the upper tooth (metacone to metastyle) and the anterior crest of the next molar behind (protoconid to paraconid). Both these pairs are longer, and more transversely orientated than in *Kuehneotherium*. A third region of shear has evolved in *Peramus*, the crest of the upper tooth which connects the paracone to the metacone acting against the cusps of the talonid of the lower molar. Although incipiently present in *Kuehneotherium*, the elaboration of this pair of edges is perhaps the single most significant development in *Peramus* and other eupantotheres from the more primitive condition, and it is a step towards the modern therian type of molars.

All the eupantotheres had teeth which were based on the same structural and functional plan as *Peramus*, but differing in the details. In the presumed primitive form *Amphitherium* (Figs 105D and 106B), the main crests are less transversely oriented, and the talonid not quite so well developed. Dryolestid pantotheres (Fig. 105C,G,H) have very highly developed transverse crests on the teeth, but the talonid is small.

Nothing is yet known of the structure of the skull and postcranial skeleton of the eupantotheres, although an as yet undescribed partial skeleton has been reported from the Upper Jurassic of Portugal (Henkel and Krebs, 1977).

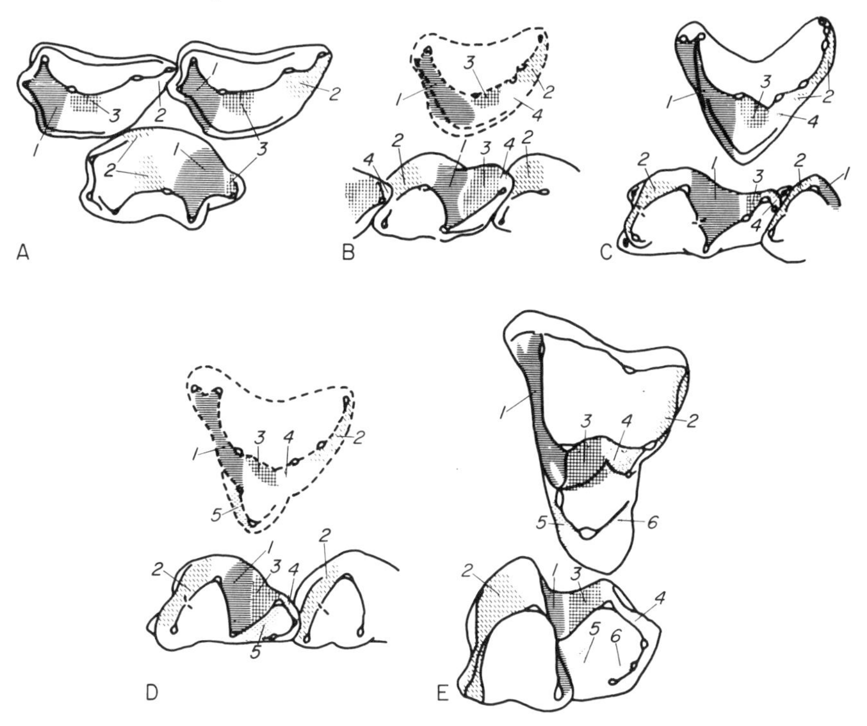


Fig. 106. The evolution of therian molars, showing corresponding pairs of wear-facets of A, *Kuehneotherium*; B, the primitive eupantothere *Amphitherium* (upper tooth hypothetical); C, the eupantothere *Peramus*; D, *Aegialodon* (upper tooth hypothetical); E, the eutherian mammal *Didelphodus*. (From Bown and Kraus, 1979.)

## *Aegialodontids*

*Aegialodon* is known from a single, incomplete lower tooth (Fig. 106D) from the Early Cretaceous Wealdon of southern England (Kermack *et al.*, 1965), and would warrant scant attention, were it not for the fact that it seems to represent an intermediate stage between the eupantothere type of molar and that of the modern therians. This lower tooth is very similar to a eupantothere such as *Peramus*, with a trigon of cusps anteriorly and a talonid posteriorly. The same wear facets are present in the two, with the exception that *Aegialodon* has an extra wear facet developed on its somewhat larger talonid. This lies anterior to the original talonid facet, and is associated with the appearance of a third talonid cusp. The only way in which a *Peramus*-like upper molar could produce such a wear facet is by developing an extra cusp on the inner side of the tooth. This is exactly the condition in the modern therians, where a new cusp called the protocone is present and acts in conjunction with the large talonid of those forms (Fig. 106E). The evolution of the protocone is the hallmark of modern therian teeth, and it appears that *Aegialodon* was the earliest and most primitive form to have gained it (Crompton, 1971). Two other very similar teeth have since been discovered, *Kermackia* from North America and *Keilantherium* (Fig. 105I) from Mongolia (Kielan-Jaworowska *et al.*, 1979b) showing that the aegialodontids were a widespread group.

These forms, to the extremely limited extent that they are known, represent the ancestral stage in tooth evolution from which both the placentals and the marsupials evolved. By the Late Cretaceous, not only had these two lines of advanced therians evolved, but there were several more which seem not to belong to either, but represent sterile side branches, evolved independently from an aegialodontid-like ancestor. Unfortunately, very little more than teeth and jaw fragments are known for these various kinds of mammals, and their interrelationships are far from established at present.

## *Marsupials and placentals*

A very close relationship between the two living groups of therian mammals, the marsupials (Metatheria) and the placentals (Eutheria), is not disputed (Marshall, 1979), and neither is the view that they diverged from a common ancestor in Early Cretaceous times (McKenna, 1969). There is a minor nomenclatural problem of what taxonomic name to apply to the monophyletic group that includes marsupials and placentals. Until relatively recently, the name Theria was applied, but this term is more usually employed for the whole stock which includes *Kuehneotherium*, symmetrodonts and eupantotheres as well. Some authors prefer the term Eutheria, which is confusing since formerly it applied exclusively to the placental mammals and is still useful in that context. Perhaps the old name Ditremata should be resurrected, since this did apply to the marsupials and placentals together, distinguishing them from the living monotreme mammals.

Fully developed modern therian teeth first appear at the end of the Early Cretaceous, the Albion, of Texas. There is some argument about whether these various forms were already diverged into marsupials and placentals (Clemens, 1971), or indeed whether some of them represent side branches of the main stock which did not give rise to either of the modern groups. The molar teeth are a type known as tribosphenic (Fig. 106E). The lower molar is comparable to that of *Aegialodon* with trigon and talonid. The trigon is relatively wider across the jaw and narrower from front to back. The two main shearing edges are therefore even closer to transversely oriented, and the faces are more vertical. The talonid is relatively larger, its occlusal surface approximately equalling the area of the occlusal surface of the trigon. The upper molar has the new protocone assumed to have been present in *Aegialodon*, but it is very large now, and extends medially, greatly increasing the transverse width of the tooth. There has even been the addition of extra cusps between the new protocone internally and the paracone and metacone externally. External to the paracone and the metacone a large stylar area of tooth persists.

Functionally, the tribosphenic type of molar teeth have added yet another pair of opposing shearing edges compared to *Aegialodon*. Instead of simply one pair of shearing edges between the protocone of the upper tooth and the talonid of the lower tooth, there are now two (Crompton, 1971; Bown and Kraus, 1979).

The next occurrence of modern therian mammals is in Late Cretaceous deposits of both North America and Central Asia (Clemens, 1971). North America is dominated by marsupials although placentals are present, while the Asian record is exclusively of placental mammals. These latter are of particular importance because they are represented by cranial and postcranial material (Fig. 107) as well as by teeth (Kielan-Jaworowska *et al.*, 1979a). By this time all the osteological features characteristic of the modern therians appear to have evolved.

Unfortunately, the virtual absence of any specimens of the more

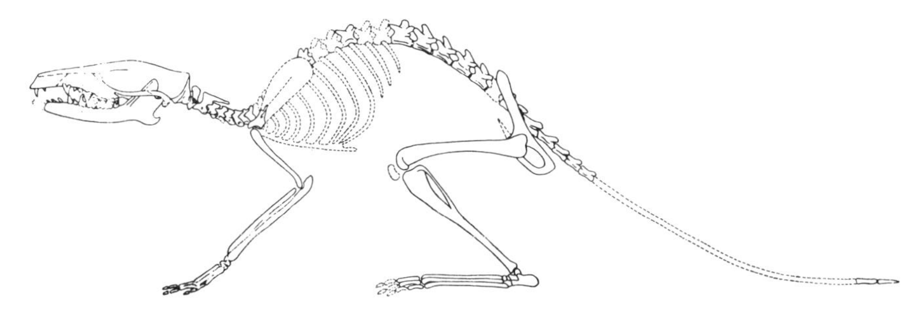

FIG. 107. The skeleton of the primitive, Late Cretaceous eutherian *Zalambdalestes*. (From Kielan-Jaworowska, 1978.)
Magnification *c*. × 0.5.

primitive therian mammals apart from teeth and jaws makes it impossible to know at what point of their evolution the modern therians acquired their characteristic features. Assuming the correctness of the proposed close relationship between morganucodontids and *Kuehneotherium*, then these characters must have evolved subsequent to the dichotomy between the therian line and morganucodontids. However, the extent to which *Kuehneotherium* itself or the eupantotheres possessed skull and postcranial features of modern therians remains unknown.

The most dramatic development of the lower jaw was the development of the accessory ear ossicles from the postdentary bones and quadrate (Fig. 93H), and therefore the complete loss of the reptilian quadrate-articular jaw hinge, at least as a functional part of the lower jaw. These bones no longer even connect to the dentary, except in the embryo. Functionally, the ear ossicles act as a lever system, improving the impedance matching between the airborne sound-waves reaching the animal and the fluid-borne sound-waves transmitted in the endolymph of the cochlea (Hopson, 1966; Webster, 1966).

The increased sensitivity of the middle ear to high-frequency sound was matched by the evolution of a coiled cochlea, which increased the length and therefore the sensitivity of the receptor organ of the inner ear, the organ of Corti.

The size of the brain of the modern therian mammals has greatly increased over that of morganucodontids, and the unique construction of the sidewall of the braincase (Fig. 103C) may be related to this. The cranial process of the squamosal becomes extensively incorporated into the sidewall, the anterior lamina of the periotic is absent, and the alisphenoid (epipterygoid) is relatively large and carries the foramina for the trigeminal nerve. Other relatively trivial changes in the skull include the loss of the ectopterygoid bone of the palate.

Compared to the primitive mammalian condition represented by morganucodontids, the postcranial skeleton of the modern therians has several new features. The separate elements of the first vertebra, the atlas, are fused to form a complete ring, which rotates around the large odontoid process of the axis, and the ribs of the rest of the neck vertebrae are reduced and fused to their respective vertebrae. These modifications may be taken as an indication that the mobility of the head on the vertebral column was increased. The lumbar ribs have been lost, which increased dorso-ventral flexibility of this part of the vertebral column. The pelvic girdle and hindlimb show no major changes from the morganucodontid pattern, except for the superposition of the astragalus on the calcaneum in the ankle. The astragalus actually loses contact with the ground completely, and the whole weight of the animal is carried by the calcaneum. The joint between the astragalus and the calcaneum increases the ability of the foot to rotate about its long axis, causing the sole to face partly medially or partly laterally. This supination and pronation of the foot must have increased the manoeuvrability of the animal. However, the most substantial modification to the postcranial skeleton of modern therians concerns the shoulder girdle and forelimb. In a curious way the forelimb appears to "catch up" with the hindlimb in an evolutionary sense, for the changes are analogous to hindlimb modifications seen as far back as the cynodonts. The interclavicle is lost, and both the coracoids are virtually absent as well, being represented by no more than a vestigial process. The scapula develops a prespinous fossa (Fig. 108C), which is a lateral facing flange extending in front of the original anterior edge of the scapula, for the supraspinatus muscle. The evidence that the prespinous fossa is indeed a reflection forwards from the inner surface of girdle is that the acromion process, for attachment of the clavicle, lies at the ventral end of the spine, which in turn lies behind the prespinous fossa. In reptiles and primitive mammals the acromion lies at the lower end of the anterior edge of the scapula blade. Also, the muscles of the more primitive forms that are associated with the anterior edge of the scapula are now found originating from the spine in modern therians. The supraspinatus muscle was derived by a dorsal migration of the supracoracoideus muscle (p. 244), and in the modern therians a further extension of the same muscle over the more posterior part of the scapula blade produced the infraspinatus muscle. This came to occupy virtually the whole of the lateral surface of the scapula behind the spine (Fig. 108C,D). The glenoid fossa for the humerus has also become modified. Although still fairly wide, the glenoid faces more or less ventrally, and it has about the same radium of curvature as does the head of the humerus. The shoulder joint is therefore a ball-and-socket joint, rather than the rolling kind of joint present in cynodonts and morganucodontids (p. 243). An important development has also occurred in the elbow joint. Here the articulating head of the ulna lies in a pulley-shaped trough, or trochlea, at the distal end of the humerus (Jenkins, 1973). This limits the movement of the lower arm to flexion and extension on the humerus, and strengthens the joint against externally applied forces tending to disarticulate it in other directions.

A great deal of misconception about how the modern therian postcranial skeleton functioned in locomotion has arisen because of undue attention to highly specialised cursorial types such as Carnivora and ungulates. The primitive version of the skeleton is represented amongst living mammals by such forms as the insectivores and the didelphid

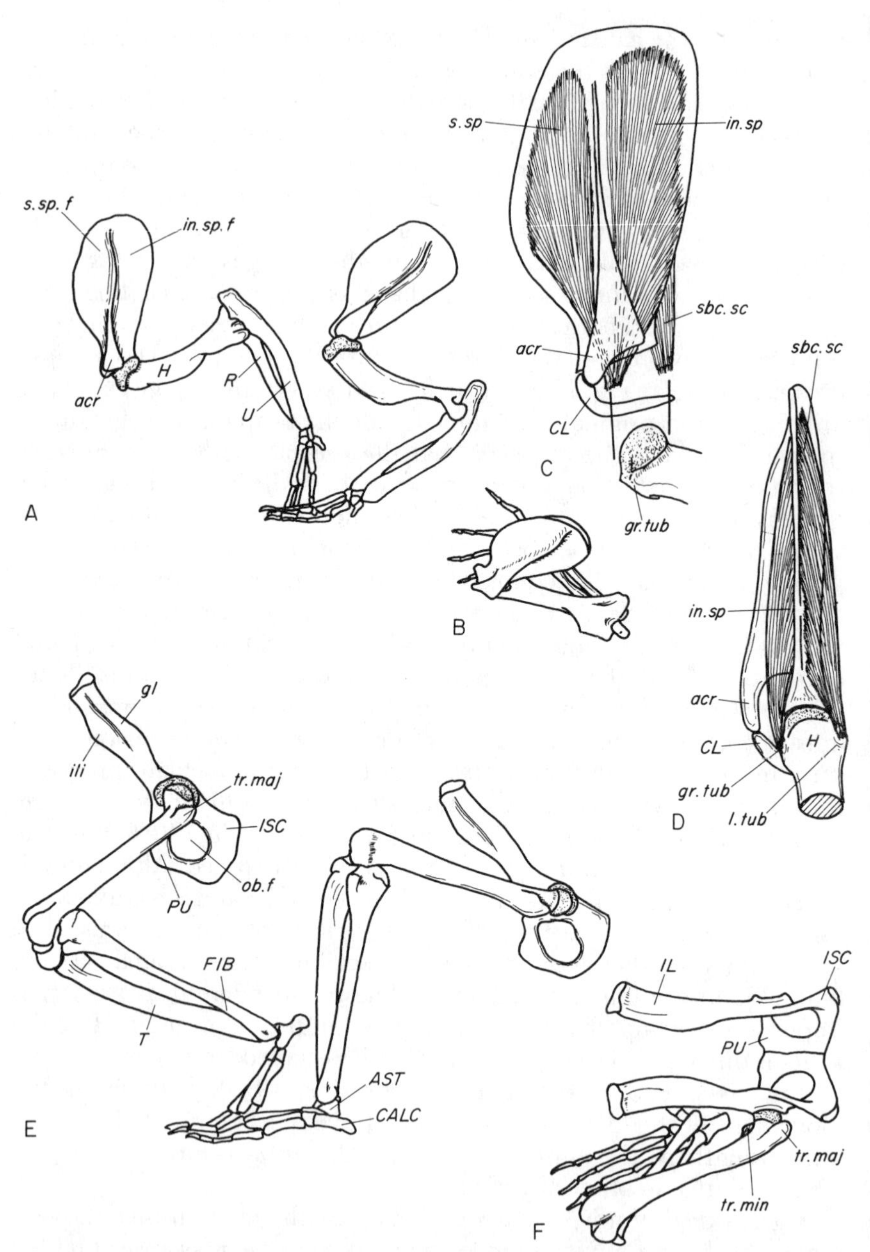


FIG. 108. Girdles and limbs of modern therians. A, lateral view of the shoulder girdle and forelimb of *Didelphis* in the retracted and protracted positions. B, dorsal view of the same, in the middle of the stride. C, lateral view of the main shoulder muscles of a therian. D, the same in posterior view. E, lateral view of the pelvis and hindlimb of

marsupials, and presumably these have also retained something like the ancestral locomotory mechanism. The humerus and femur do not approach the vertical orientation seen in advanced cursorial mammals, but operate close to the horizontal. Jenkins (1971b), in his important radiofluorographic analysis of locomotion in *Didelphis* and other small, non-cursorial mammals, has shown that at the commencement of a stride, the humerus extends posteriorly and slightly laterally from the glenoid, and its distal end is only slightly depressed so that the bone makes an angle of about 30° to the horizontal (Fig. 108A,B). The radius and ulna together slope forwards and downwards at about 30–35° to the horizontal. They also slope slightly inwards, so that the forefoot lies below the shoulder girdle. The actual stride consists of an elevation of the distal end of the humerus until the bone lies about 15° above the horizontal. This occurs partly by movement of the humerus at the shoulder joint, and partly by a rotation of the whole of the scapula, relative to the rib cage. At the same time, the elbow joint extends, the wrist joint extends, and the sole of the forefoot comes off the ground. The result is that the radius and ulna become vertically oriented, and at the end of the stride actually extend somewhat posteriorly. The position of the foot relative to the body is important. During the stride, the shoulder region moves slightly sideways and therefore the foot lies very close to a position below the midline of the body. The principal difference between this modern therian mode of operation of the forelimb and that of the cynodonts and probably primitive mammals such as the morganucodontids concerns the role of the humerus. In the primitive forms, most of the stride resulted from the retraction of the humerus in a horizontal plane, from transverse to posteriorly oriented. In the therians, the humerus moves relatively little, and most of the stride comes from the action of the radius and ulna, and forefoot, along with movement of the shoulder girdle on the ribcage.

The pelvis and hindlimb of modern therians shows far less change from a primitive, morganucodontid-like condition. Indeed the two types are closely similar in structure and presumably therefore in function. In *Didelphis* and other primitive living therians (Fig. 108E,F)

*Didelphis* in the retracted and protracted positions. F, the same in dorsal view, in the middle of a stride. (A, B, E and F modified after Jenkins, 1971b.)

*acr*, acromion; *AST*, astragalus; *CALC*, calcaneum; *CL*, clavicle; *FIB*, fibula; *gl*, gluteal (ilio-femoralis) muscle; *gr.tub*, greater tuberosity; *H*, humerus; *ili*, iliacus (pubo-ischio-femoralis internus) muscle; *in.sp*, infraspinatus muscle; *in.sp.f*, infra-spinatus fossa; *ISC*, ischium; *l.tub*, lesser tuberosity; *ob.f*, obturator fenestra; *PU*, pubis; *R*, radius; *s.sp*, supraspinatus muscle; *s.sp.f*, supraspinatus fossa; *tr.min*, trochanter minor; *tr.maj*, trochanter major; *U*, ulna.

the femur operates in an essentially similar although rather more active manner to the humerus.

At the beginning of a stride, the femur extends anteriorly and laterally from the acetabulum. It thus lies in a horizontal plane, or even slightly above. The tibia and fibula are approximately vertical and the foot lies completely on the ground. The actual stride consists of the knee joint flexing at the same time as the back of the foot comes off the ground, which causes the tibia and fibula to slope posteriorly from the knee and the animal to move forwards. This is followed by a second phase of movement, when the distal end of the femur depresses and the knee now extends. The foot comes further off the ground. During the stride, the hips shift towards the side and the pelvis thus lies over the active foot, much as in the case of the forelimb.

Traditionally it has been believed that the evolutionary changes in the postcranial skeleton of modern therians were designed to increase both the efficiency and the speed of locomotion. Unfortunately neither of these seems to be true of the primitive modern therians. Measurements of the cost of locomotion show that such mammals are no better off than similar sized lizards, while the top speed of the two kinds of animal are about the same (Bakker, 1975). The mammals have a much greater endurance, being able to maintain a high speed for longer periods, but this has nothing to do with the limb anatomy, but with the higher rate of aerobic metabolism. The advantage of modern therian locomotion is probably to do with manoeuvrability of the animal. The limbs, both front and back, are designed to operate in such a way that the feet always lie close to the mid-sagittal plane of the body. The inherent instability which comes from having the feet so close to one another can allow the animal to change direction rapidly and cope with very uneven terrain, providing adequate neuro-muscular control is available. Clearly one of the main purposes of having a large, complex brain is to provide just this level of control. The theme of increasing agility has run right through the mammal-like reptiles, and has merely reached its maximum expression in the modern therians.

## Non-therian mammals

There is a variety of other kinds of mammals whose molar teeth do not conform to the triangular pattern of cusps seen in the therian line. These include the Upper Triassic morganucodontids, several other Mesozoic groups, and possibly the living monotremes. Variously referred to as the non-therian, the atherian (Kermack *et al.*, 1973) and prototherian (Hopson, 1970) mammals, it is frequently accepted that they form a monophyletic group, recognised by the structure of the side wall of the braincase. As has been seen in the case of morganucodontids (Fig. 103B), this involves an anterior lamina of the periotic bone with foramina for the maxillary and mandibular branches of the trigeminal nerve. The alisphenoid does not form a large part of the sidewall, and neither does the squamosal. The second character that unites the non-therian mammals is the possession of molar teeth based on a linear series of cusps after the style of morganucodontids, or at least having teeth derivable from such a condition. However, both the braincase and the molar structure of non-therian mammals are probably simply ancestral mammal features (Kemp, in preparation) and cannot be used to define a monophyletic group. An alternative interpretation of the braincase structure in modern mammals (p. 291) indicates that there is much less real difference between non-therians (monotremes) and therians than supposed in this regard, and that the non-therian type can be accepted as ancestral to the therian type. As for the cusp arrangement, several groups of cynodonts have a linear arrangement, resembling that of the non-therian mammals.

### *Triconodonts*

Triconodonts (Fig. 109A) are known from the Middle Jurassic through to the Early Cretaceous, as teeth, jaws and a few fragments of the skull. A complete specimen has been reported from North America, but not yet described (Jenkins and Crompton, 1979). The molar teeth are basically similar to those of the Upper Triassic morganucodontids, and these earlier forms are actually included in the Triconodonta by several authors.

Triconodonts tended to be rather larger than other Mesozoic mammals, and the Upper Jurassic *Triconodon* for example was about as large as a cat. The molar teeth have a linear row of three main cusps which in the typical forms were all of about the same size. A cingulum is present internally and also externally on the upper teeth, but is not prominent. Upper and lower molars sheared against one another and the triconodonts were possibly carnivorous rather than insectivorous, a view supported by the relatively large canines present.

This group is one of the few where the sidewall of the braincase is known (Kermack, 1963), and it undoubtedly possessed an anterior lamina of the periotic. There is also a well-known cast of the internal surface of the bones of the skull roof, showing something of the size and structure of the brain. This appears to have been considerably smaller

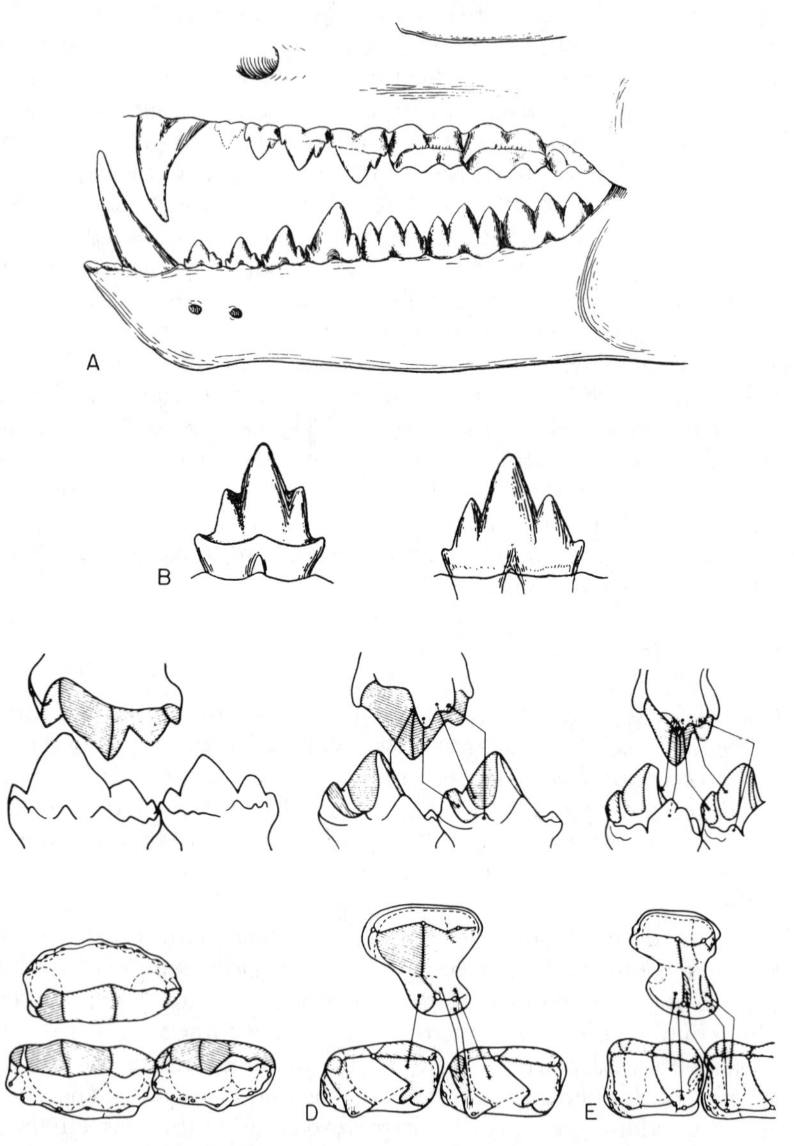


FIG. 109. Triconodonts and docodonts. A, lateral view of the jaws of *Triconodon*. B, lower molar of *Amphilestes* in internal and external views. C, internal and occlusal views of the molars of *Eozostrodon* for comparison with docodonts. D, internal and occlusal views of the molars of the primitive docodont *Haldanodon*. E, internal and occlusal views of *Docodon*. Corresponding wear facets of D and E are indicated by the lines. (A and B from Simpson, 1928; C–E from Hopson and Crompton, 1969.)

Magnifications: A, *c*. × 2.4; B, *c*. × 12.

than the brains of similar sized modern therian mammals (Hopson, 1979).

The amphilestines are a very poorly known group, resembling triconodonts but having the central cusp of the molar teeth larger than the anterior or posterior cusps (Fig. 109B). Possibly these are not closely related to the typical triconodonts (Kermack, 1967).

## *Docodonts*

These non-therian mammals occur exclusively in the Upper Jurassic and are known only from teeth and jaw fragments (Kron, 1979). The best known form is *Docodon* (Fig. 109E) whose molar teeth are expanded across the line of the jaw. They are nevertheless quite closely comparable to the equivalent teeth of morganucodontids (Crompton and Jenkins, 1968), and appear to have arisen by an expansion of the internal cingulum of the upper teeth and the development of new pairs of shearing crests between uppers and lowers. The process was, in general, analogous to the development of new shearing crests in the therian mammals. The earlier form *Haldanodon* (Fig. 109D) from Portugal shows an intermediate stage in the supposed transition of morganucodontid-like teeth to *Docodon*.

## *Multituberculates*

The multituberculates were a highly distinctive group of herbivorous or possibly omnivorous mammals (Clemens and Kielan-Jaworowska, 1979). They are first known from the Upper Jurassic, underwent a wide radiation in the Cretaceous, and actually survived into the Tertiary to become extinct at the end of the Eocene. In general adaptations they were rodent-like, and this comparison extends not only to their anatomy, but also to the large number of individuals and of species that occurred, on a world-wide basis. Their eventual extinction is usually attributed to the rise of the rodents in the early Tertiary (Hopson, 1967).

The skull of multituberculates is flat, short and wide (Fig. 110), with powerful zygomatic arches and lateral-facing eyes. A remarkable specialisation is the loss of the jugal bone. The braincase is built on the non-therian pattern, with a very large anterior lamina of the periotic and reduced alisphenoid. No postdentary bones have been found in association with the dentary, although whether or not they had been transformed into ear ossicles is unknown. It is certainly possible. The dentition is highly specialised for the herbivorous diet. Only one pair of enlarged lower incisors is present, although there are up to three pairs

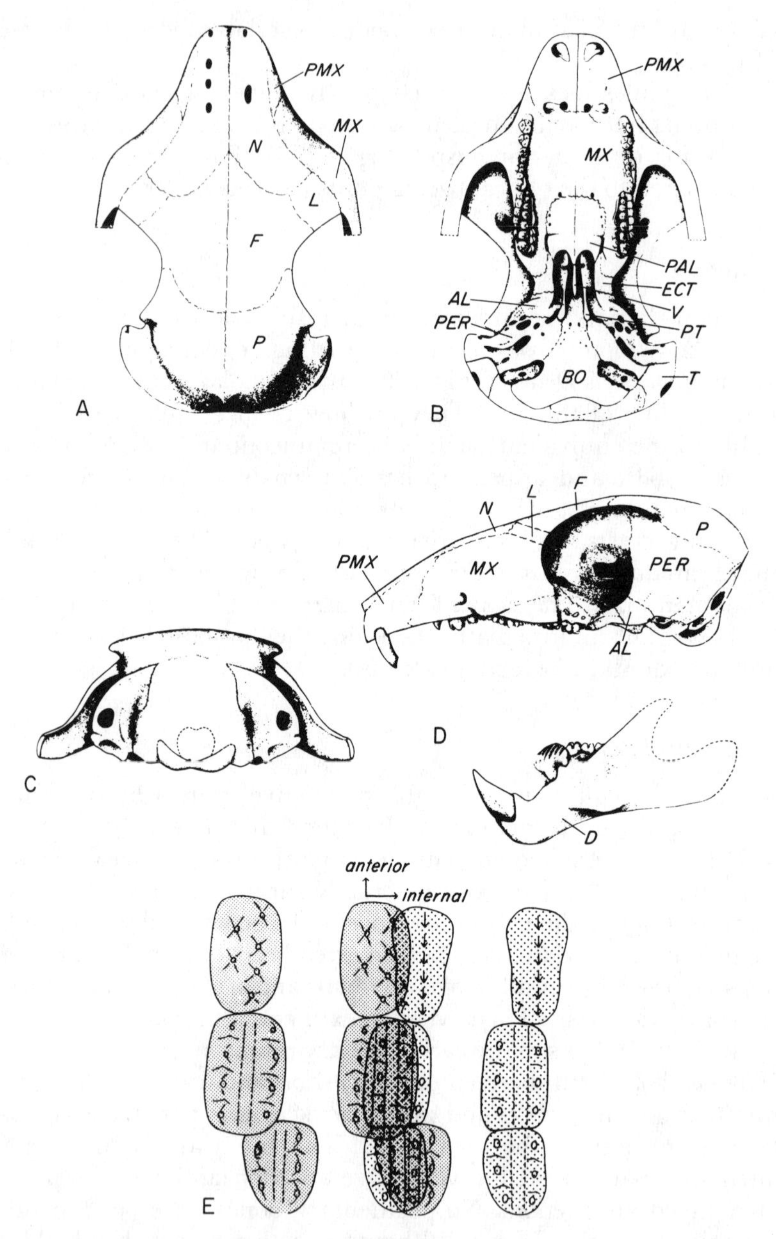


Fig. 110. The multituberculate skull. Skull of *Kamptobaatar*: A, dorsal view. B, ventral view. C, posterior view. D, lateral view with lower jaw. E, occlusion in a plagiaulacid multituberculate, upper dentition finely stippled, lower dentition coarsely stippled. (From Clemens and Kielan-Jaworowska, 1979.)

of upper incisors. Canines are absent and there is a long gap, or diastema between the incisors and the first of the postcanine teeth. In the majority of forms, the last one or two lower premolars have been transformed into large, blade-like teeth, heavily striated and used against the upper premolars for chopping up tough, fibrous vegetation. The molars bear two longitudinal rows of blunt cusps. The lower molars occlude with the uppers by the rows of cusps fitting into the valleys between the cusp rows, creating a large grinding surface (Fig. 110E).

The relationships of the multituberculates are unknown. The molar teeth are often compared with haramiyid teeth (p. 266) from the Upper Triassic. Both have two longitudinal rows of cusps, and both have a method of tooth occlusion that involves a posterior movement of the lower jaw during the bite. However, until haramiyids are better known, the possible relationship between the two is very tentative.

## Monotremes

The monotremes are the primitive living mammals of Australasia, the platypus *Ornithorhynchus* and the echidnas *Tachyglossus* and *Zaglossus*. Apart from Pleistocene representatives, the only fossil form which can be attributed to this group is *Obdurodon* (Fig. 111C), known by two isolated teeth from the Miocene of Australia (Woodburne and Tedford, 1975) and which resemble the short-lived teeth of the juvenile platypus.

The soft anatomy, physiology and general biology of the monotremes is very similar to those of the marsupial and placental mammals. They share such features as endothermic temperature physiology, hair, a completely divided heart, brain structure, mammary glands, diaphragm and many details of the biochemistry and endocrine glands (Griffiths, 1978). Only in the reproductive system are monotremes clearly more primitive than other modern mammals, for they lay eggs and the mammary glands lack nipples. Also a reptile-like cloaca, which is a pocket on the ventral side receiving the genital and urinary ducts, and housing the anus is present. Certain points of the musculature are also probably primitive compared to the modern therian mammals, such as the detrahens muscle for jaw-opening, instead of the digastric

*AL*, alisphenoid; *BO*, basioccipital; *D*, dentary; *ECT*, ectopterygoid; *F*, frontal; *L*, lachrymal; *MX*, maxilla; *N*, nasal; *P*, parietal; *PAL*, palatine; *PER*, periotic; *PMX*, premaxilla; *PT*, pterygoid; *T*, tabular; *V*, vomer.

Magnification A–D, *c*. × 2.6.

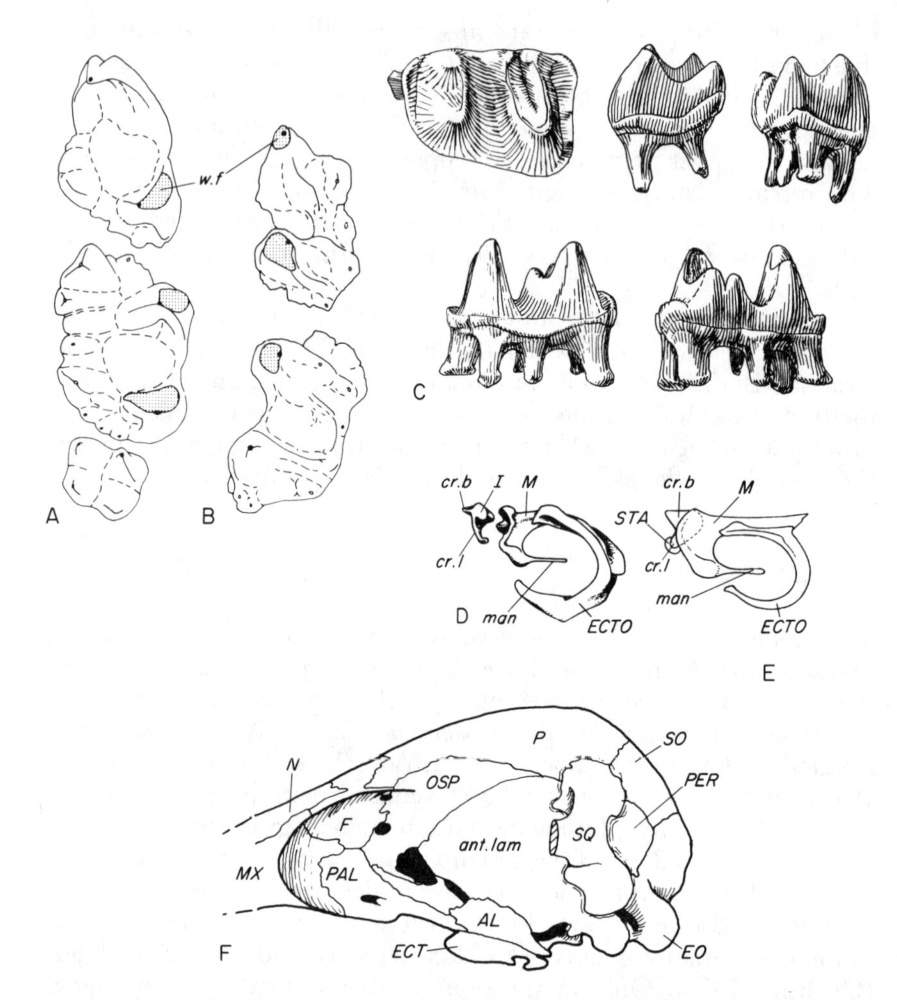


FIG. 111. Monotremes. A, right lower molars of *Ornithorhynchus* juvenile. B, left upper molars of *Ornithorhynchus* juvenile. C, isolated tooth of *Obdurodon* in occlusal, anterior and posterior views (top), and in internal and external views (bottom). D, ear ossicles of a therian mammal. E, ear ossicles of a monotreme. F, lateral view of the sidewall of the braincase of *Ornithorhynchus*. (A and B from Hopson and Crompton, 1969; C from Clemens, 1979; D and E from Allin, 1975; F redrawn after Kermack and Kielan-Jaworowska, 1971.)

*AL*, alisphenoid; *ant.lam*, anterior lamina of the periotic; *cr.b*, crus brevis of the incus, *cr.l*, crus longus of the incus; *ECT*, ectopterygoid; *ECTO*, ectotympanic; *EO*, exoccipital; *F*, frontal; *I*, incus; *M*, malleus; *man*, manubrium of the malleus; *MX*, maxilla; *N*, nasal; *OSP*, orbitosphenoid; *P*, parietal; *PAL*, palatine; *PER*, periotic; *SO*, supraoccipital; *SQ*, squamosal; *STA*, stapes; *w.f*, wear facet.

Magnification of C, *c*. × 6.

(p. 230), and the absence of an infraspinatus muscle in the shoulder (p. 281).

However, these characters of the soft anatomy are of no help in elucidating the relationships of the monotremes to fossil mammal groups. Indeed, the relationships of the monotremes is a matter of contention (Clemens, 1979b). The most widely held view at present is that they are related to the non-therian mammals of the Mesozoic and together with them form a group Prototheria, or Atheria. The two characters used to define the Prototheria in this sense are molar teeth based on a linear series of main cusps rather than the triangulated tribosphenic molars of the therians, and the presence of a large anterior lamina of the periotic in the side wall of the braincase (Kermack, 1967; Hopson and Crompton, 1969; Hopson, 1970; Compton and Jenkins, 1979). The braincase of monotremes, as interpreted so influentially by Watson (1916), consists of a very large anterior lamina of the periotic through which both the maxillary ($V_2$) and the mandibular ($V_3$) branches of the trigeminal nerve run. The alisphenoid is reduced and displaced forwards (Fig. 103A). Monotreme teeth are degenerate, for the echidnas have completely lost them, while the platypus possesses only a number of juvenile teeth (Fig. 111A,B) which are lost soon after leaving the burrow, and replaced by horny pads. The molar teeth of the platypus have been compared with the molars of morganucodontid non-therians (Hopson and Crompton, 1969), a comparison which is, to say the least, rather vague.

Among the particular non-therian groups, a more specific relationship between the multituberculates and the monotremes has been proposed (Kielan-Jaworowska, 1971), based not only upon the braincase structure, but also the absence of a jugal and certain points of similarity of the palates of the two groups.

The concept of a monophyletic taxon Prototheria to include all the non-therian groups of mammals is, however, of dubious validity. The linear arrangement of the main cusps of the molar teeth probably represents the ancestral mammalian condition and is therefore of little value in deciding relationships. More importantly, an alternative interpretation of the braincase structure is possible, whereby the primitive therians such as *Kuehneotherium* and the eupantotheres could have possessed the non-therian type of braincase with an anterior lamina exactly like morganucodontids. In other words, the so-called non-therian braincase may be the ancestral mammalian braincase and not a specialisation (Presley and Steel, 1976). In monotremes, the anterior lamina actually ossifies as an independent membrane bone in the membranous side wall of the braincase, and only later fuses with the periotic

(Fig. 103E) (Griffiths, 1978). In the modern therians (Fig. 103H), the supposed alisphenoid consists largely of membrane bone that ossifies in the same membranous sidewall of the braincase as in monotremes. The only difference from the monotremes is that here the membrane bone fuses with the ventral part of the alisphenoid instead of with the periotic. On this interpretation, the monotreme and the therian do not differ in the constituents of the braincase, but only in which particular bone the anterior lamina fuses with. There is no difficulty in evolving either condition from a morganucodontid braincase (Fig. 103F), and therefore in assuming that early, primitive therians had a morganucodontid-like braincase as well. The reason why the anterior lamina of therians later came to fuse ventrally with the remnant of the alisphenoid may be related to the expansion of the squamosal component of the sidewall of the braincase, which effectively isolated the anterior lamina from the periotic. This could well have been a late development in therian evolution, not occurring until the appearance of the modern therians in the Cretaceous. It must, of course, be stressed again that the braincase structure is not yet known in any of the primitive therians, and until it is such a hypothesis must remain speculative (Kemp, in preparation).

Nevertheless, it must be concluded that the structure of the braincase does not provide a satisfactory basis for widely separating monotremes from therians. The alleged similarity between platypus teeth and morganucodontid teeth is no more reliable a character. In view of the enormous divergence of molar structure that occurred during the evolution of many modern therian groups from the primitive tribosphenic molar, no weight can be placed on such a vague comparison. It is just as reasonable to suppose that the platypus dentition could be a major modification from a primitive therian one. It is certainly no more bizarre a suggestion than the evolution of, say, rodent teeth. A major difficulty with the hypothesis of a non-therian origin of monotremes is their possession of ear ossicles derived from the quadrate and post-dentary bones that closely resemble the ossicles of modern therians (Fig. 111D,E). This is usually regarded as parallel evolution, justified by the weak argument that such independent acquisition of so obviously a useful feature is to be expected. Slight differences in the structure of the ossicles are also invoked as evidence of parallelism. The tympanic bone in particular differs in not becoming fused to the otic capsule and in vibrating as part of the typanic membrane (Allin, 1976). However, this tends to disregard the otherwise very similar structure in the two groups.

If it is assumed that neither the braincase nor the dentition can be used to separate monotremes widely from the therians, then it ceases to be necessary to assume that the ear ossicles evolved independently. The exact stage at which the therian ossicles evolved is unknown. *Kuehneotherium*, the earliest and most primitive therian must have lacked them, for a groove to house the post-dentary bones is still present on the inner face of the dentary (Kermack *et al.*, 1968). The same is true of the earlier eupantotheres, but the Early Cretaceous eupantotheres probably did not have postdentary bones, and could therefore have possessed ear ossicles (Krebs, 1971). It follows that the monotremes could have diverged from the therian stock at or soon after the eupantothere level (Fig. 112), before the characteristic modern therian features evolved. This hypothesis of a therian origin of the monotremes is certainly more attractive than the non-therian hypothesis, although confirmation must await the description of more complete material of early therians.

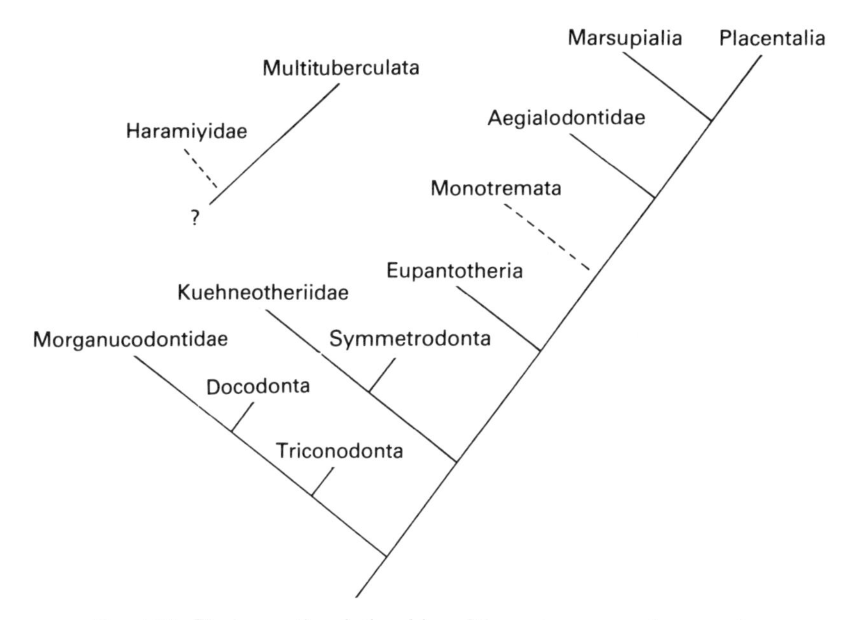


FIG. 112. Phylogenetic relationships of the main groups of mammals.

## The definition of a mammal

It has been taken for granted so far that the groups discussed in this chapter are properly regarded as mammals.

A great deal has been written about the problem of what exactly is a mammal. Until about fifteen years ago the difficulty was compounded by the then prevalent belief that the mammals were a polyphyletic grade of organisms, evolved several times independently from therapsid ancestors. The discovery of the Triassic mammals changed this view and it is now widely held that the various groups of mammals can all be traced to a single, hypothetical ancestor (Fig. 112) that had itself achieved the mammalian organisation. To justify this conclusion that the mammals form a monophyletic group, they must all be shown to possess (or at least be derived from forms which possessed) certain derived characters not present in other groups. These particular characters, or synapomorphies, will then be the basis for the definition of the taxon Mammalia. Unfortunately this ideal state of affairs is upset by the incompleteness of many of the fossil specimens, as a result of which there is not one single synapomorphy known for certain to exist in all the groups generally felt to be mammalian. A second, less regrettable complication arises from the close approach to mammalian structure found in certain groups traditionally labelled as mammal-like reptiles.

Two extreme attempts to solve the problem of the mammalian definition are purely semantic and completely evade the issue. MacIntyre (1967) proposed that only the therian mammals be regarded as mammals, which can then be defined on the tooth structure and possibly the braincase. On the other hand, several authors in the past have suggested that at least the cynodonts, and possibly all the mammal-like reptiles should be included in the Mammalia, whence the taxon Mammalia becomes a synonym of the pre-existing taxon Theropsida.

The traditional definition of the mammals has been based upon the structure of the jaw articulation. Mammals have lost the reptilian articular-quadrate hinge and gained the new dentary-squamosal hinge (e.g. Simpson, 1960). However, the array of forms now known to have possessed both hinges simultaneously has rather reduced the clarity of this character. If the gain of the dentary-squamosal hinge is used, then both the therian and the non-therian mammals would be included, but also the tritheledontids and possibly tritylodontids and *Probainognathus*. If, on the other hand, the loss of the articular-quadrate hinge is used, Mammalia would be restricted to the modern therians, the monotremes, and probably the multituberculates, while the obvious relatives of these groups in the Mesozoic, such as *Kuehneotherium* and the morganucodontids, would be excluded.

A more recent definition of mammals is based on the mode of action of the molar teeth (Crompton and Jenkins, 1973). In the two main Triassic groups, *Kuehneotherium* and morganucodontids, the lower molar moves slightly medially as it occludes with the upper molars. Therefore during the jaw opening phase of mastication, the lower tooth must move laterally ready for the next bite. This triangular orbit of movement of the lower teeth, and therefore the lower jaw as a whole is highly characteristic of these forms and of most of their relatives, but is unknown in any cynodont, tritylodontid or tritheledontid. Unfortunately, neither the haramiyids nor the multituberculates show this feature and their lower jaws seem to have had an antero-posterior shift instead. Both must be omitted from the Mammalia, unless it is argued that they descended by modification from a form that did have the triangular orbit of movement of the jaw. Although there is no direct evidence for this, the multituberculates at least have other features showing a relationship to the rest of the mammals, particularly a braincase with a large anterior lamina of the periotic, and a characteristically mammalian postcranial skeleton.

There is necessarily a subjective decision to be made about which groups should be included as mammals. If the tentative phylogeny of the origin of mammals proposed here (Fig. 104) is accepted, it remains to be decided which of the various hypothetical ancestors (i.e. which branching point on the phylogeny) should be taken as the first mammal. Thus both the tritylodontids and the tritheledontids could be included. The resulting taxon would be monophyletic and mammals would be defined as having lost the prefrontal and postorbital bones, possibly gained a dentary-squamosal contact, and having numerous new postcranial features such as the trochanter minor on the femur. An alternative would be to include the tritheledontids, but not the tritylodontids, in which case the mammals would be defined as having prismatic enamel on the teeth. Thirdly, the common ancestor of morganucodontids and *Kuehneotherium* could be selected as the first mammal, and the group defined as having the triangular orbit of movement of the lower jaw, an actual condyle on the dentary, diphyodont tooth replacement, and an enlarged anterior lamina of the periotic. In every one of these three possible schemes, at least one of the groups in question is not sufficiently well known to be certain that the definitive characters suggested are actually present, and this itself is a measure of the existing doubt about these interrelationships. In practical terms, the relationships of the tritheledontids are the most insecurely determined and therefore it is best at the present time to exclude them by accepting the third possible scheme and associated mammalian definition. Even so, the picture will not be clear until more is known of *Kuehneotherium* and the haramiyids, as well as the more primitive groups.

# 12 The Origin of Mammals

THE EVOLUTION of the mammal-like reptiles from their earliest appearance in the fossil record until the time of the first mammals occupied some 130 million years. During this period of time a complex radiation occurred producing a great variety of different types of animals, differing both in habitat and in degree of development of mammalian features. Only one of the evolutionary lines led ultimately to the mammals, and there is no reason to suppose that this particular line differed in any general way from all the other lines. Nevertheless it is of the greatest interest to pick out the sequence of known fossil forms that lie closest to the hypothetical set of ancestors from the primitive pelycosaurs to the mammals (Fig. 113). This sequence illustrates features of the evolutionary transition from one vertebrate class, the reptiles, to another, the mammals, and indeed this is the only such major transition in the animal kingdom that is anything like well documented by an actual fossil record.

The first step in elucidating the origin of the mammals is to attempt conceptual reconstructions of the various ancestral stages, using the nature of the fossil forms that were most closely related to them. It then becomes possible to consider more deeply the significance of the anatomical changes that are shown to have occurred.

## The acquisition of mammalian characters

In the course of the previous chapters the relationships of the various mammal-like reptiles and early mammals to one another have been assessed on the basis of the possession of shared derived characters. It is now possible to select those particular fossil forms that branched directly off the hypothetical line leading to the mammals and to ignore all the others. Each one may be placed in its supposed phylogenetic position relative to the hypothetical line. The resulting cladogram (Fig. 113) illustrates a series of branching points, each one of which represents a hypothetical ancestral form, named A to V. The nature of each of the

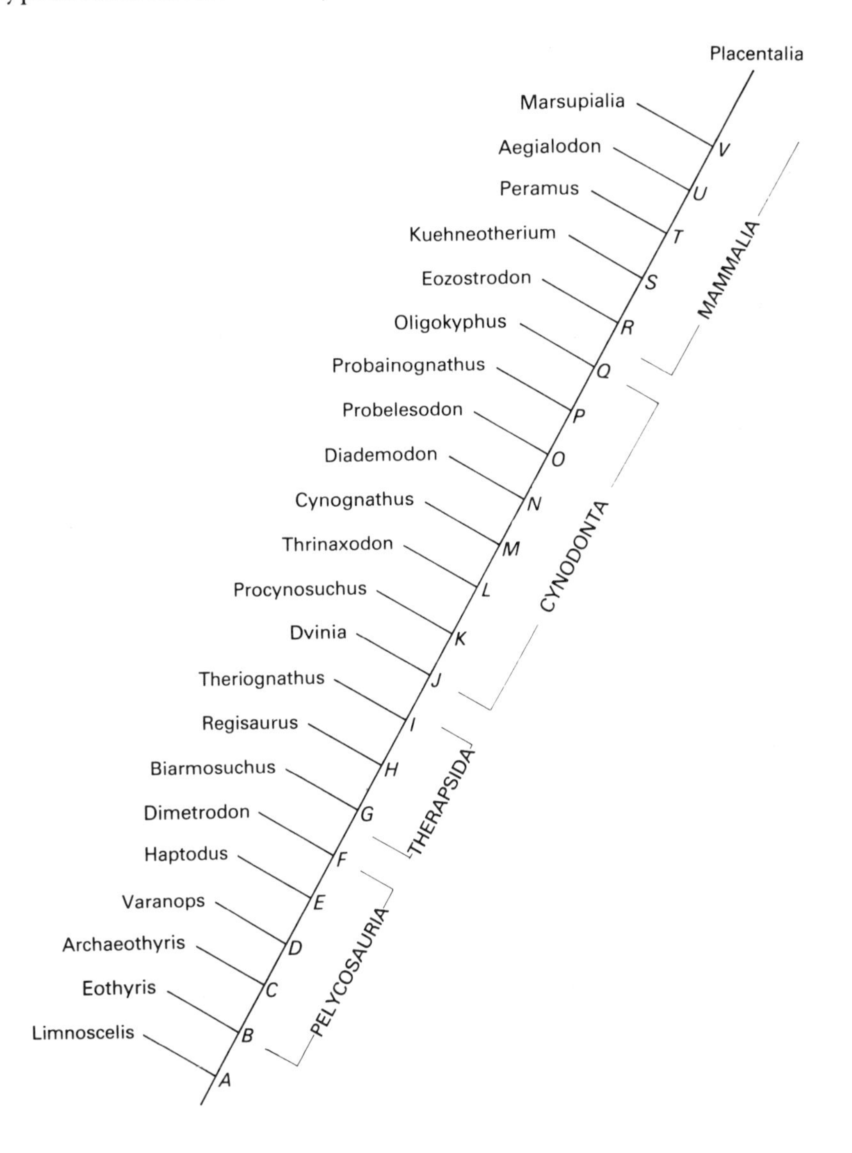


FIG. 113. A cladogram illustrating the relationships of the genera most closely related to the hypothetical lineage from stem-reptiles to modern therian mammals.

hypothetical ancestors can be deduced. To take as an example, hypothetical ancestor G, it will possess those new derived characters common to both the particular known fossil descended from it, *Biarmosuchus*, and the later forms, therapsids, cynodonts and mammals. It will also possess those characters already deduced to have been present in its own immediate ancestor, F. The characters which G will not possess are those particular, unique specialisations of *Biarmosuchus* alone.

The strength of this particular approach to the origin of the mammals is that it expresses perfectly clearly what is known of the sequence in which mammal characters were gained, and discards complications due to specialisations of particular fossil forms. There are of course certain areas of doubt. The resulting lineage of hypothetical ancestors depends entirely on the accuracy with which characters have been recognised respectively as uniquely derived (synapomorphies), ancestral (plesiomorphies) or specialised (autapomorphies). The cladogram can therefore be no better than is permitted by the methods of so assessing the various characters. Thus, a possible source of error would be the independent development of certain characters in a particular fossil form on the one hand and the later mammal-like reptiles on the other. The hypothetical common ancestor at that level would not itself have possessed the character in question although it might be assumed that it had. Similarly, the loss of a character in either the fossil form or in the later forms might lead to the view that the hypothetical common ancestor did not possess it, when in fact it did. These, however, are simply problems of systematics in general, and have been avoided as best possible by the methods already discussed (p. 12). Another inevitable limitation is a complete ignorance of other intermediate stages, lying between those we know about. Only further fossil forms will enable these gaps to be filled.

With reference to the cladogram, the principal features of each of the reconstructed hypothetical ancestors can be briefly considered.

*A. "Limnoscelis" level (p. 21)* This hypothetical ancestor was a very primitive tetrapod, lacking a temporal fenestra and still having the remnant of the lateral hinge line separating the cheek region from the skull roof. The teeth were all the same size and the jaw muscles small. Locomotion was by a slow, clumsy sprawling gait. A spiral-shaped glenoid caused restriction and predetermination of the movements of the massive humerus. The ilium was small and a Y-shaped system of trochanters occupied the ventral side of the femur. The feet were broad and the wrist and ankle held permanently off the ground. The metabolic rate was very low and the animal completely ectothermic.

*B. "Eothyris" level (p. 31)* A small temporal fenestra was present in place of the hinge line, bounded above by the still large supratemporal bone meeting the postorbital bone. The occiput had become plate-like. The jaw muscles were enlarged and the teeth in the canine region had slightly increased in size. The postcranial skeleton of *Eothyris* is unknown. In the related *Oedaleops*, the ilium was probably slightly enlarged.

*C. "Archaeothyris" level (p. 18)* The jaws were elongated, both by increasing the relative length of the snout and by sloping the cheek region backwards to give a posteriorly placed jaw hinge. The supratemporal bone was reduced. The neural spines of the vertebrae were well developed.

*D. "Varanops" level (p. 33)* The posterior part of the skull was broad, with more or less vertical cheeks. The temporal fenestra had enlarged and the occiput became more strongly attached to the rest of the skull. The lower jaw had a slight coronoid eminence, and a retroarticular process for the jaw-opening musculature. The body was narrower and deeper, and the vertebral column more strongly constructed. The limbs were relatively slender and long, the ilium further enlarged, and a prominent fourth trochanter for the caudi femoralis musculature was present on the underside of the femur. The animal was more mobile and active.

*E. "Haptodus" level (p. 34)* The facial part of the skull was high, with a convex dorsal profile. This curvature continued posteriorly to the more ventrally positioned jaw articulation. The size of the posterior part of the skull was increased, and massive paroccipital and supraoccipital processes buttressed the occiput to the sides of the skull. The coronoid eminence of the lower jaw was more developed, and the downturning of the articular region of the jaw led to the development of the angular notch. The jaw musculature was more powerful. There was some reduction in the number of teeth, and individually the teeth were laterally compressed. In the postcranial skeleton, the centra of the vertebrae were larger, and the zygapophyseal articulations no longer horizontal but inclined to the sagittal plane. A third sacral vertebra was present. The limbs were very well developed.

*F. "Dimetrodon" level (p. 34)* The canine teeth were reduced to one in each jaw and were enlarged. The lachrymal bone had shortened, to allow enlargement of the maxilla to accommodate the roots of the canine teeth.

*G. "Biarmosuchus" level (p. 102)* This, the hypothetical ancestor of all the therapsids, shows one of the largest accumulations of new characters. The temporal fenestra was enlarged and occupied virtually the whole of the cheek. The occipital plate was vertically orientated, and still more strongly attached by the paroccipital processes to the rest of the skull. The jaw hinge was relatively anterior in position and the upper part of the hinge was formed from the quadrate in intimate association with the reduced quadratojugal. The supratemporal bone was lost. The reflected lamina of the angular bone of the lower jaw had expanded as a very large, thin sheet attached only along its anterior edge to the rest of the angular. The jaw musculature was greatly increased. The canine teeth were larger still, and dominated the dentition. The basipterygoid articulation between the braincase and the primary upper jaw had become immoveable. The stapes was reduced in size.

Several new features of the postcranial skeleton indicate a radical alteration in the method of locomotion. The scapula blade was narrow and the whole shoulder girdle mobile, while the glenoid had lost the screw-shape of more primitive forms. This, coupled with the more bulbous head of the humerus, allowed the forelimb much more freedom of movement. The sternum was ossified. In the hindlimb, the ilium was much enlarged and the pubis reduced. The femur was S-shaped, and had developed a distinct trochanter major behind the articulating head of the bone. The internal trochanter had shifted to the middle of the underside of the bone, and the fourth trochanter was virtually lost. The digits of both the fore and the hind feet were approaching equality of length by the reduction and loss of phalanges and the feet were plantigrade. Trunk intercentra were absent from all but the neck and tail vertebrae. In locomotion, the hindlimb had developed the ability to work in a more erect fashion in addition to sprawling. The metabolic rate had increased somewhat and the animal was an inertial homeotherm.

*H. "Regisaurus" (therocephalian) level (p. 167)* The temporal fenestra was enlarged medially, giving rise to a narrow intertemporal bar and associated with a reduction of the postorbital bone. A coronoid process of the dentary had developed on the lower jaw. The jaw musculature was enlarged, and the external adductor muscle had started to differentiate into a medial temporalis and a lateral masseter component. The quadrate was reduced in size and moveable. The epipterygoid was slightly widened, and the processus cultriformis of the parasphenoid reduced. In the postcranial skeleton, the limbs were long and slender, and the ilium extended forwards.

*I. "Whaitsiid" level (p. 165)* The significant feature at this level was the wide expansion of the epipterygoid, which developed a deep notch in its posterior boundary for the emerging trigeminal nerve.

*J. "Dvinia" level (p. 182)* It was at this level that the characteristics of the cynodonts appeared. The temporal fenestra increased still further, expanding backwards to produce a posteriorly flared squamosal, and also laterally to give rise to a bowed zygomatic arch. The intertemporal region, or sagittal crest, was deep and narrow. The dentary bone had increased in relative size and the coronoid process was broad, with a small adductor fossa on the lateral face. The postdentary bones were somewhat reduced, and the reflected lamina of the angular was smaller. The quadrate complex was also reduced to some extent. The post-canine teeth had developed a cingulum, bearing cuspules, on the inner side of the crowns of both the upper and the lower teeth. On the skull roof, the nasal bones had expanded and made contact with the lachrymal. On the palate, an incomplete secondary palate was formed and the suborbital vacuities were closed. In the braincase the expanded epipterygoid made a more extensive contact with the prootic, the braincase floor was thinner, and a lateral process of the prootic was developed. The occipital surface showed a characteristic development, the narrowing of the supraoccipital and widening of the tabular bones, and a double occipital condyle was present. The brain was enlarged. The postcranial skeleton of *Dvinia* is virtually unknown and it is not therefore certain whether the typical cynodont features found at the next level had already evolved by this one. The metabolic rate was probably increased and the animal was at last approaching endothermic temperature physiology.

*K. "Procynosuchus" level (p. 185)* There were few new developments of the skull at this stage. The vomers had fused, and the supraoccipital was even narrower. The number of lower incisor teeth was reduced to four, from the six found at the *Dvinia* level.

Several important developments of the postcranial skeleton had occurred at this stage, although they may have been present at the preceding stage. The first two vertebrae, atlas and axis, were modified to increase the flexion–extension and rotation movements of the head. The vertebral column showed incipient functional division into thoracic and lumbar regions, with a reduction of the size of the lumbar ribs. In the limb girdles, reduction of the corocoids of the shoulder girdle had commenced, while enlargement of the ilium and further reduction of the pubis had also occurred. A diaphragm may have been present.

*L. "Thrinaxodon" level (p. 187)* A further development of several of the cynodont features seen already marked this level. The temporal fenestra had continued to expand and the zygomatic arch bowed dorsally as well as laterally. The posterior end of this arch was separated from the occipital plate by a deep incisure. The dentary had enlarged further, and the adductor fossa on its lateral surface occupied the entire posterior part of the bone. The coronoid process rose higher into the temporal fenestra, almost to the level of the skull roof. The postdentary bones were correspondingly reduced, including the reflected lamina of the angular. The jaw musculature was enlarged, particularly the masseter. The quadrate complex was smaller, and had developed a more complex attachment to the squamosal, allowing it to move to and fro. An incipient secondary jaw articulation was present in the form of a ligamentous connection between the surangular of the lower jaw and the squamosal and quadratojugal of the skull. The small stapes came to abut distally against the inner side of the quadrate. The number of incisor teeth was reduced to four uppers and three lowers, and maxillary precanine teeth were absent. The postcanine teeth had become more complex, with the addition of relatively large anterior and posterior accessory cusps derived from the cingulum. The ectopterygoid bone and the lateral pterygoid processes of the palate were reduced, and the interpterygoid vacuity closed. The secondary palate was complete. The most important postcranial skeleton developments concerned the vertebral column and ribs. The vertebrae had accessory zygapophyseal articulations, and the diapophyseal articulation for the ribs were shared between adjacent centra. The ribs themselves evolved large, overlapping costal plates. The scapula blade was deeply concave and an acromium process for articulation with the clavicle was present, indicating the development of a supraspinatus muscle. The corocoids were further reduced. The ilium was slightly inreased again, and the trochanter major of the femur was of increased prominence. The hindlimb was probably incapable of a sprawling gait by now, and the whole locomotion was more agile.

*M. "Cynognathus" level (p. 207)* This represents the advanced cynodont level, where the temporal fenestra had enlarged yet further. A deep external auditory meatus occupied the postero-lateral part of the occiput. The dentary was relatively even larger, and formed virtually the whole of the muscle-bearing part of the lower jaw. The coronoid process reached high into the skull, and a ventral angular process was present. The postdentary bones had become reduced to a compound rod, lying in a trough on the medial side of the dentary. A bony secondary jaw articulation was present between the squamosal and the surangular bone. The jaw muscles were further enlarged, and were now balanced in such a way as to practically remove the reaction stresses at the jaw hinge. The postcanine teeth met in true occlusion, wear-facets forming on both the upper and the lower teeth. The size of the sub-temporal fossa was increased by a medial shift of the quadrate rami of the pterygoids alongside the braincase. In the postcranial skeleton, the costal plates of the more anterior thoracic ribs were lost, and those of the lumbar ribs simplified, although still large. The acromion process of the scapula was enlarged and more turned outwards.

*N. "Diademodon" level (p. 195)* The temporal fenestra was enlarged still more. The ectopterygoid bones and lateral pterygoid processes of the palate were further reduced. The rate of tooth replacement had decreased.

*O. "Probelesodon" level (p. 208)* The postorbital bar was slender, and the sagittal crest long and low. The dentary had developed a definite articular process, extending posteriorly above the postdentary bones, although not quite contacting the squamosal of the skull. The secondary palate had lengthened by increase of the palatine component. The costal plates of all the ribs were lost, as were the secondary zygapophyseal articulations between the vertebrae. There had been some reduction of the posterior process of the ilium, and the trochanter major extended further proximally on the femur.

*P. "Probainognathus" level* (p. 212) The animal was of reduced overall size. The zygomatic arch had become slender, and originated from low down the side of the occiput. In the lower jaw, the dentary had possibly just made contact with the squamosal, forming the mammalian secondary jaw hinge alongside the reptilian hinge. The canine teeth were smaller. The postcanines possessed a principal cusp which was not very much larger than the anterior and posterior accessory cusps.

*Q. "Oligokyphus" (tritylodontid) level (p. 203)* The prefrontal and postorbital bones were lost, and the orbit and temporal fenestra became confluent. The teeth were multi-rooted. In the braincase the anterior lamina of the periotic was slightly enlarged. The external auditory meatus was less prominent on the back of the skull. Major new features of the postcranial skeleton had evolved, including platycoelous (flat-faced) vertebrae. The acromion process of the scapula was very large and permitted the development of a large supraspinatus muscle. The glenoid was widened and shortened and the humerus had become very slender. The forelimb was now capable of operating in a more erect fashion. The posterior process of the ilium was lost, while the long anterior process was divided into upper and lower parts by an external ridge. The pubis had turned posteriorly and the ischium was reduced and horizontal. The femur had developed a mammalian trochanter minor, and lost the reptilian internal trochanter. Both the musculature and the locomotion were virtually fully mammalian.

*R. "Eozostrodon" level (p. 253)* This was a very small animal and may be regarded as the hypothetical ancestor of all mammals. The zygomatic arch was very slender and low. On the palate, the lateral pterygoid flanges were finally lost, although a small ectopterygoid remained. The side wall of the braincase was modified by anterior extension of the anterior lamina of the periotic, and the alisphenoid (epipterygoid) was displaced forwards. The number of lower incisors was increased to four, and the postcanines were differentiated both anatomically and in terms of tooth replacement into premolars and molars. Diphyodont tooth replacement occurred. The molar teeth consisted of a large, transversely flattened main cusp and well-developed anterior and posterior accessory cusps. The cingula of both upper and lower molars were well developed. During tooth occlusion, the lower teeth moved medially and therefore the lower jaw moved through a characteristic triangular orbit during the feeding cycle, only one side of the dentition being used at a time. The articular process of the dentary had developed a swollen condyle which articulated with the glenoid of the squamosal, this being the dominant jaw articulation. The quadrate-articular hinge was reduced.

*S. "Kuehneotherium" level (p. 263)* The main cusp and accessory cusps of the molars were arranged in a triangle, upper and lower teeth tending to interlock. Nothing is yet known of the skull and postcranial skeleton of this level, or of the next two. It is therefore impossible to say at what level the various characters of the modern therian skeleton arose.

*T. "Peramus" (eupantothere) level (p. 276)* The talonid of the lower molar teeth was developed. The skull and postcranial skeleton of forms at this level are as yet completely unknown.

*U. "Aegialodon" level (p. 277)* The talonid was formed as a full basin, the wear on which indicates that a major new cusp, the protocone, had evolved on the upper molars. It has been suggested that the monotremes diverged at about this level (p. 291). If so, then the ear ossicles and fused atlas ring developed.

*V. "Modern therian" level (p. 278)* This form represents the latest ancestor of both the marsupials and the placentals, and as such had those characters that distinguish modern therians from all other mammals. However, as so little is yet known of the more primitive therians it is impossible to say at what point in therian evolution these various features actually appeared. There is no good reason to believe that they all evolved together.

The dominant feature of the skull was a considerable enlargement of the brain. The side wall of the braincase now consisted of an expanded alisphenoid anteriorly and a large cranial process of the squamosal posteriorly. The anterior lamina of the periotic was absent. In the palate, the ectopterygoid was lost, and the parasphenoid bone was very greatly reduced. The postdentary bones had lost their contact with the dentary and formed the accessory ear ossicles of the middle ear. The cochlea was coiled.

The most significant changes in the postcranial skeleton concern the shoulder girdle, which had lost the interclavicle and developed a lateral-facing supraspinatus fossa. The shoulder joint was a ball-and-socket. In the axial skeleton, the neck ribs were fused to their vertebrae, and the lumbar ribs were lost. The hind foot was narrowed, and the astragalus was superimposed upon the calcaneum.

The two functional systems most clearly illustrated by the fossil record are feeding and locomotion. The evolution of each of these to the ultimate mammalian condition is clearly seen to have been gradual. From the very earliest stage representing the ancestral pelycosaur, features of mammals began to appear in incipient form, and at various subsequent stages other features of mammals were added, or those already present became more mammal-like. For example, the size of the adductor musculature of the jaw increased by increments at stages *B,D,E,G,H,J,L,M,N,O*, while the complexity of the dentition increased at *B,F,G,J,L,M,P,Q,R,S,T,U,V*. In the same way modification of the postcranial skeleton and therefore of locomotion is detectable at stages *B,D,G,H,K,L,M,O,Q,V*. Other functional systems are less adequately reflected in the osteology and therefore their evolution is not so well illustrated. Where there is evidence, however, it suggests a similar pattern of gradual acquisition of the mammalian condition to that of feeding and locomotion. Improvement of sound reception, as judged from the nature of the stapes, occurred at *G,J* and, if Allin's theory of sound conduction by the cynodont postdentary bone is accepted, at *L,M,P,R,V* as well. The brain probably increased in relative size at *J,Q,R,V*. The rate of tooth replacement decreased at stage *N* and again at *R*. Other mammal characters which appeared seem to have evolved more or less at random in the sequence, such as the diaphragm at level *J* or *K*, and turbinals in the nose at *F*.

It is not possible to know how large a change in any character or system could occur between an ancestral and its immediate descendant species because of the absence of fossil specimens at levels between those stages which are represented by known forms. Thus we cannot tell whether the evolution of these various systems was a series of step-wise quantum jumps in morphology, or whether it was a gradual accumulation of minute changes. Overall, however, the evolution of each system can be described as gradual at the particular phylogenetic resolution presently available. It follows that there is an apparent correlation between the evolution of the separate systems, changes in one tending to be accompanied by changes in the others. This correlation is not absolute because at certain stages major changes in one system are not accompaied by large changes in others. For example, at stage *H* a substantial enlargement of the adductor musculature was associated with only a very minor modification of the locomotor system, as was probably also true at stage *J*.

There are other mammalian characteristics which are not represented at all in the fossil forms, or at best only by highly tentative associations with skeletal characters. Such things as the structure of the heart and double circulation, the kidney and its specialised physiology, hair, lactation and temperature physiology are of fundamental importance in understanding the nature of the origin of mammals. By analogy with the osteological features, it is likely that these soft structures also evolved gradually through the mammal-like reptiles and early mammals, but in order to investigate such a hypothesis further, the way in which all the various characters relate functionally to one another and to the environment must be considered.

## The evolution of mammalness

There are three great problems (and several lesser ones) facing an animal living on land. The first is the wide temperature variation, both diurnally and seasonally. The second is the tendency to lose water because of the huge water gradient between the air and the animal's tissues, and the resulting difficulties of maintaining osmotic, and ionic balances. The third is the gravitational problem arising from the absence of buoyancy in air. Various simple biological solutions to these problems are possible, but unfortunatly the solution of one problem frequently tends to aggravate one or both of the others. A simple increase in body size reduces the surface area to volume ratio and therefore ameliorates the temperature and water-loss problems, but aggravates the locomotory and support problems associated with gravity. The evolution of an impermeable skin reduces water loss, but increases the temperature problem because it reduces the animal's ability to use evaporative cooling when overheated. Endothermy is a means of maintaining a constant temperature in the face of a fluctuating environmental temperature, but it has an adverse effect on the rate of water loss because of the higher body temperature and higher rate of $O_2/CO_2$ gas exchange involved. It also has an indirect effect on the gravity problem because the greater amount of food required to be collected makes increased demands on the locomotory system. An improvement of the design of the locomotory system to combat gravity increases both the temperature and the water-loss problem because of the higher heat production associated with increased locomotion.

It is perfectly clear therefore that progress towards perfecting terrestrial adaptation must involve an integrated solution to these three problems. They cannot be overcome one at a time in piecemeal fashion. Mammalian organisation can be decribed very largely in terms of just such an integrated solution.

## *Temperature control*

Mammals maintain a constant, relatively high body temperature. It is achieved by endothermy, which is internal heat production from a high cellular metabolic rate, typically some seven times that of a similar sized ectothermic reptile. There is also insulation of the body surface by hair, and a variety of mechanisms for finely adjusting the balance between heat production and heat loss, such as variable blood flow to the superficial blood vessel, sweat glands in the skin, and involuntary heat production by the muscles.

It is frequently claimed that the adaptive significance of endothermic temperature physiology in mammals is that the high body temperature raises the level of activity by increasing the rate of its metabolic processes. In fact, it is probably not the high temperature itself which is important. Although the rate of most chemical reactions rises with temperature, this is not true in a simple sense of enzyme-controlled reactions. The activity of an enzyme has a maximum at a particular optimum temperature, and is reduced both above and below that temperature. The value of the optimum temperature of the enzymes in different organisms differs, and the set of enzymes in any particular organism could in theory be adapted to operate maximally at any particular temperature (Heinrich, 1977). Mammalian endothermy must be seen as a highly successful, if somewhat extravagant, means simply of maintaining a constant body temperature within fine limits and under a wide range of environmental conditions. Given that the objective is to keep the temperature constant, the question arises of what should be the thermostat setting of the animal. The answer is a compromise between two conflicting requirements. On the one hand, the cost in terms of food requirements rises as the proposed thermostat setting rises. On the other hand, the difficulty of dispersing the excess heat produced by the muscles when the animal is fully active requires that the thermostat is not set too low, and therefore a reasonable temperature gradient from the animal's body to cooler surroundings exists. Amongst modern mammals there are differences in body temperature which support the view that it is constancy rather than absolute value that is important. Primitively nocturnal forms such as the Insectivora and didelphid marsupials which are normally active under cool conditions have a body temperature round about 30°C, and the same is true of very inactive forms such as the sloths. Basically diurnal forms, which operate in warmer surroundings, and larger forms which are more liable to problems of metabolic heat load, have temperatures maintained around 38°C (Crompton *et al.*, 1978). The clear advantage of the endothermic method of temperature regulation is that it guarantees constancy of body temperature under a wide variety of ambient temperatures and levels of activity, while freeing the animal of the behavioural obligations and noctural (or diurnal) cessation of activity of ectotherms.

The adaptive value of a permanently constant body temperature itself comes back to the action of enzymes. All the enzymes of the body can be arranged to work at their optimal levels throughout the life of the animal. This means that complex multi-enzyme systems can evolve and function, because the action of each individual enzyme within the system is constant, and therefore predictable. If the enzymes did not behave in this way, then the probability of a complex system malfunctioning would be high. It would not be exaggerating to claim that the constancy of body temperature is the *sine qua non* of the highly complex biological organisation that characterises the mammals compared to the reptiles of today.

Against this benefit of endothermy there is, however, a considerable cost. The direct cost is in the greater amount of food that has to be collected and assimilated. Another, more subtle cost which may be termed the evolutionary cost is the large amount of evolutionary change necessary in order to achieve endothermy. The rise in metabolic rate seems to involve little more than an increase in the number of mitochondria in the cells, and presumably minor alterations to the pattern of hormonal control of metabolism. In addition, however, the fine control mechanisms of temperature regulation are necessary, so that neither alterations in the rate of metabolic heat output during differing levels of activity, nor variations in ambient temperature are allowed to cause a change in body temperature. Thus hair, sweat glands and specialised skin blood vessels must evolve. More indirectly, but equally important in the functioning of endothermy are several other aspects of the biology of mammals. The locomotory apparatus must become capable of carrying the animal about in search of its some tenfold increase in food requirements. The feeding apparatus has to ingest at this greater rate, and also assist in the breakdown of the food, a process which would be far too slow if left solely to the intestinal processes. The diaphragm is needed for the greater rate of external gas exchange that occurs. The potential increase in water loss that would result from the higher temperature and greater breathing rate must be combated by the kidney, and finally the sense organs and central nervous system must be designed to organise and control all these activities.

### *Chemical control*

The second great terrestrial problem solved by mammals is the tendency to lose water from the tissues, with the attendant alteration in osmotic pressure and ionic balances. The means by which the chemical internal environment is maintained constant is the kidney, which is more elaborate than in any other vertebrate. The blood pressure in the renal artery supplying the kidneys is high and the number of kidney tubules is large. The first point about the mammalian kidney, therefore, is that there is a very high ultrafiltration rate of the blood. The second point is the very long loop of Henle, which is associated with the production of a concentrated, hypertonic urine, the main means of water conservation. The third point of importance is that by producing hypertonic urine, sufficient water is conserved that the animal can afford to excrete liquid. There is therefore a flow of aqueous solution passing out of the body which gives the opportunity for very fine regulation of the plasma levels of ions and other soluble substances. By appropriate rates of secretion into or reabsorbtion from the fluid flowing through the kidney tubules, the level of each ion or molecule can be maintained constant in the blood.

As in the case of temperature regulation, so the advantage to mammals of fine chemical regulation relates to the activities of enzymes. Rate of enzyme action is sensitive to changes in the concentration of ions, including particularly hydrogen ions, and osmotic pressure. In order to integrate large numbers of enzyme-controlled reactions, the conditions under which they occur must be maintained constant. Environmentally, the regulation mechanisms free the animal from dependency upon excessive external water supplies or specialised diets. Again, however, there is an evolutionary cost of developing the finely tuned chemostatic system. The heart and circulatory system must be designed to produce the high blood pressure needed by the kidney. There must also be a complex endocrine system in order to detect the level of each of the substances controlled, and to initiate appropriate rates of secretion and reabsorbtion in the kidney tubules.

### *Spatial control*

The problem associated with gravity on land is the least obviously recognisable as a physiological problem requiring control mechanisms in the animal's biological organisation. The land surface is highly irregular, with huge variations in slope and texture and an infinite variety of obstacles, while the collection of food and avoidance of predators pose further locomotory problems. By analogy with both temperature and water problems, the mammals can be regarded as having a locomotory system capable of operating in a wide variety of conditions, thereby extending the range of environments within which they live.

Mammalian limbs have a wide range of amplitudes and angles through which they can be moved, and they can cope therefore with very irregular ground, surmounting or circumventing obstacles. The limbs are long but slender giving a potentially rapid locomotion, and the feet are placed on the ground close to the animal's midline, which makes the animal highly manoevrable or agile. As well as these geometrical properties of the locomotory system, mammals have muscles capable of sustained aerobic activity at a high rate. Thus, although they are not noticeably faster or more efficient at moving, they can maintain rapid movement for much longer periods than equivalent-sized reptiles.

Once again there is an evolutionary cost of the solution to the gravity problem in mammals. As well as the obvious changes in the limbs, the locomotory system requires an increased food and oxygen supply, a complex sensory and central nervous system for control, and possibly a higher metabolic rate for the sustained efforts.

*Homeostasis*

Temperature control, chemical control and spatial control are three aspects of the overall concept of homeostasis, the ability to withstand external fluctuations of the environment by means of regulation of the animal's internal environment. This is a property of all life to a greater or lesser extent, but the mammals possess it to a greater degree than any other forms. The interrelationships between the three main aspects of homeostasis will have been noted, and they can be expressed diagrammatically (Fig. 114). Each arrow indicates a structure or a function which is necessary for the performance of another structure or function. Although naturally an oversimplification of mammalian biology, this figure makes clear the functional interrelationship between the different processes and structures. No one of the individual components of homeostasis can function without the others and one can only really speak of a single, integrated homeostatic mechanism.

An area of mammalian biology not yet considered is reproduction. Juvenile mammals face the same problems as the adults in a potentially more severe form, partly because of their smaller size and consequently greater surface area to volume ratio, making them more subject to temperature fluctuations and water loss. Also, the high degree of complexity of mammals means that a relatively long period is required for development and complete expression of the various regulatory mechanisms of homeostasis. During this period the young are particularly susceptible to environmental variations. The solution to the problem of the young mammals is for the parents to maintain a more or less constant external environment in which the juveniles develop, which is the significance of the various aspects of parental care which have evolved. A reasonably constant external temperature was maintained initially by a nest or burrow, as in the monotremes, and ultimately by means of viviparity. Adequate fluid is provided by lactation. The provision of food by lactation, and the protection by the parents against predators remove the need for a fully developed locomotory system. It will be seen from the figure that this aspect of mammalian life can be fitted into the overall scheme of homeostasis. Without the pre-existing homeostatic mechanisms, the characteristics of mammalian reproduction would not be possible. Equally, if the mammals were not such finely tuned homeostatic organisms, such reproduction would be unnecessary.

The adaptive significance of homeostasis of mammals lies in the buffering provided against spatial and temporal variations in the environment. Therefore they can exist in a wider geographical range, and remain active during both diurnal and seasonal cycles to a greater extent than other organisms. In this sense, the habitats and niches potentially available to mammals are increased. On the other hand, the

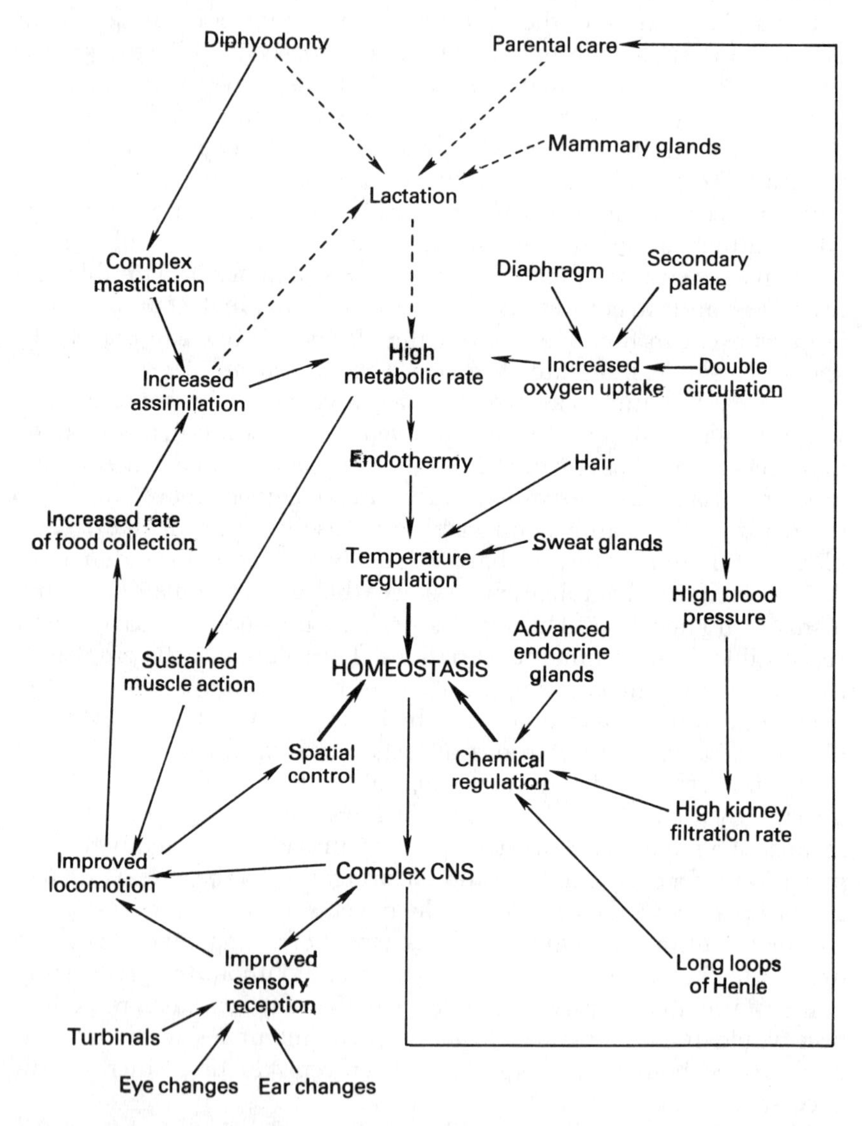


FIG. 114. The main structures and processes of mammals, illustrating the functional interrelationship between them. The arrows indicate how each such structure or function depends on others for its existence, and the whole network centres upon homeostasis. Dashed arrows relate to the special problems of the juvenile mammal.

high metabolic cost of mammalian biology has a limiting effect, for they are either restricted to areas of high food availability, or alternatively can exist only at a relatively low population density.

To conclude, the essence of mammalian biology is the very high degree of complexity and internal integration of the various structures and functional processes. By understanding this concept, it becomes possible to predict certain likely features of the evolution of mammalness.

1. No single characteristic could evolve very far towards the mammalian condition unless it was accompanied by appropriate progression of all the other characteristics. However, the likelihood of simultaneous change in all the systems is infinitessimally small. Therefore only a small advance in any one system could occur, after which that system would have to await the accumulation of small changes in all the other systems, before evolving a further step towards the mammalian condition. The result would be a gradual acquisition of increasingly mammalian characteristics, each system appearing to evolve in a loose correlation with all the other systems.
2. The fundamental biological organisation of mammals is suitable for virtually any adaptive type of terrestrial animal, herbivore, carnivore, insectivore, etc. Therefore the presence of mammalian characteristics, whether incipiently or fully developed, will not be correlated with any particular environmental circumstances.
3. Mammalian biology is not an all-or-nothing adaptation, and the modern mammals merely represent a particular degree of homeostatic control and environmental independence. A lesser degree of homeostasis is still of value. However, in general the greater the refinement of the homeostatic mechanisms, the greater the advantage to animals living in the mammalian habitat, and therefore there will be an orthogenetic trend towards increasing mammalness. At any particular level of this development, adaptive radiations can occur under suitable circumstances, each radiation consisting of animals at about the same level of organisation.

It only needs to be added that these three predicted features of the evolution of the mammals are those actually demonstrated by the fossil record.

# 13 The Macroevolutionary Pattern

THE MAMMAL-LIKE REPTILES show a complex pattern of evolution through time, from their earliest appearance in the Carboniferous through to the end of the Triassic and the appearance of the first mammals. There are numerous gaps evident in the fossil record, which is to say the known species do not grade into one another and therefore it must be assumed that large numbers of intermediate species are missing. The resolution of this particular fossil record is, in fact, lower than that known for certain other groups of animals. In its favour though is the enormous morphological range of forms that it includes, in terms both of progress from a primitive reptilian grade to the mammalian grade, and of the extent of adaptive radiation at various levels. Thus while mammal-like reptiles cannot be used to illustrate the evolutionary processes at the species or genus level very well, they do allow interesting observations to be made about the origin and evolution of higher taxa such as families, orders and even a class.

The evolutionary history of the mammal-like reptiles is reviewed briefly first, and then the more general features of the pattern are noted. Finally an attempt is made to identify possible causes underlying the pattern.

## Review of synapsid evolution

Figure 15 shows the phylogenetic relationships of the main groups of mammal-like reptiles. The earliest record of pelycosaurs dates from Westphalian B which is early Upper Carboniferous (i.e. Lower Pennsylvanian of North America). By Westphalian D times, at least two pelycosaur families had differentiated, the ophiacodontids and the sphenacodontids, and before the close of the period (Stephanian) the archaic carnivorous eothyridids and the herbivorous edaphosaurids were also present. Indeed, the only important pelycosaurs not known from the Carboniferous are the herbivorous caseids, which are not found until the Early Permian. During the Early Permian of North America, the pelycosaurs formed the dominant part of the terrestrial fauna, sharing their habitat only with a variety of small insectivorous captorhinomorphs, and semi-aquatic amphibians. They radiated mainly at the species and at most genus level, some members of each of the main groups achieving large size.

The end of the Early Permian saw the extinction of both the ophiacodontids and the edaphosaurids, although the more advanced forms, sphenacodontids and caseids, lingered into the lowest part of the Late Permian, represented by the North American Guadalupian and the almost contemporaneous Zone I of the Russian Kazanian. Thus these later surviving pelycosaurs briefly overlapped the earliest therapsids, which had evolved as a single lineage from a hypothetical sphenacodontid. Even at their earliest appearance therapsids had already diverged into three of the main groups. The majority were large dinocephalians, both primitive carnivorous brithopodids and specialised estemmenosuchids. The second group, the eotitanosuchids were medium-sized carnivores, and the third, the anomodonts were represented by the small, herbivorous venjukovoids. The rest of the fauna associated with these early therapsids again consisted of captorhinomorphs, such as the North American *Rothia* which was probably a small herbivore. In the Russian record, primitive procolophonoids are present, very small insectivorous or possibly herbivorous relatives of the captohinomorphs. Amphibians were again present.

The next phase of Late Permian evolution is represented by the *Tapinocephalus*-zone fauna of the southern African Karroo. A major radiation of the huge dinocephalians occurred, with carnivorous anteosaurids and titanosuchids, and herbivorous tapinocephalids. They coexisted with the earliest members of the more advanced therapsid groups, small herbivorous dicynodonts, fairly small carnivorous gorgonopsids, and therocephalians. Many of the latter were small, insectivorous forms but one group, the pristerognathids, were large carnivores. Other tetrapods continued to form a very minor part of the fauna. The most important were the pareiasaurs, large herbivorous reptiles related to the procolophonoids, but otherwise there were only a few amphibians such as *Rhinesuchus*, and small eosuchian diapsids.

The dinocephalians disappeared completely at the end of the *Tapinocephalus*-zone, along with the pristerognathid therocephalians. The final phase of Late Permian evolution is represented by the rich *Cistecephalus*-zone and *Daptocephalus*-zone of southern Africa. It was marked by a

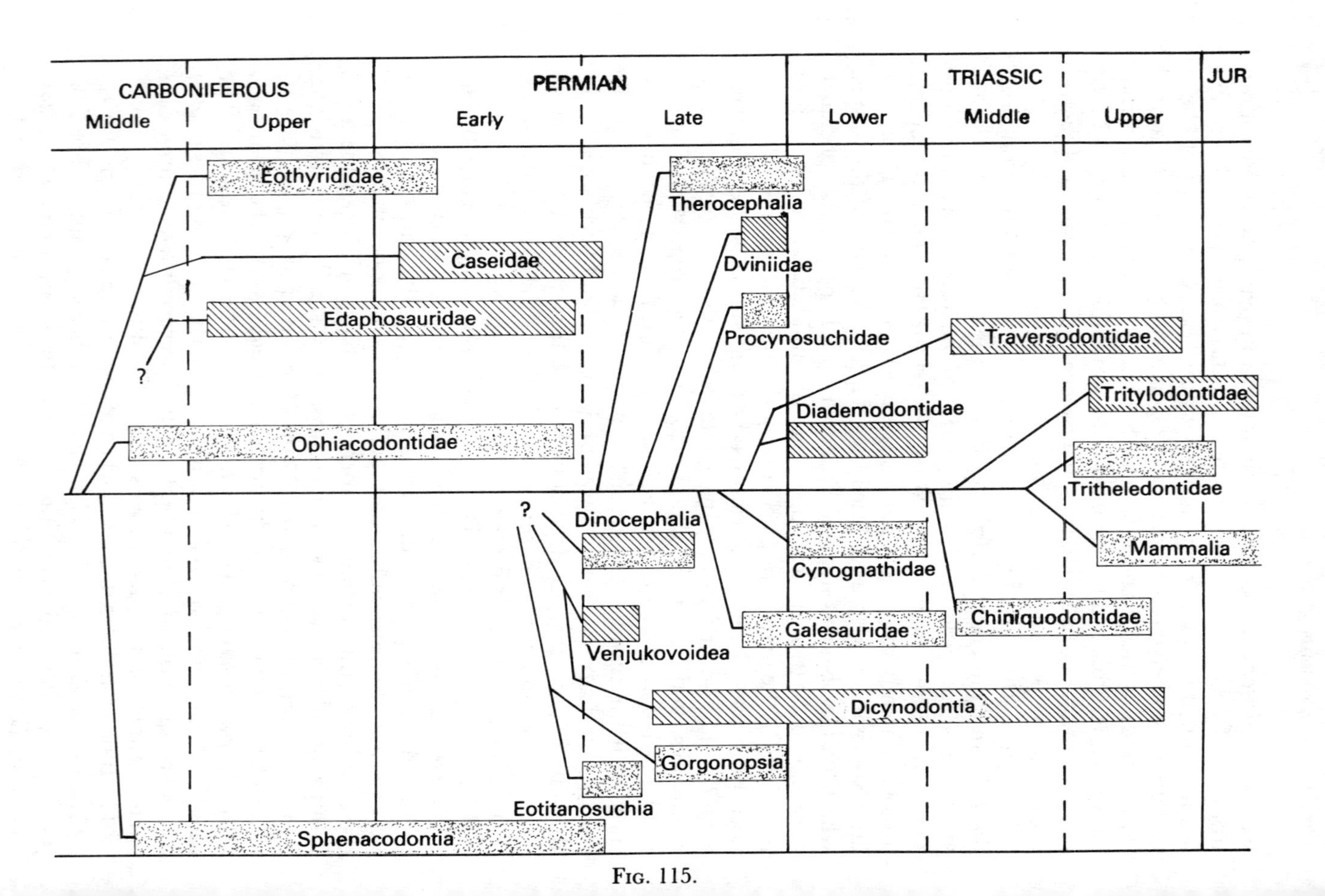
CARBONIFEROUS
Middle
Upper
PERMIAN
Early
Late
TRIASSIC
Lower
Middle
Upper
JUR
Eothyrididae
Caseidae
Edaphosauridae
?
Ophiacodontidae
Sphenacodontia
Therocephalia
Dviniidae
Procynosuchidae
Traversodontidae
Diademodontidae
Tritylodontidae
Tritheledontidae
Mammalia
?
Dinocephalia
Venjukovoidea
Dicynodontia
Gorgonopsia
Eotitanosuchia
Cynognathidae
Galesauridae
Chiniquodontidae

FIG. 115.

flourishing of dicynodonts, gorgonopsids and therocephalians. Several genera of dicynodonts and gorgonopsids achieved large size, although none were as gigantic as the dinocephalians had been. Most of the therocephalians remained small, often insectivorous but one group, the whaitsiids, were large carnivores paralleling the gorgonopsids. This period also saw the first appearance of the cynodonts, which were to become the next great dominant group of mammal-like reptiles. At this time, however, they were small and primitive, having evolved from a therocephalian-like ancestor. The associated fauna, still greatly overshadowed by the therapids, consisted of pareiasaurs, which had survived from the *Tapinocephalus*-zone, a few small eosuchians, and amphibians.

The end of the *Daptocephalus*-zone marks the close of the Permian, and saw another major phase of extinction. All the gorgonopsids and the great majority of the dicynodonts and therocephalians disappeared. Indeed, the therapsid fauna of the lowest Triassic *Lystrosaurus*-zone of southern Africa and several other parts of the world is taxonomically quite impoverished. The dominant form was the specialised dicynodont *Lystrosaurus*, which was probably semi-aquatic. The only other dicynodont was the very small *Myosaurus*. For the first time there were no large carnivorous therapsids although a number of small carnivorous forms were present, including therocephalians and galesaurid cynodonts. The non-synapsid part of the *Lystrosaurus*-zone fauna contained amphibians and small eosuchians as before. The pareiasaurs had disappeared but small herbivorous procolphonids were relatively abundant. But the most important new appearance was that of the archosaur reptiles. *Proterosuchus* was a semi-aquatic thecodont up to about one metre in length.

The next appearance of mammal-like reptiles is towards the top of the Lower Triassic, in the *Cynognathus*-zone of southern Africa. By this time the cynodonts had radiated and formed the dominant part of the synapsid fauna. Large carnivores, the cynognathids, and fairly large herbivorous diademodontids had diverged from a common ancestor derived from a galesaurid-like form. The therocephalians were reduced to the small, herbivorous lineage of the bauriids, and the dicynodonts were represented by the large herbivorous kannemeyeriids, relatives of *Lystrosaurus*, and the single, small and very rare *Kombuisia*. More advanced terrestrial archosaurs had also appeared such as *Euparkeria*, along with several groups of small lepidosaurian diapsids and amphibians.

FIG. 115. Phylogenetic relationships of the main groups of mammal-like reptiles, indicating the particular duration of each. Predominantly carnivorous groups are stippled, and predominantly herbivorous groups are cross-hatched.

By the Middle Triassic formations of Africa and South America, more advanced cynodonts had evolved from the same linage as *Cynognathus* and *Diademodon*. Carnivores, both large and small, were represented by the chiniquodontids, while the herbivorous traversodontids had undergone quite a considerable radiation of mostly modest-sized forms. The large herbivorous dicynodont line related to *Kannemeyeria* also persisted. By this time, the archosaur radiation was well under way, with a variety of advanced thecodontian carnivores. A new group of fairly large herbivores had also evolved. These were the rhynchosaurs, derived from the primitive lepidosaurs of the Lower Triassic. The main amphibians were the large, more or less fully aquatic capitosaurs.

The early part of the Upper Triassic saw little change in the synapsid fauna, as indicated in the Ischigualasto Formation of Argentina. Large traversodontids such as *Exaeretodon* and a few chiniquodontids were present, along with large kannemeyeriid dicynodonts. However, during the later part of the period, the large cynodonts and dicynodonts both disappeared, leaving only the very mammal-like herbivorous tritylodontids, and insectivorous tritheledontids of the Red Beds of the southern African Karroo. The earliest mammals also make their appearance right at the end of the Upper Triassic in the form of the morganucodontids, kuehneotheriids and haramiyids. In contrast to this decline in mammal-like reptiles, the archosaur radiation was well under way. Both saurischian and ornithischian dinosaurs had evolved from earlier thecodonts, and crocodiles were present in some numbers.

With the exception of the tritylodontid *Stereognathus* of the Middle Jurassic, no therapsids are known to have survived beyond the Triassic. Only the mammals continued, but for the entire Jurassic and Cretaceous periods they failed to evolve into large forms, and remained as insectivores, small carnivores and small herbivores.

The question of how incomplete this fossil record is cannot be answered with certainty. The more important collecting areas such as the Early Permian of mid-western USA and the Late Permian and Lower Triassic Karroo of South Africa have been searched intensively for a century. It is unlikely that any major taxon, such as a family, of mammal-like reptiles existed in them but remains unrecorded. If there was such a group, then certainly it was so rare that it could not have formed a very significant part of the community. However, the great majority of the fossil-bearing localities were lowland basins and it is possible that some quite different kinds of mammal-like reptiles existed in other regions, such as the uplands. Only two cases of upland faunas are known for certain, both consisting of bones contained in the material filling fissures in limestone. The earlier is the Fort Sill locality (p. 67) of Oklahoma, which contains a variety of very small pelycosaurs of Early Permian age. Although these are unique genera, they all belong to families well known from the lowlands. The second case is the series of Upper Triassic-basal Jurassic localities of South Wales and Somerset (p. 253), which contains the tritylodontid *Oligokyphus* and *Eozostrodon*, both members of families known elsewhere from lowland deposits. Two other forms, *Kuehneotherium* and the haramiyids, belong to early mammal families which do not occur in contemporaneous lowlands. However, in the case of *Kuehneotherium* certainly and possibly also the haramiyids, related forms do appear in lowland deposits later on. Thus on the limited evidence available concerning upland faunas, it seems unlikely that forms were present there which are completely unrepresented in the lowlands at the family level.

Gaps at a lower taxonomic level, species and genera, are practically universal in the fossil record of the mammal-like reptiles. In no single adequately documented case is it possible to trace a transition, species by species, from one genus to another. This phenomenon is found to occur in the great majority of fossil records of all groups of organisms and has been accounted for by the punctuated equilibrium model of evolution (Eldridge and Gould, 1972; Stanley, 1979). On this hypothesis, most evolutionary change is associated with the formation of new species. Speciation is a result of the geographical and therefore reproductive isolation of small parts of the population of an existing species, and is most likely to occur at the periphery of the species range. This peripheral isolate, as it is termed, will be a small, perhaps extremely small population. It will be subjected to rather different selective forces to the main population, while at the same time being freed from direct competition with individuals of the main populaton. Very rapid evolution may therefore occur. The probability of discovering fossils of such small, geographically remote populations is negligible and therefore the new forms are not manifested in the fossil record until, for one reason or another, they spread into the area formerly occupied by the ancestral species. By this time, so much evolutionary change has occurred that the new species is radically different from its ancestral species and a morphological gap separates successive species in the record.

## A model adaptive radiation

The pattern of evolution of the mammal-like reptiles can be considered to consist of three phases. The first was the adaptive radiation of the pelycosaurs in the Upper Carboniferous and Early Permian. This was superseded by the radiation of the non-cynodont therapsids in the Late Permian. The third phase was the Triassic radiation of the cynodonts. A fourth radiation, that of the Jurassic and Cretaceous mammals, followed the cynodont radiation, but was rather different because the synapsids had ceased to be the dominant terrestrial fauna, a role played by the dinosaurs. Each of the three mammal-like reptile radiations commenced from a single lineage that had originated during the previous phase, and followed a broadly comparable course. The fact that many features of the evolutionary pattern occurred independently three times in succession in the overall fossil record implies that there may be certain underlying rules governing the process of macroevolution; the pattern is not simply random and does not depend on any very particular environmental factors.

The principal features of a generalised adaptive radiation, derived from the features seen in common in all three individual radiations may be summarised as follows (Fig. 116).

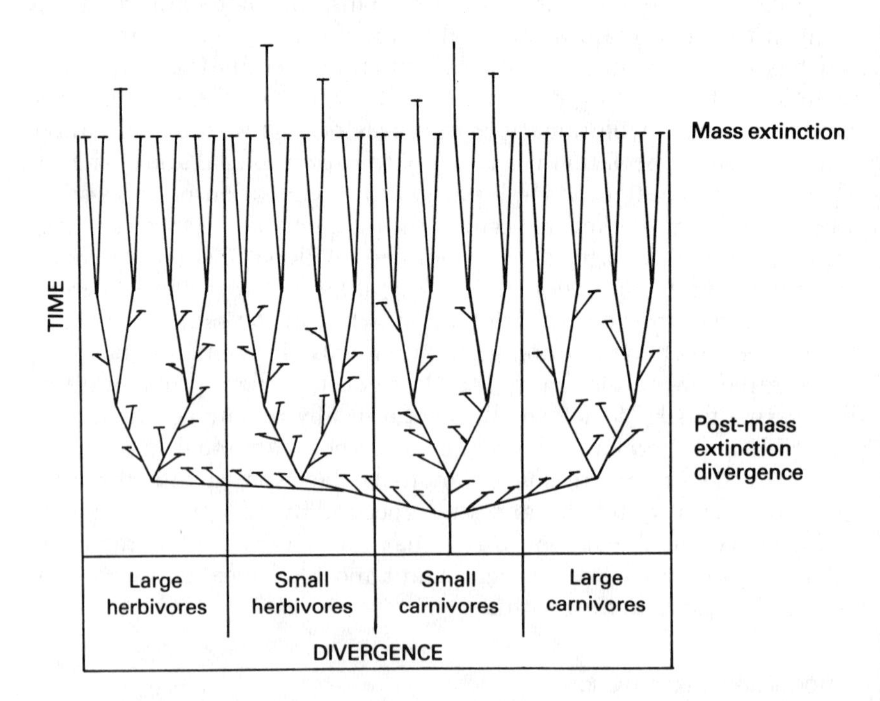


Fig. 116. Some of the important features of the model adaptive radiation. Branchings indicate speciation events, and horizontal lines indicate extinctions of species. The time scale is not to be taken as linear.

1. The radiation commences after a sudden mass-extinction has removed most although not all of the lineages of the previous radiation.
2. The radiation commences from a single lineage of small carnivore-insectivore which was part of the previous radiation and survived the mass extinction. (It is not known for certain in the case of the non-cynodont therapsid radiation that the ancestor was small, but it is likely.)
3. The members of the radiation are more progressive, that is, more mammal-like than the members of the previous radiation.
4. The radiation produces lineages of insectivores, large carnivores, and both small and large herbivores.
5. The new lineages are more or less fully developed at their first appearance in the fossil record, and thereafter they remain relatively conservative.
6. Within each of the lineages, individual species and genera disappear and new ones appear continuously, which is the phenomenon of species turnover. Only the lineage, or higher taxon, as a whole persists for the full course of the radiation.
7. During the radiation, minor phases of mass extinction occur, with the simultaneous loss of several or many members of the separate lineages.
8. The termination of the radiation is marked by the simultaneous loss of all the species composing it. A few lineages survive, as manifested by the appearance of new species of these lineages at the commencement of the next radiation. The surviving lineages consist mainly of small animals.

## Explanations

Three general factors must have been involved in determining the course of evolution of the mammal-like reptiles, and in so far as these can be described, thus far can the features of the model adaptive radiation be accounted for.

i. *The rate of extinction and the rate of speciation within a lineage.* If the extinction rate is greater, the lineage will become extinct, but if the speciation rate is greater an adaptive radiation will occur.
ii. *The rate of morphological change within a lineage.* This will be a function of both the average size of the change between an ancestral and a descendent species, and also of the rate of species turnover of the lineage.
iii. *The kind of morphological change.* This will depend on the sorts of

niches potentially available to a new morphotype, and also the constraints on change arising from the need for the organism to remain integrated at the genetic, developmental and phenotypic levels (p. 15).

### *Rates of extinction and speciation*

Species turnover, the effect whereby a lineage is seen over a period of time to consist of continually replacing species, indicates that the evolution of a lineage is a highly dynamic process (Stanley, 1979). There is a rough balance between the rate at which existing species become extinct and the rate at which new ones appear. If, however, species disappear more rapidly than new ones appear, the number of species constituting the lineage declines and the whole lineage becomes extinct. Conversely, an adaptive radiation, or increase in species number in the lineage will result from the formation of new species more rapidly than the existing ones disappear.

The actual extinction of a species is presumed to be related to an alteration to the environment. The adaptation of the species is reduced, for example subjecting it to excessive predation or competition, reducing its food source, or perhaps reducing its capacity to reproduce or develop properly. The exact cause of extinction cannot be discovered from the fossil record, and neither can the particular environmental change initiating the extinction. It is probable that a really very small change is all that is often required, and in the course of the history of life on earth, minor environmental fluctuations have been extremely frequent (Pearson, 1978). Examples of changes in temperature, rainfall, seasonality and geomorphology are abundant, and even if factors such as these do not affect a particular species directly, they can have profound effects on that species' biotic environment. It is reasonable to believe that the continual extinction of species reflects no more than this ever-changing environment, and that a somewhat greater environmental perturbation than usual will cause more widespread extinction.

The current view of speciation is that it is usually, if not always, associated with the geographic isolation of a small part of the species population, i.e. allopatric speciation. The small population formed in this way possesses its own, separate gene pool, which can be affected by a number of random genetic processes. The Founder Effect refers to the chance initial differences between the newly isolated gene pool and the main gene pool, while random mutations and recombinations lead to genetic divergence. At the same time, different selective pressures will apply to the new population because it will almost certainly live in a somewhat different environment to the ancestral population from which it arose. Thus increasing genetic divergence occurs until the new population is sufficiently distinct from the old to be a new species.

Any number of isolated populations can occur around the margins of the range of the ancestral species, and therefore more than one new species can arise virtually simultaneously. The overall rate at which speciation occurs will depend upon the number of opportunities which exist for geographic isolation. In areas where the geography is more varied, with hills, extensive water barriers, steep climatic gradients and so on, a higher rate of speciation is to be expected. In contrast, a more monotonous habitat will tend to be associated with a lower speciation rate.

The respective rates of extinction and speciation may also be partly dependent upon one another. If the extinction rate rises, it is possible that more niches become available for new species to invade from the peripheral areas, and therefore a greater number of the new species become incorporated in the main ecosystem. Conversely, if speciation rate rises, there could be more competition between the old species and the invading new species, causing a rise in extinction rate. There is therefore at least a tendency towards a negative feedback system, which will keep the numbers of extant species of a lineage approximately constant. This might be important in explaining why a lineage is so often far longer-lived than any of its constituent species.

At any event, during the course of evolution of a lineage, variations in the ratio of extinction rate to speciation rate do occur. The most important is the higher extinction rate compared to speciation rate that marks the periods of mass extinction, which play such a central role in the evolution of the mammal-like reptiles. As indicated in the model adaptive radiation, a mass extinction separates successive phases of adaptive radiation, and is characterised by the simultaneous extinction of several lineages. However, the extinction is not complete for in every case certain lineages survive, although the particular species of these lineages present after the mass extinction differ from those present before. The differential extinction of lineages is important because it indicates that certain types of lineage are more resistant to total extinction than others, which is to say that even under the same environmental conditions, the ratio of extinction rate to speciation rate of different types of organisms differs. If the main lineages of mammal-like reptiles are considered to be composed of four simplified ecotypes, large and small carnivores and herbivores respectively, then it becomes possible to see how a mass extinction could affect different lineages in different ways.

Large carnivore lineages are likely to be the most susceptible of all to mass extinction. Each species has a relatively small population size in a given geographic area because of its high trophic level and the size of the individuals. Therefore there is a high chance that an environmental change will cause reduction of the population below a viable level, from which it will be unable to recover its numbers. Also there are relatively few species of large carnivore at any one time, and therefore less chance of one of the species happening by good fortune to be suitably adapted to survive the particular alteration to the environment. On the other hand, large carnivores are less likely to speciate. The existing species are highly mobile and potentially disperse widely, which reduces the likelihood of isolated populations forming at the periphery of the species range. Also, relatively large geographical barriers are necessary for isolation of parts of the population of large animals.

At a time of environmental change, therefore, the extinction rate of large carnivores will rise but the speciation rate remain too low to replace the lost species, and the lineage will rapidly become extinct. The one exception to this general rule among the mammal-like reptiles concerns the sphenacodontid pelycosaurs. These survived for a short period into the Late Permian, where they very briefly co-existed with the therapsid radiation. However, these forms were apparently closely associated with water from which they must have got much of their food as fish. Semi-aquatic animals tend not to have the same degree of mobility as fully terrestrial forms, since their migrations are tied to waterways. They therefore have a greater potential to form isolated populations and hence new species. There will be a consequent increased resistance to mass extinction, as seen in other semi-aquatic groups such as crocodiles and fresh water chelonians. Other than sphenacodontids, all the large carnivore groups suffered prompt extinction during the main mass extinction phases, and also during more minor mass extinction. The carnivorous dinocephalians and the pristerognathid therocephalians disappeared during the course of the therapsid radiation, the gorgonopsids and whaitsiid therocephalians at the end of the Permian, the cynognathid cynodonts in the Lower Triassic, and the chiniquodontids in the Upper Triassic.

Like large carnivores, the large herbivore lineages tend to have low speciation rates because of their high mobility and consequently low probability of forming peripheral isolates. However, being at a low trophic level, they tend to have larger species populations and more species than equivalent carnivores, and therefore have a slightly greater chance of avoiding total extinction. It might be expected that lineages of large herbivores would be a little less subject to total mass extinction.

Although many lineages did become extinct at the same time as the large carnivores, such as the herbivorous dinocephalians during the Late Permian and most of the large dicynodonts at the close of that period, others survived. The caseid pelycosaurs continued just into the Late Permian to overlap with the early therapsids, and the lineage resulting in the Triassic dicynodonts survived the Permian–Triassic boundary. The Triassic dicynodonts themselves were a particularly long-lived lineage of large herbivores.

Small herbivore lineages will be expected to have a relatively high resistance to mass extinction because of the large size of species populations and the relatively large number of species that can occur. The probability of complete extinction of such a lineage will tend to be low. At the same time, small animals generally have less mobility than large ones, and therefore accidentally isolated parts of the population will tend to remain isolated for long enough to speciate. Only relatively small geographical barriers are necessary to cause this isolation. Therefore the rate of speciation will tend to be high. The fossil record supports this view, since several small herbivore lineages are seen to survive mass extinction. The dicynodonts present in the lower Late Permain were small, and survived the extinction of the dinocephalians, while the presence of *Myosaurus* and the unrelated *Kombuisia* in the Lower Triassic indicates that at least two small dicynodont lineages survived the Permian–Triassic extinction. The traversodontid cynodonts were mostly small and this lineage lasted for most of the second half of the Triassic, while the smaller of the very mammal-like tritylodontids lasted into the Jurassic.

The final ecotype to consider is the small carnivore, or insectivore. Like the small herbivores, they too tend to have large populations and a high number of species. Thus they tend to have low extinction rates. Again, because of their small size and low mobility, they can potentially speciate at a high rate. Small carnivores may have an additional advantage over small herbivores, which is their generally unspecialised nature and therefore wide niches. To deal efficiently with plant food requires many specialised features of the masticatory and digestive systems, which, when fully expressed, tend to limit the organism to its chosen specialised diet. In contrast, small carnivore-insectivores remain unspecialised and can cope with a more catholic diet. With an environmental change, there is a higher chance of a carnivore still possessing an adequate food source compared to a herbivore.

Certainly numerous lineages of small carnivores survived the various extinction phases. The archaic little pelycosaur *Eothyris* was an Early Permian survivor of a predominantly Carboniferous group, and at the end of the Early Permian a presumed lineage of small carnivorous pelycosaurs survived to give rise to the therapsids. During the early part of the Late Permian, the gorgonopsids were relatively small, and survived the dinocephalian extinction along with a variety of small therocephalians. Insectivorous therocephalians also survived into the Lower Triassic, as did the small galesaurid cynodonts. The presence of small chiniquodontids in the Middle Triassic indicates that this lineage survived the period when the cynognathid cynodontids became extinct, while the presence of tritheledontids and the first mammals in the uppermost Triassic and beyond demonstrates further the viability of insectivores.

It is likely that the phases of mass extinction that periodically interrupted the history of the mammal-like reptiles were caused by relatively small environmental changes that altered the ratio of extinction rate to speciation rate of the various lineages, some more than others. Although on a geological time scale these events appear instantaneous, the resolution of the fossil record is too crude to determine how long it took a lineage to decline on a biological time scale. It could well have been many thousands of years in fact. There is certainly no need to postulate large sudden catastrophes of the kind frequently invoked to account for mass extinctions in general. Indeed the prime difficulty with such explanations is that they fail to account for the survival of the lineages which did not become extinct. One recent hypothesis by Bakker (1977) has the merit of stressing a reduction in speciation rate as well as possible increase in extinction rate, and also accounts well for the differential extinction between lineages of small and large animals. He notes a coincidence between mass extinctions and reductions of mountain-building (orogenic) activity. With continuing erosion of the land, the effect was to reduce the variation in height of the upland, inland areas and produce a monotonous landscape surrounding the main lowland basins. Thus the opportunities for geographical isolation of parts of the species populations were reduced and the rate of formation of new species declined. This effect would be most marked for lineages of large animals, which require larger geographic barriers anyway. Appropriate reductions in orogeny have been documented for the middle of the Permian, to a lesser extent during the Late Permian, and quite dramatically at the boundary between the Permian and the Triassic, all of which coincide with mass extinctions. Other lower levels of mountain building are noted in the Lower and the Upper Triassic. Bakker further documents major regressions of the sea at these points in time, which must have caused modifications to the environment sufficient to increase the rate of extinction of the various lineages occupying the lowland basins.

Much more detailed studies are required to test this hypothesis of Bakker, but even if it is not fully correct, it is the kind of explanation that most satisfactorily accounts for the biological features of a phase of mass extinction.

A final question concerning rates of extinction is whether direct competition between similar ecotypes is significant. Certainly it is not necessary. The gorgonopsids and whaitsiids for example became extinct at the close of the Permian, long before the newer version of large carnivores, the cynognathid cynodonts, appeared in the Lower Triassic. In other cases, however, similar ecotypes from successive radiations do overlap and could theoretically compete. The sphenocodontid and caseid pelycosaurs for example co-existed with the dinocephalian therapsids for a brief time in the Late Permian. The extinction of the pelycosaurs soon after may have been caused by these therapsids, although it does not seem possible to test such a possibility.

### *Rate of morphological change*

The apparent rate of morphological change in the main lineages of the mammal-like reptiles varies. The sudden appearance of new higher taxa, families and even orders, immediately after a mass extinction, with all the features more or less developed, implies a very rapid evolution. This is invariably followed by a much slower rate of morphological change of the lineage, usually no more than generic, or at most sub-family level. It is possible that this observation is an artifact, and that the new taxa had long histories before they appeared in the fossil record, during which they gradually acquired their characteristic features. However, in no case is such a long history known by even a single specimen, and therefore it is much more reasonable to accept that very high rates of morphological evolution characteristically occur following a mass extinction.

Three factors theoretically control the rate of morphological evolution within a lineage. These are the magnitude of the change that occurs when a new species is formed, the subsequent rate of evolution of the new species, and the rate at which species replace one another over a period of time. There is a dispute about the relative importance of the first two factors, abbreviated as quantum speciation and phyletic gradualism respectively and reviewed by Stanley (1979). For many years now it has been believed that evolution occurs by slow changes in gene frequency and the accumulation of beneficial mutations by natural selection, which produces slow, gradual morphological change. On this model, speciation is also seen to be gradual, due to the slow genetic divergence of reproductively isolated populations. The evidence behind the gradualistic view is in the field of population genetics, where just such phenomena can be observed in natural populations. However, a recurrent heterodoxy in biology is the view, expressed with varying degrees of extremeness, that much larger morphological changes can occur suddenly, between an ancestral species and its descendant. The advantage of the concept of quantum speciation, as it is currently termed, is that it helps to explain the apparently very high rates of change involved in the origin of major new taxa. No fossil record is likely to permit testing of the hypothesis of quantum speciation because of the possibility that the record is incomplete. Certainly this is true of the mammal-like reptiles. Therefore acceptance of quantum speciation as part of the explanation of macroevolutionary events depends on whether the biological properties of living organisms are such that it can be reasonably supposed to occur often enough. Evidence is actually growing that a possible mechanism for quantum change does exist. The hierarchieal arrangement of genes, with regulator genes affecting the integrated action of numbers of structural genes implies that mutations of such regulator genes could affect the overall genome in a cascading fashion, bringing about a quite large change but not upsetting the integration of the genome as a whole. At the developmental level, a substantial alteration to one part of the organism could induce appropriate modifications to other parts so that the resulting phenotype is still viable. Assuming that quantum speciation steps are theoretically possible, it is interesting to speculate on the environmental conditions under which they are most likely to produce species that become successfully incorporated into the community. A new species which differs fairly radically from its immediate ancestor as a result of quantum change is unlikely to be well adapted to any particular niche. If it is isolated from the ancestral species population, however, it has a high chance of surviving even in its relatively ill-adaptive state, because it will not suffer from competition. It thus has the time for a more gradual improvement by natural selection, a process corresponding to the non-adaptive phase of major evolutionary transition of Simpson (1953).

The fossil record is no better for testing the hypothesis that the rate of species turnover was greater immediately after a mass extinction phase, but it can be argued from first principles that this is likely. The following model speculates on possible evolutionary events immediately following a phase of mass extinction.

1. The mass extinction removes all the species occupying the main arena, but of all the peripheral, isolated populations that had recently

speciated some are potentially suitably adapted to survive in the new conditions of this main arena.
2. These particular new species invade the main arena and survive by occupying the vacated niches. They face no competition and can be relatively poorly adapted at this time.
3. In due course these new species give rise to new peripheral isolates which speciate, possibly by quantum speciation. Among this second generation of new species there is a high probability that by chance some of them have changed in such a way that they are potentially better adapted for the conditions in the main arena than the present occupiers. When they accidentally invade the main arena they will consequently displace the present occupiers.
4. The process will be repeated, the second generation of new species giving rise to a third generation of new species by the isolation of small peripheral populations. Since the second generation of species is better adapted than the first generation, then there is a reduced but still reasonable probability that the third generation will include better adapted forms than the second. So the process will continue, but the probability of successive generations of new species including better adapted forms than present occupiers of the main arena will decline. Therefore the rate at which species are replaced in the main arena will slow down until all the occupiers of the main arena are so well adapted that the chances of them being replaced by new species is low. This will be the condition of stasis, or greatly reduced rate of morphological change in the lineage.

It is impossible to estimate how long the initial phase of this evolutionary model might take, when species turnover and rate of morphological change is very high. If it only lasted for perhaps a few thousand years, it would not be surprising if it did not show in the fossil record of the mammal-like reptiles. Thus the model accounts for the fact that major new taxa appear abruptly, and are more or less fully expressed in the record.

### *Direction of morphological change*

The factors which determined exactly what kinds of morphological change occurred in the course of evolution of the mammal-like reptiles fall into two general categories. The most obvious are the environmental opportunities available. These opportunities will be in the form of unoccupied niches, or occupied niches for which new species can successfully compete by causing the extinction of the present occupier, or by causing a division of the niche to permit co-existence of two erstwhile competitors. The particular morphological changes that occur will be appropriate adaptations for the niche in question, and in earlier chapters much of the anatomy of mammal-like reptiles has been interpreted by implication along these lines. The broad similarity of ecotypes produced in the separate phases of adaptive radiation of the mammal-like reptiles and expressed in the model adaptive radiation as large and small carnivores and herbivores hardly needs comment. Allowing for the oversimplification of these terms, such categories refer simply to the main environmental opportunities which can exist in the terrestrial context.

Of considerably greater theoretical interest are the other factors controlling morphological change, which are the internal biological features of organisms that channel evolution along certain possible paths, but prevent change in other directions. The first question to consider from this point of view is why each of the main three radiations commences from a small carnivore or insectivore. Reasons for the tendency of small carnivore lineages to survive mass extinction have been suggested, but it was noted that small herbivore lineages have almost the same resistance. This selection of the small carnivores as progenitors of the next phase is all the more remarkable because they give rise to a radiation which includes herbivores. This widely known phenomenon has not been satisfactorily explained. It is generally held that herbivores are more specialised than the carnivores, and are therefore unable to evolve into new ecotypes. After a mass extinction, the small surviving carnivores have an obvious advantage as far as the vacant carnivore niches are concerned and it is not surprising that they evolved to fill them. However, on the face of it the same should be true of the surviving herbivores with respect to the vacant herbivore niches. There is no clear reason why the caseids should not have radiated in the Late Permian rather than the dicynodonts and herbivorous dinocephalians. Similarly, after the Permian–Triassic extinction, why did not the dicynodonts rather than the cynodonts produce the small and medium-sized herbivores, instead of being restricted to the large herbivore niches of the Triassic? One possible factor involved is the temporary absence of large carnivores immediately after these mass extinctions. Natural selection would cease to favour the evolution of devices for escape among the herbivores, and such things as locomotory ability may have declined. Indeed the reduced metabolic requirements of a reduced locomotory ability may have been positively selected for. The same would not have been true of the small carnivores which would continue to require speed and agility for catching insects. When larger carnivores did appear, the herbivore lineages would be susceptible to extinction from excessive predation. Yet again these are speculative events that must have occurred in the dark, brief period of evolution immediately after the mass extinction, and the fossil record offers no direct evidence.

The second question regarding internal factors and morphological channelling concerns the integration between the various structures and functions of the individual, which has to be maintained at all times during the course of evolutionary change. This was discussed at length in the previous chapter, with reference to the origin of mammals. It was noted that the fossil record supports the view that evolution towards mammalian levels of homeostasis involved practically all aspects of the organism simultaneously. No single structure or function could evolve very far without being accompanied by appropriate changes in all the other features. Thus a long-term trend appears, manifested by the gradual, progressive evolution of all the systems more or less in parallel, from the primitive pelycosaur level to the mammals themselves. This progressive evolution of increasing homeostasis, or mammalness, is the simple explanation behind the observation that each main phase of radiation consists of members more advanced than the previous radiation's members. The later members of a radiation are the ones which have progressed furthest down this path, and the high rate of evolution of the lineages surviving the mass extinction no doubt gave a further boost towards mammalness at the start of the new radiation phase.

This kind of correlated progression mechanism, as it is termed, must also play a part in the divergent evolution of the particular specialised lineages. In these cases, the features involved are not related to any general biological characteristic such as homeostasis, but to particular adaptations for particular habitats. To take as an example dicynodonts, their herbivorous specialisation requires the replacement of the teeth by horny tooth plates, reorientation of the jaw musculature, changes in the form of the jaw hinge, and an extensive remodelling of the shape of the skull and lower jaws. Also suitable locomotion, and central nervous programming and behaviour are needed. No one of these features has much adaptive value unless accompanied by the others, and therefore the evolution of the dicynodont type of organism must have followed a correlated progression, each feature evolving gradually and in association with changes in all the other features. Once this particular evolutionary pathway had commenced, it is most unlikely that any result could follow except that which did. The morphological changes were in this sense channelled.

The concept of correlated progression has implications for two commonly expressed ideas about mammal-like reptile evolution. The first is based on the view that the origin of any major new taxon, or adaptive type of animal involves a "key adaptation", by which is meant a critical new feature which finally frees the evolving line to invade a new adaptive zone. Barghusen (1968) for example, believed that the key innovation in the evolution of the cynodonts was the development of adductor musculature on the lateral side of the lower jaw. Such views must be criticised on the grounds that many other changes necessarily accompanied the key innovation, in order for the latter to be an integrated part of the cynodont's overall biology. It is meaningless to select any particular one of these changes and regard it as more significant than the rest. The key innovation of cynodonts was the whole complex of changes that brought about the basic cynodont organisation.

The second commonly expressed idea about the evolution of mammmal-like reptiles which is inconsistent with correlated progressions is that parallel evolution was widespread. On the face of it, the tendency to channel morphological change towards a particular end, such as homeostasis, which is of benefit to all ecotypes might be expected to lead to parallel evolution in separate lineages. Against this, however, it must be pointed out that most lineages do not survive a mass extinction, immediately after which is the main period of morphological change, and therefore the main time when significant advances in homeostasis must be expected. Furthermore, the complexity of a correlated progression, involving as it does so many parts of the organism, makes it extremely improbable that more than one lineage would achieve the same degree of homeostasis at the same time. The more improved one would be expected to out-compete the less improved one, and this is probably the very reason why, of the several small carnivore lineages that may survive a mass extinction, only one of them goes on to generate the new radiation. Looking at the actual fossil record, there are several cases of parallel evolution of a very trivial kind, such as secondary palates in dicynodonts, cynodonts and some therocephalians, or reduction of the postcanine teeth in gorgonopsids, pristerognathid therocephalians and whaitsiid therocephalians. A somewhat more marked but still superficial case is the parallel acquisition of multicusped teeth and secondary palate in cynodonts and bauriid therocephalians of the Lower Triassic. This example is interesting because these are two of the lineages which survived the mass extinction at the end of the Permian. It may be the one case known where two lineages did in fact follow in parallel the trend towards increased homeostasis. At any event, the bauriids shortly became extinct.

It is within the cynodonts themselves that parallel evolution has been most widely invoked, with the suggestion that most of the main groups evolved separately from a primitive, galesaurid level. As has been argued elsewhere, this view rests on the assumption that the wide variation in the structure of the post-canine teeth reflects a long period of separation between the various families. Given this, it is then assumed that the enlarged dentary, temporal fenestra, occluding teeth and many postcranial features must have evolved in parallel. On the macroevolutionary model presented here, which best fits the known facts of mammal-like reptile evolution, parallel evolution of cynodonts is most improbable. First there is no reason to suppose that postcanine teeth could not evolve very rapidly, indeed it is very likely. Second, it is simply very unlikely that the correlated progression that must have been involved in the development of all the advanced cynodont features, features found in all ecotypes, could have evolved more than once in exactly the same manner at exactly the same rate.

# References

ALLIN, E. F. (1975). Evolution of the mammalian middle ear. *J. Morph.* **147**, 403–438.

ANDERSON, J. M. and CRUICKSHANK, A. R. I. (1978). The biostratigraphy of the Permian and Triassic Part 5. A review of the classification and distribution of Permo-Triassic tetrapods. *Palaeont. afr.* **21**, 15–44.

BAIRD, D. and CARROLL, R. L. (1967). *Romeriscus*, the oldest known reptile. *Science, N.Y.* **157**, 56–59.

BAKKER, R. T. (1974). Experimental and fossil evidence for the evolution of tetrapod bioenergetics. *In* "Perspectives in Biophysical Ecology" (Eds D. Gates and R. Schmerl), pp. 365–399. Springer-Verlag, New York.

BAKKER, R. T. (1977). Tetrapod mass extinctions—a model of the regulation of speciation rates and immigration by cycles of topographic diversity. *In* "Patterns of Evolution as Illustrated by the Fossil Record" (Ed. A. Hallam), pp. 439–468. Elsevier Scientific, Amsterdam, Oxford and New York.

BARGHUSEN, H. R. (1968). The lower jaw of cynodonts (Reptilia, Therapsida) and the evolutionary origin of mammal-like adductor jaw musculature. *Postilla* No. 116, 1–49.

BARGHUSEN, H. R. (1973). The adductor jaw musculature of *Dimetrodon* (Reptilia, Pelycosauria). *J. Paleont.* **47**, 823–834.

BARGHUSEN, H. R. (1975). A review of fighting adaptations in dinocephalians (Reptilia, Therapsida). *Paleobiology* **1**, 295–311.

BARGHUSEN, H. R. (1976). Notes on the adductor jaw musculature of *Venjukovia*, a primitive anomodont therapsid from the Permian of the U.S.S.R. *Ann. S. Afr. Mus.* **69**, 249–260.

BARRY, T. H. (1968). Sound conduction in the fossil anomodont *Lystrosaurus*. *Ann. S. Afr. Mus.* **50**, 275–281.

BARRY, T. H. (1974). A new dicynodont ancestor from the Upper Ecca (lower Middle Permian) of South Africa. *Ann. S. Afr. Mus.* **64**, 117–136.

BERMAN, D. S. (1977). A new species of *Dimetrodon* (Reptilia, Pelycosauria) from a non-deltaic facies in the Lower Permian of north-central New Mexico. *J. Paleont.* **51**, 108–115.

BOLT, J. R. (1969). Lissamphibian origins: possible protolissamphibian from the Lower Permian of Oklahoma. *Science, N.Y.* **166**, 888–891.

BONAPARTE, J. F. (1962). Descripción del cráneo y mandíbula de *Exaeretodon frenguellii*, Cabrera. y su comparación con Diademodontidae, Tritylodontidae y los cinodontes sudamericanos. *Publnes. Mus. Munic. Nat. tradic, Mar del Plata* **1**, 135–202.

BONAPARTE, J. F. (1963a). La familia Traversodontidae (Terapsida–Cynodontia). *Acta geol. lilloana* **4**, 163–194.

BONAPARTE, J. F. (1963b). Descripción del esqueleto postcraneano de *Exaeretodon* (Cynodontia–Traversodontidae). *Acta geol. lilloana* **4**, 5–51.

Bonaparte, J. F. (1963c). Descripción de *Ischignathus sudamericanus* n.gen.n.sp., nueuo cinodonte gonfodonte del Triásico medio superior de San Juan, Argentina (Cynodontia–Traversodontidae). *Acta geol. lilloana* **4**, 111–128.

Bonaparte, J. F. (1970). Annotated list of the South American Triassic tetrapods. *Second Gondwana symposium*, pp. 665–682. International Union of geological sciences commission on stratigraphy. C.S.I.R. Pretoria.

Bonaparte, J. F. and Barbarena, M. C. (1975). A possible mammalian ancestor from the Middle Triassic of Brazil (Therapsida–Cynodontia). *J. Paleont.* **49**, 931–936.

Boonstra, L. D. (1934). A contribution to the morphology of the Gorgonopsia. *Ann. S. Afr. Mus.* **31**, 137–174.

Boonstra, L. D. (1936). The cranial morphology of some titanosuchid dinocephalians. *Bull. Am. Mus. nat. Hist.* **72**, 99–116.

Boonstra, L. D. (1954). The cranial structure of the titanosuchian: *Anteosaurus*. *Ann. S. Afr. Mus.* **42**, 108–148.

Boonstra, L. D. (1955). The girdles and limbs of the South African Deinocephalia. *Ann. S. Afr. Mus.* **42**, 185–326.

Boonstra, L. D. (1962). The dentition of the titanosuchian dinocephalians. *Ann. S. Afr. Mus.* **46**, 57–112.

Boonstra, L. D. (1963). Early dichotomies in the therapsids. *S. Afr. J. Sci.* **59**, 176–195.

Boonstra, L. D. (1965a). The Russian dinocephalian *Deuterosaurus*. *Ann. S. Afr. Mus.* **48**, 233–236.

Boonstra, L. D. (1965b). The skull of *Struthiocephalus kitchingi*. *Ann. S. Afr. Mus.* **48**, 251–265.

Boonstra, L. D. (1965c). The girdles and limbs of the Gorgonopsia of the *Tapinocephalus* zone. *Ann. S. Afr. Mus.* **48**, 237–249.

Boonstra, L. D. (1966). The dinocephalian manus and pes. *Ann. S. Afr. Mus.* **50**, 13–26.

Boonstra, L. D. (1968). The braincase, basicranial axis and median septum in the Dinocephalia. *Ann. S. Afr. Mus.* **50**, 195–273.

Boonstra, L. D. (1969). The fauna of the *Tapinocephalus* Zone (Beaufort beds of the Karroo). *Ann. S. Afr. Mus.* **56**, 1–73.

Boonstra, L. D. (1971). The early therapsids. *Ann. S. Afr. Mus.* **59**, 17–46.

Boonstra, L. D. (1972). Discard the names Theriodonta and Anomodonta: a new classification of the Therapsida. *Ann. S. Afr. Mus.* **59**, 315–338.

Bown, T. M. and Kraus, M. J. (1979). Origin of the tribosphenic molar and metatherian and eutherian dental formulae. *In* "Mesozoic Mammals: the first two-thirds of mammalian evolution" (Eds J. A. Lillegraven, Z. Kielan-Jaworowska and W. A. Clemens), pp. 172–181. University of California Press, Berkeley, Los Angeles and London.

Bramble, D. M. (1978). Origin of the mammalian feeding complex: models and mechanisms. *Paleobiology* **4**, 271–301.

Bramwell, C. D. and Fellgett, P. B. (1973). Thermal regulation in sail lizards. *Nature, Lond.* **242**, 203–205.

Brink, A. S. (1956a). Speculations on some advanced mammalian characteristics in the higher mammal-like reptiles. *Palaeont. afr.* **4**, 77–96.

Brink, A. S. (1956b). On *Aneugomphius ictidoceps* Broom and Robinson. *Palaeont. afr.* **4**, 97–115.

Brink, A. S. (1963a). On *Bauria cynops* Broom. *Paleont. afr.* **8**, 39–56.

Brink, A. S. (1963b). Two cynodonts from the Ntawere Formation in the Luangwa Valley of Northern Rhodesia. *Palaeont. afr.* **8**, 77–96.

Brink, A. S. (1965). A new ictidosuchid (Scaloposauria) from the *Lystrosaurus*-zone. *Palaeont.afr.* **9**, 129–138.
Broili, F. and Schröder, J. (1934a). Über den Cynodontier *Tribolodon frerensis* Seeley. *Sitz. Bayer. Akad. Wiss.* for 1934, 163–177.
Broili, F. and Schröder, J. (1934b). Zur Osteologie des Kopfes von *Cynognathus. Sitz. Bayer. Akad. Wiss.* for 1934, 95–128.
Broili, F. and Schröder, J. (1935a). Über die Skelettrests eines Gorgonopsiers aus den unteren Beaufort-Schichten. *Sitz. Bayer. Akad. Wiss.* for 1935, 279–330.
Broili, F. and Schröder, J. (1935b). Über den Schädel von *Gomphognathus. Sitz. Bayer. Akad. Wiss.* for 1935, 115–182.
Broom, R. (1910). A comparison of the Permian reptiles of North America with those of South Africa. *Bull. Am. Mus. nat. Hist.* **28**, 197–234.
Broom, R. (1912). On a new type of cynodont from the Stormberg. *Ann. S. Afr. Mus.* **7**, 334–336.
Broom, R. (1932). "The Mammal-like Reptiles of South Africa and the Origin of Mammals", 376 pp. H. F. and G. Witherby, London.
Camp, C. L. (1956). Triassic dicynodont reptiles, Part I. The North American genus *Placerias. Mem. Univ. Calif.* **13**, 255–304.
Carroll, R. L. (1964). The earliest reptiles. *Zool. J. Linn. Soc.* **45**, 61–83.
Carroll, R. L. (1969a). A Middle Pennsylvanian captorhinomorph, and the interrelationships of primitive reptiles. *J. Paleont.* **43**, 151–170.
Carroll, R. L. (1969b). Problems of the origin of reptiles. *Biol. Rev.* **44**, 393–432.
Carroll, R. L. (1970). The ancestry of reptiles. *Phil. Trans. R. Soc.* **B257**, 267–308.
Cassiliano, M. L. and Clemens, W. A. (1979). Symmetrodonta. *In* "Mesozoic Mammals: the first two-thirds of mammalian history" (Eds J. A. Lillegraven, Z. Kielan-Jaworowska and W. A. Clemens), pp. 150–161. University of California Press, Berkeley, Los Angeles and London.
Charig, A. J. (1972). The evolution of the archosaur pelvis and hindlimb: an explanation in functional terms. *In* "Studies in Vertebrate Evolution" (Eds K. A. Joysey and T. S. Kemp), pp. 121–155. Oliver and Boyd, Edinburgh.
Chudinov, P. K. (1960). Upper Permian therapsids of the Ezhovo locality. *Palaeont. Zh. Palaeont. Inst. Acad. Sci. U.S.S.R.* **4**, 81–94. (In Russian.)
Chudinov, P. K. (1965). New facts about the fauna of the Upper Permian of the U.S.S.R. *J. Geol.* **73**, 117–140.
Clemens, W. A. (1971). Mammalian evolution in the Cretaceous. *In* "Early Mammals" (Eds D. M. Kermack and K. A. Kermack), pp. 165–180. *Zool. J. Linn. Soc.* **50**, suppl. 1.
Clemens, W. A. (1979a). A problem in morganucodontid taxonomy (Mammalia). *Zool. J. Linn. Soc.* **66**, 1–14.
Clemens, W. A. (1979b). Notes on the Monotremata. *In* "Mesozoic Mammals: the first two-thirds of mammalian history" (Eds J. A. Lillegraven, Z. Kielan-Jaworowska and W. A. Clemens), pp. 309–311. University of California Press, Berkeley, Los Angeles and London.
Clemens, W. A. and Kielan-Jaworowska, Z. (1979). Multituberculata. *In* "Mesozoic Mammals: the first two-thirds of mammalian history." (Eds J. A. Lillegraven, Z. Kielan-Jaworowska and W. A. Clemens), pp. 99–149. University of California Press, Berkeley, Los Angeles and London.
Clemens, W. A. and Mills, J. R. E. (1971). Review of *Peramus tenuirostris* Owen (Eupantotheria, Mammalia). *Bull. Br. Mus. nat. Hist. (Geol.)* **20**, 87–113.
Cluver, M. A. (1971). The cranial morphology of the dicynodont genus *Lystrosaurus. Ann. S. Afr. Mus.* **56**, 155–274.

Cluver, M. A. (1974a). The skull and mandible of a new cistecephalid dicynodont. *Ann. S. Afr. Mus.* **64**, 137–156.

Cluver, M. A. (1974b). The cranial morphology of the Lower Triassic dicynodont *Myosaurus gracilis. Ann. S. Afr. Mus.* **64**, 35–54.

Cluver, M. A. (1975). A new dicynodont reptile from the *Tapinocephalus* zone (Karoo System, Beaufort Series) of South Africa, with evidence of the jaw adductor musculature. *Ann. S. Afr. Mus.* **67**, 7–23.

Cluver, M. A. (1978). The skeleton of the mammal-like reptile *Cistecephalus* with evidence for a fossorial mode of life. *Ann. S. Afr. Mus.* **76**, 213–246.

Cluver, M. A. and Hotton, N. (1981). The Genera *Dicynodon* and *Diictodon* and their bearing on the classification of the Dicynodontia (Reptilia, Therapsida). *Ann. S. Afr. Mus.*, **83**, 99–146.

Colbert, E. H. (1948). The mammal-like reptile *Lycaenops. Bull. Am. Mus. nat. Hist.* **89**, 357–404.

Colbert, E. H. and Kitching, J. W. (1977). Triassic cynodont reptiles from Antarctica. *Am. Mus. Novit.* No. 2611, 1–30.

Cott, H. B. (1961). Scientific results of an enquiry into the ecology and economic status of the Nile crocodile (*Crocodilus niloticus*) in Uganda and Northern Rhodesia. *Trans. zool. Soc. Lond.* **29**, 211–356.

Cox, C. B. (1959). On the anatomy of a new dicynodont genus with evidence of the position of the tympanum. *Proc. zool. Soc. Lond.* **132**, 321–367.

Cox, C. B. (1964). On the palate, dentition and classification of the fossil reptile *Endothiodon* and related genera. *Am. Mus. Novit.* No. 2171, pp. 1–25.

Cox, C. B. (1965). New Triassic dicynodonts from South America, their origins and relationships. *Phil. Trans. R. Soc. Lond.* **B248**, 457–516.

Cox, C. B. (1972). A new digging dicynodont from the Upper Permian of Tanzania. *In* "Studies in Vertebrate Evolution" (Eds K. A. Joysey and T. S. Kemp), pp. 173–189. Oliver and Boyd, Edinburgh.

Crompton, A. W. (1955a). A revision of the Scaloposauridae with special reference to kinetism in this family. *Navors nas. Mus., Bloemfontein* **1**, 149–183.

Crompton, A. W. (1955b). On some Triassic cynodonts from Tanganyika. *Proc. zool. Soc. Lond.* **125**, 617–669.

Crompton, A. W. (1958). The cranial morphology of a new genus and species of ictidosaurian. *Proc. zool. Soc. Lond.* **130**, 183–216.

Crompton, A. W. (1963). On the lower jaw of *Diarthrognathus* and the origin of the mammalian lower jaw. *Proc. zool. Soc. Lond.* **140**, 697–753.

Crompton, A. W. (1964). On the skull of *Oligokyphus. Bull. Br. Mus. nat. Hist. (Geol.)* **9**, 69–82.

Crompton, A. W. (1971). The origin of the tribosphenic molar. *In* "Early Mammals" (Eds D. M. Kermack and K. A. Kermack), pp. 65–87. *Zool. J. Linn. Soc.* **50**, suppl.1.

Crompton, A. W. (1972a). The evolution of the jaw articulation of cynodonts. *In* "Studies in Vertebrate Evolution" (Eds K. A. Joysey and T. S. Kemp), pp. 231–253. Oliver and Boyd, Edinburgh.

Crompton, A. W. (1972b). Postcanine occlusion in cynodonts and tritylodontids. *Bull. Br. Mus. nat. Hist. (Geol.)* **21**, 29–71.

Crompton, A. W. (1974). The dentitions and relationships of the Southern African Triassic mammals, *Erythrotherium parringtoni* and *Megazostrodon rudnerae*. *Bull. Br. Mus. nat. Hist. (Geol.)* **24**, 399–437.

Crompton, A. W. and Ellenberger, F. (1957). On a new cynodont from the Molteno Beds and the origin of the tritylodontids. *Ann. S. Afr. Mus.* **44**, 1–14.

Crompton, A. W. and Hotton, N. III (1967). Functional morphology of the masticatory apparatus of two dicynodonts (Reptilia, Therapsida). *Postilla*, No. 109, pp. 1–51.

Crompton, A. W. and Jenkins, F. A. (1968). Molar occlusion in late Triassic mammals. *Biol. Rev.* **43**, 427–458.

Crompton, A. W. and Jenkins, F. A. (1973). Mammals from reptiles: a review of mammalian origins. *Ann. Rev. Earth Planet. Sci.* **1**, 131–155.

Crompton, A. W. and Jenkins, F. A. (1979). Origin of mammals. *In* "Mesozoic Mammals: the first two-thirds of mammalian history" (Eds J. A. Lillegraven, Z. Kielan-Jaworowska and W. A. Clemens), pp. 59–73. University of California Press, Berkeley, Los Angeles and London.

Crompton, A. W. and Parker, P. (1978). Evolution of the mammalian masticatory apparatus. *Am. Scient.* **66**, 192–201.

Crompton, A. W., Taylor, C. R. and Jagger, J. A. (1978). Evolution of homeothermy in mammals. *Nature, Lond.* **272**, 333–336.

Cruickshank, A. R. I. (1967). A new dicynodont genus from the Manda Formation of Tanzania (Tanganyika). *J. Zool. Lond.* **153**, 163–208.

Cruickshank, A. R. I. (1978). Feeding adaptations in Triassic dicynodonts. *Palaeont. afr.* **21**, 121–132.

Currie, P. J. (1977). A new haptodontine sphenacodont (Reptilia: Pelycosauria) from the Upper Pennsylvanian of North America. *J. Paleont.* **51**, 927–942.

Currie, P. J. (1979). The osteology of haptodontine sphenacodonts (Reptilia: Pelycosauria). *Palaeontographica Abt. A* **163**, 130–188.

DeMar, R. (1970). A primitive pelycosaur from the Pennsylvanian of Illinois. *J. Paleont.* **44**, 154–163.

Efremov, I. A. (1940). *Ulemosaurus svijagensis* Riab., ein Dinocephale aus den Ablangungen des Perm der U.S.S.R. *Nova Acta Acad. Caesar. Leop. Carol.* **9**, 12–205.

Efremov, I. A. (1954). Fauna of terrestrial vertebrates from the Permian copper sandstones of the Western cis-Urals. *Trudi Palaeont. Inst. Acad. Sci. U.S.S.R.* **54**, 1–416. (In Russian.)

Efremov, I. A. and Vjushkov, B. P. (1955). Catalogue of localities of Permian and Triassic terrestrial vertebrates in the territories of the U.S.S.R. *Trudi. Palaeont. Inst. Acad. Sci. U.S.S.R.* **46**, 1–185. (In Russian.)

Eldridge, N. and Gould, S. J. (1972). Punctuated equilibria: an alternative to phyletic gradualism. *In* "Models in Paleobiology" (Ed. T. J. M. Schopf), pp. 82–115. Freeman, Cooper and Co., San Francisco.

Estes, R. (1961). Cranial anatomy of the cynodont reptile *Thrinaxodon liorhinus*. *Bull. Mus. comp. Zool. Harv.* **125**, 165–180.

Ewer, R. F. (1961). The anatomy of the anomodont *Daptocephalus leoniceps* (Owen). *Proc. zool. Soc. Lond.* **136**, 375–402.

Fourie, S. (1962). Notes on a new tritylodontid from the Cave Sandstone of South Africa. *Researches nas. Mus. Bloemfontein* **2**, 7–19.

Fourie, S. (1968). The jaw articulation of *Tritylodontoides maximus*. *S. Afr. J. Sci.* **64**, 255–265.

Fourie, S. (1974). The cranial morphology of *Thrinaxodon liorhinus* Seeley. *Ann. S. Afr. Mus.* **65**, 337–400.

Fox, R. C. (1962). Two new pelycosaurs from the Lower Permian of Oklahoma. *Univ. Kansas Publ. Mus. nat. Hist.* **12**, 297–307.

Frazzetta, T. H. (1969). Adaptive problems and possibilities in the temporal fenestration of tetrapod skulls. *J. Morph.* **125**, 145–158.

GEIST, V. (1971). An ecological and behavioural explanation of mammalian characteristics, and their implication to therapsid evolution. *Z. Säugetierk.* **37**, 1–15.

GOW, C. E. (1978). The advent of herbivory in certain reptilian lineages during the Triassic. *Palaeont. afr.* **21**, 133–141.

GOW, C. E. (1980). The dentitions of the Tritheledontidae (Therapsida: Cynodontia) *Proc. R. Soc. Lond.* **B208**, 461–481.

GREGORY, W. K. (1926). The skeleton of *Moschops capensis* Broom, a dinocephalian reptile from the Permian of South Africa. *Bull. Am. Mus. nat. Hist.* **56**, 179–251.

GRIFFITHS, M. (1978). "The Biology of the Monotremes," 365 pp. Academic Press, New York, San Francisco and London.

GRINE, F. E., MITCHELL, D., GOW, C. E., KITCHING, J. W. and TURNER, B. R. (1979). Evidence for salt glands in the Triassic reptile *Diademodon* (Therapsida; Cynodontia). *Palaeont. afr.* **22**, 35–39.

GRINE, F. E. and VRBA, E. S. (1980). Prismatic enamel: a preadaptation for mammalian diphyodonty. *S. Afr. J. Sci.* **76**, 134–141.

HAINES, R. W. (1942). The tetrapod knee joint. *J. Anat.* **76**, 270–301.

HAINES, R. W. (1946). A revision of the movement of the forearm in tetrapods. *J. Anat.* **80**, 1–11.

HAUGHTON, S. H. and BRINK, A. S. (1954). A bibliographic list of Reptilia from the Karroo beds of Africa. *Palaeont. afr.* **2**, 1–187.

HEINRICH, B. (1977). Why have some animals evolved to regulate a high body temperature? *Am. Nat.* **111**, 623–640.

HENKEL, S. and KREBS, B. (1977). Der erst Fund eines Säugetier-Skelettes aus der Jura-Zeit. *Umschau* **77**, 217–218.

HOPSON, J. A. (1964). The braincase of the advanced mammal-like reptile *Bienotherium*. *Postilla* No. 87, 1–30.

HOPSON, J. A. (1966). The origin of the mammalian middle ear. *Am. Zool.* **6**, 437–450.

HOPSON, J. A. (1976). Comments on the competitive inferiority of the multituberculates. *Syst. Zool.* **16**, 352–355.

HOPSON, J. A. (1970). The classification of non-therian mammals. *J. Mammalogy* **51**, 1–9.

HOPSON, J. A. (1971). Postcanine replacement in the gomphodont cynodont *Diademodon*. *In* "Early mammals" (Eds D. M. Kermack and K. A. Kermack), pp. 1–21. *Zool. J. Linn. Soc.* **50**, Suppl. 1.

HOPSON, J. A. (1973). Endothermy, small size, and the origin of mammalian reproduction. *Am. Nat.* **107**, 446–452.

HOPSON, J. A. (1979). Paleoneurology. *In* "Biology of the Reptilia" (Ed. C. Gans), Vol. 9, pp. 39–146. Academic Press, London, New York and San Francisco.

HOPSON, J. A. and CROMPTON, A. W. (1969). Origin of mammals. *Evol. Biol.* **3**, 15–72.

HOPSON, J. A. and KITCHING, J. W. (1972). A revised classification of cynodonts (Reptilia: Therapsida). *Palaeont. afr.* **14**, 71–85.

HOTTON, N. III (1967). Stratigraphy and sedimentation in the Beaufort Series (Permian-Triassic), South Africa. *In* "Essays in Paleontology and Stratigraphy" (Eds C. Teichert and E. L. Yochelson), pp. 390–428. Dept. Univ. Kans. Spec. Publi. **2**, University of Kansas Press.

HOTTON, N. III (1974). A new dicynodont (Reptilia, Therapsida) from *Cynognathus* zone deposits of South Africa. *Ann. S. Afr. Mus.* **64**, 157–166.

JENKINS, F. A. (1970). The Chañares (Argentina) Triassic reptile fauna VII. The postcranial skeleton of the traversodontid *Massetognathus pascuali* (Therapsida, Cynodontia). *Breviora*, No. 352, 1–28.

Jenkins, F. A. (1971a). The postcranial skeleton of African cynodonts. *Bull. Peabody Mus. nat. Hist.* No. 36, 1–216.

Jenkins, F. A. (1971b). Limb posture and locomotion in the Virginia opposum *(Didelphis marsupialis)* and in other non-cursorial mammals. *J. Zool. Lond.* **165**, 303–315.

Jenkins, F. A. (1973). The functional anatomy and evolution of the mammalian humero-ulnar articulation. *Am. J. Anat.* **137**, 281–298.

Jenkins, F. A. and Crompton, A. W. (1979). Triconodonta. *In* "Mesozoic Mammals: the first two-thirds of mammalian history" (Eds J. A. Lillegraven, Z. Kielan-Jaworowska and W. A. Clemens), pp. 74–90. University of California Press, Berkeley, Los Angeles and London.

Jenkins, F. A. and Parrington, F. R. (1976). The postcranial skeleton of the Triassic mammals *Eozostrodon*, *Megazostrodon* and *Erythrotherium*. *Phil. Trans. R. Soc.* **B273**, 387–431.

Jerison, H. J. (1973). "Evolution of the Brain and Intelligence," pp. 482. Academic Press, New York, San Francisco and London.

Kemp, T. S. (1969a). On the functional morphology of the gorgonopsid skull. *Phil. Trans. R., Soc.* **B256**, 1–83.

Kemp, T. S. (1969b). The atlas-axis complex of the mammal-like reptiles. *J. Zool., Lond.* **159**, 223–248.

Kemp, T. S. (1972a). Whaitsiid Therocephalia and the origin of cynodonts. *Phil. Trans. R. Soc.* **B264**, 1–54.

Kemp, T. S. (1972b). The jaw articulation and musculature of the whaitsiid Therocephalia. *In* "Studies in Vertebrate Evolution" (Eds K. A. Joysey and T. S. Kemp), pp. 213–230. Oliver and Boyd, Edinburgh.

Kemp, T. S. (1978). Stance and gait in the hindlimb of a therocephalian mammal-like reptile. *J. Zool., Lond.* **186**, 143–161.

Kemp, T. S. (1979). The primitive cynodont *Procynosuchus*: functional anatomy of the skull and relationships. *Phil. Trans. R. Soc.* **B285**, 73–122.

Kemp, T. S. (1980a). Origin of the mammal-like reptiles. *Nature, Lond.* **283**, 378–380.

Kemp, T. S. (1980b). The primitive cynodont *Procynosuchus*: structure, function and evolution of the postcranial skeleton. *Phil. Trans. R. Soc.* **B288**, 217–258.

Kemp, T. S. (1980c). Aspect of the structure and functional anatomy of the Middle Triassic cynodont *Luangwa*. *J. Zool. Lond.* **191**, 193–239.

Kermack, K. A. (1963). The cranial structure of the triconodonts. *Phil. Trans. R. Soc.* **B246**, 83–103.

Kermack, K. A. (1967). The interrelations of early mammals. *J. Linn. Soc. (Zool.)* **47**, 241–249.

Kermack, D. M., Kermack, K. A., and Mussett, F. (1968). The Welsh pantothere *Kuehneotherium praecursoris*. *J. Linn. Soc. (Zool.)* **47**, 407–423.

Kermack, K. A. and Kielan-Jaworowska, Z. (1971). Therian and non-therian mammals. *In* "Early Mammals' (Eds. D. M. Kermack and K. A. Kermack). *Zool. J. Linn. Soc. Lond.* **50**, Suppl. 1. pp. 103–115.

Kermack, K. A., Lees, P. M. and Mussett, F. (1965). *Aegialodon dawsoni*, a new trituberculosectorial tooth from the lower Wealdon. *Proc. R. Soc. Lond.* **B162**, 535–554.

Kermack, K. A., Mussett, F. and Rigney, H. W. (1973). The lower jaw of *Morganucodon*. *Zool. J. Linn. Soc.* **53**, 87–175.

Kermack, K. A., Mussett, F. and Rigney, H. W. (1981). The skull of *Morganucodon*. *Zool. J. Linn. Soc.* **71**, 1–158.

Keyser, A. W. (1975). A re-evaluation of the cranial morphology and systematics of some tuskless Anomodontia. *Mem. geol. Surv. Rep. S. Afr.* **67**, 1–110.

Keyser, A. W. and Cruickshank, A. R. I. (1979). The origins and classification of Triassic dicynodonts. *Trans. geol. Soc. S. Afr.* **82**, 81–108.

Kielan-Jaworowska, Z. (1971). Results of the Polish-Mongolian Palaeontological Expeditions. Part III. Skull structure and affinities of the Multituberculata. *Palaeont. pol.* No. 25, 5–41.

Kielan-Jaworowska, Z. (1978). Results of the Polish-Mongolian Palaeontological Expeditions. Part VIII: Evolution of the therian mammals in the Late Cretaceous of Asia. Part III. Postcranial skeleton in the Zalambdalestidae. *Palaeont. pol.* No. 38, 5–41.

Kielan-Jaworowska, Z., Bown, T. M. and Lillegraven, J. A. (1979a). Eutheria. *In* "Mesozoic Mammals: the first two-thirds of mammalian history" (Eds J. A. Lillegraven, Z. Kielan-Jaworowska and W. A. Clemens), pp. 221–258. University of California Press, Berkeley, Los Angeles and London.

Kielan-Jaworowska, Z., Eaton, J. G. and Bown, T. M. (1979b). Theria of the metatherian-eutherian grade. *In* "Mesozoic Mammals: the first two-thirds of mammalian history", pp. 182–191. University of California Press, Berkeley, Los Angeles and London.

King, G. M. ms (1979). Permian dicynodonts from Zambia. D. Phil. Thesis. University of Oxford.

King, G. M. (1981a). The postcranial skeleton of *Robertia broomiana*, an early dicynodont (Reptilia, Therapsida) from the South African Karoo. *Ann. S. Afr. Mus.* **84**, 203–231.

King, G. M. (1981b). The functional anatomy of a Permian dicynodont. *Phil. Trans. R. Soc.* **B291**, 243–322.

King, G. M. and Cluver, M. A. (in preparation). A reassesssment of the relationships within the Permian Dicynodontia (Reptilia, Therapsida) and a new classification of dicynodonts. *Ann. S. Afr. Mus.*

Kitching, J. W. (1977). The distribution of the Karroo vertebrate fauna. *Mem. No. 1.* Bernard Price Institute for Palaeontological Research. University of Witwatersrand. pp. 131.

Kraus, M. J. (1979) Eupantotheria. *In* "Mesozoic Mammals: the first two-thirds of mammalian history" (Eds J. A. Lillegraven, Z. Kielan-Jaworowska and W. A. Clemens), pp. 162–171. University of California Press, Berkeley, Los Angeles and London.

Krebs, B. (1971). Evolution of the mandible and lower dentition in dryolestids (Pantotheria, Mammalia). *In* "Early Mammals" (Eds D. M. Kermack and K. A. Kermack), pp. 89–102. *Zool. J. Linn. Soc.* **50**, Suppl. 1.

Kron, D. G. (1979). Docodonta. *In* "Mesozoic Mammals: the first two-thirds of mammalian history" (Eds J. A. Lillegraven, Z. Kielan-Jaworowska and W.A. Clemens), pp. 91–98. University of California Press, Berkeley, Los Angeles and London.

Kühne, W. G. (1956). "The Liassic Therapsid Oligokyphus," pp. 149. British Museum (Natural History), London.

Kutty, T. S. (1972). Permian reptilian fauna from India. *Nature, Lond.* **237**, 462–463.

Langston, W. J. (1965). *Oedaleops campi* (Reptilia: Pelycosauria) new genus and species from the Lower Permian of New Mexico, and the family Eothyrididae. *Tex. Mem. Mus. Bull.* **9**, 5–47.

Lillegraven, J. A., Kielan-Jaworowska, Z. and Clemens, W. A. (Eds) (1979). "Mesozoic Mammals: the first two thirds of mammalian history," pp.311. University of California Press, Berkeley, Los Angeles and London.

Lombard, R. E. and Bolt, J. R. (1978) Evolution of the tetrapod ear: an analysis and reinterpretation. *Biol. J. Linn. Soc.* **11**, 19–76.

MacIntyre, G. T. (1967). Foramen pseudovale and quasi-mammals. *Evolution* **21**, 834–841.

McKenna, M. C. (1969). The origin and early differentiation of therian mammals. *Ann. N. Y. Acad. Sci.* **167**, 217–240.

McNab, B. K. (1978). The evolution of endothermy in the phylogeny of mammals. *Am. Nat.* **112**, 1–21.

Manley, G. A. (1972). A review of some current concepts of the functional evolution of the ear in terrestrial vertebrates. *Evolution* **26**, 608–621.

Marshall, L. G. (1979). Evolution of metatherian and eutherian (mammalian) characters: a review based on cladistic methodology. *Zool. J. Linn. Soc.* **66**, 369–410.

Mendrez, Ch. H. (1972). On the skull of *Regisaurus jacobi*, a new genus and species of Bauriamorpha Watson and Romer 1956 (=Scaloposauria Boonstra 1953), from the *Lystrosaurus*-zone of South Africa. *In* "Studies in Vertebrate Evolution" (Eds K. A. Joysey and T. S. Kemp), pp. 191–212. Oliver and Boyd, Edinburgh.

Mendrez, Ch. H. (1974a). Etude du crâne d'un jeune specimen de *Moschorhinus kitchingi* Broom, 1920 (? *Tigrisuchus simus* Owen, 1876), Therocephalia Pristerosauria Moschorhinidae d'Afrique australe. *Ann. S. Afr. Mus.* **64**, 71–115.

Mendrez, Ch. H. (1974b). A new specimen of *Promoschorhynchus platyrhinus* Brink 1954 (Moschorhinidae) from the *Daptocephalus*-zone (Upper Permian) of South Africa. *Palaeont. afr.* **17**, 69–85.

Mendrez, Ch. H. (1975). Principales variations du palais chez les thérocéphales Sud-Africains (Pristerosauria et Scaloposauria) au cours du Permien Supérieur et du Trias Inférieur. *Colloque International C.N.R.S. No. 218 (Paris). Problèmes actuels de paléontologie–évolution des vertébrés.* pp. 379–408.

Mendrez-Carroll, Ch. H. (1979). Nouvelle étude du crâne du type de *Scaloposaurus constrictus* Owen, 1876, de la zone à *Cistecephalus* (Permien supérieur) d'Afrique australe. *Bull. Mus. Natn. Hist. nat., Paris*, 4e sér., **1**, *section C*, No. 3, 155–201.

Mills, J. R. E. (1971). The dentition of *Morganucodon*. *In* "Early Mammals" (Eds D. M. Kermack and K. A. Kermack.) pp. 29–63. *Zool. J. Linn. Soc. Lond.* **50**, Suppl. 1.

Milner, A. R. and Panchen, A. L. (1973). Geographical variation in the tetrapod faunas of the Upper Carboniferous and Lower Permian. *In* "Implications of Continental Drift to the Earth Sciences" (Eds D. H. Tarling and S. K. Runcorn), Vol. I, pp. 353–368. Academic Press, New York, San Francisco, London.

Olson, E. C. (1944). The origin of mammals based upon the cranial morphology of the therapsid suborders. *Spec. Pap. Geol. Soc. Am* . **55**, 1–136.

Olson, E. C. (1962). Late Permian terrestrial vertebrates, U.S.A. and U.S.S.R. *Trans. Am. Phil. Soc.* **52**, 3–224.

Olson, E. C. (1965). Vertebrates from the Chickasha Formation Permian of Oklahoma. *Okla. Geol. Survey*, Circular 70, p. 70.

Olson, E. C. (1968). The family Caseidae. *Fieldiana Geol.* **17**, 225–349.

Olson, E. C. (1971). "Vertebrate Paleozoology," pp. 839. Wiley–Interscience, New York, London, Sydney, Toronto.

Olson, E. C. (1974). On the source of the therapsids. *Ann. S. Afr. Mus.* **64**, 27–46.

Olson, E. C. (1975). Permo-Carboniferous paleoecology and morphotypic series. *Am. Zool.* **15**, 371–389.

Olson, E. C. and Vaughn, P. P. (1970). The changes of terrestrial vertebrates and climates during the Permian of North America. *Forma et Functio* **3**, 113–138.

Orlov, Y. A. (1958). The carnivorous dinocephalians of the Isheevo fauna (titanosuchians). *Trudi. Paleon. Inst. Acad. Sci. U.S.S.R.* **72**, 3–113. (In Russian.)

Osborn, J. W. and Crompton, A. W. (1973). The evolution of mammalian from reptilian dentitions. *Breviora*, No. 399, 1–18.

Panchen, A. L. (1972). The interrelationships of the earliest tetrapods. *In* "Studies in Vertebrate Evolution" (Eds K. A. Joysey and T. S. Kemp), pp. 65–87. Oliver and Boyd, Edinburgh.

Panchen, A. L. (1977). Geographical and ecological distribution of the earliest tetrapods. *In* (Eds M. K. Hecht, P. C. Goody and B. M. Hecht), pp. 723–738. "Major Patterns in Vertebrate Evolution," Plenum, New York.

Parrington, F. R. (1934). On the cynodont genus *Galesaurus*, with a note on the functional significance of the changes in the evolution of the theriodont skull. *Ann. Mag. nat. Hist.* ser. 10, **13**, 38–67.

Parrington, F. R. (1936). On the tooth-replacement in the theriodont reptiles. *Phil. Trans. R. Soc.* **226B**, 121–142.

Parrington, F. R. (1939). On the digital formulae of theriodont reptiles. *Ann. Mag. nat. Hist.* Ser. 11, **3**, 209–214.

Parrington, F. R. (1945). On the middle ear of the Anomodontia. *Ann. Mag. nat. Hist.* Ser. 11, **12**, 625–631.

Parrington, F. R. (1946). On the cranial anatomy of cynodonts. *Proc. zool. Soc. Lond.* **116**, 181–197.

Parrington, F. R. (1947). On a collection of Rhaetic mammalian teeth. *Proc. zool. Soc. Lond.* **116**, 707–728.

Parrington, F. R. (1948). Labyrinthodonts from South Africa. *Proc. zool. Soc. Lond.* **118**, 426–445.

Parrington, F. R. (1955). On the cranial anatomy of some gorgonopsids and the synapsid middle ear. *Proc. zool. Soc. Lond.* **125**, 1–40.

Parrington, F. R. (1971). On the Upper Triassic mammals. *Phil. Trans. R. Soc.* **261B**, 231–272.

Parrington, F. R. (1973). The dentitions of the earliest mammals. *Zool. J. Linn. Soc. Lond.* **52**, 85–95.

Parrington, F. R. (1978). A further account of the Triassic mammals. *Phil. Trans. R. Soc.* **282B**, 177–204.

Parrington, F. R. (1979). The evolution of the mammalian middle and outer ears: a personal review. *Biol. Rev.* **54**, 369–387.

Patterson, B. and Olson, E. C. (1961). A triconodontid mammal from the Triassic of Yunnan. *In* "International Colloquium on the Evolution of Lower and non Specialised mammals" (Ed. G. Vandebroek), Part 1, pp. 129–191. Kon. Vlaamse Acad. Wetensch., Lett. Schone Kunsten Belgie, Brussels.

Peabody, F. E. (1957). Pennsylvanian reptiles of Garnett, Kansas: edaphosaurs. *J. Paleont.* **31**, 947–949.

Pearson, R. (1978). "Climate and Evolution," pp. 274. Academic Press, London, New York and San Francisco.

Pearson, H. S. (1924). A dicynodont reptile reconstructed. *Proc. zool. Soc. Lond.* for 1924. pp. 827–855.

Plumstead, E. P. (1973). The late Palaeozoic *Glossopteris* flora. *In* "Atlas of Palaeobiogeography" (Ed. A. Hallam), pp. 186–205. Elsevier, Amsterdam, London and New York.

Presley, R. and Steel, F. L. D. (1976). On the homology of the alisphenoid. *J. anat.* **121**, 441–459.

Reisz, R. (1972). Pelycosaurian reptiles from the Middle Pennsylvanian of North America. *Bull. Mus. comp. Zool. Harv.* **144**, 27–62.

REISZ, R. (1975). Pennsylvanian pelycosaurs from Linton, Ohio and Nýřany, Czechoslovakia. *J. Paleont.* **49**, 522-527.

REISZ, R. (1981). The Pelycosauria: a review of phylogenetic relationships. *In* "The Terrestrial Environment and the Origin of Land Vertebrates" (Ed. A. L. Panchen), pp. 553–591 The Systematics Association Special Volume No. 15.

RICQLÈS, A. de (1974). Evolution of endothermy. *Evolutionary Theory*, **1**, 51–80.

ROBINSON, P. L. (1971). A problem of faunal replacement on Permo-Triassic continents. *Palaeontology* **14**, 131–153.

ROBINSON, P. L. (1973). Palaeoclimatology and continental drift. *In* "Implications of Continental Drift to the Earth Sciences" (Eds D. H. Tarling and S. K. Runcorn), Vol. 1, pp. 449–474 Academic Press, London and New York.

ROMER, A. S. (1922). The locomotor apparatus of certain primitive and mammal-like reptiles. *Bull. Am. Mus. nat. Hist.* **46**, 517–606.

ROMER, A. S. (1966). "Vertebrate Paleontology," 3rd Ed., pp. 468. University of Chicago Press, Chicago and London.

ROMER, A. S. (1967). The Chañares (Argentina) Triassic reptile fauna III. Two new gomphodonts, *Massetognathus pascuali* and *M. teruggii*. *Breviora*, No. 264, 1–25.

ROMER, A. S. (1969a). The Chañares (Argentina) Triassic reptile fauna V. A new chiniquodont cynodont *Probelesodon lewisi*—cynodont ancestry. *Breviora*, No. 333, 1–24.

ROMER, A. S. (1969b). The Brazilian cynodont reptiles *Belesodon* and *Chiniquodon*. *Breviora*, No. 332, 1–16.

ROMER, A. S. (1970). The Chañares (Argentina) Triassic reptile fauna VI. A chiniquodontid cynodont with an incipient squamosal-dentary jaw articulation. *Breviora* No. 344, 1–18.

ROMER, A. S. (1973). The Chañares (Argentina) Triassic reptile fauna XVIII. *Probelesodon minor*, a new species of carnivorous cynodont; family Probainognathidae nov. *Breviora* No. 401, 1–4.

ROMER, A. S. and LEWIS, A. D. (1973). The Chañares (Argentina) Triassic reptile fauna 19. Postcranial material of the cynodonts *Probelesodon* and *Probainognathus*. *Breviora* No. 407, 1–26.

ROMER, A. S. and PRICE, L. W. (1940). Review of the Pelycosauria. *Spec. Pap. Geol. Soc. Am.* No. 28, 1–538.

RUSSELL, D., RUSSELL, D. and WOUTERS, G. (1976). Une dent d'aspect mammalian en provenance du Rhetien Francais. *Geobios* **4**, 377–392.

SCHAEFFER, B. (1941). The morphological and functional evolution of the tarsus in amphibians and reptiles. *Bull. Am. Mus. nat. Hist.* **78**, 395–472.

SIGOGNEAU, D. (1968). On the classification of the Gorgonopsia. *Palaeont. afr.* **11**, 33–46.

SIGOGNEAU, D. (1970a). Révision systématique des gorgonopsiens. *Editions du Centre National de la Recherche Scientifique*, Paris., pp. 1–416.

SIGOGNEAU, D. (1970b). Contribution à la connaissance des ictidorhinidés (*Gorgonopsia*). *Palaeont. afr.* **13**, 25–38.

SIGOGNEAU, D. and CHUDINOV, P. K. (1972). Reflections on some Russian eotheriodonts. *Palaeovertebrata* **5**, 79–109.

SIGOGNEAU-RUSSELL, D. and RUSSELL, D. E. (1974). Étude du premier Caséidé (Reptilia, Pelycosauria) d'Europe occidentale. *Bulletin du Muséum National d'Histoire Naturelle*. 3rd Series No. 230, Sciences de la Terre 38.

SIMPSON, G. G. (1928) "A Catalogue of the Mesozoic Mammalia in the Geological Department of the British Museum," pp. 215. British Museum (Natural History), London.

Simpson, G. G. (1929). American Mesozoic Mammalia. *Peabody Mus. Mem.* **3**, 1–171.
Simpson, G. G. (1949). "The Meaning of Evolution," 364 pp. Yale University Press, New Haven, Connecticut.
Simpson, G. G. (1953). "The Major Features of Evolution," 434 pp. Columbia University Press, New York.
Simpson, G. G. (1960). Diagnosis of the classes Reptilia and Mammalia. *Evolution, Lancaster, Pa.* **14**, 388–392.
Stanley, S. M. (1979). "Macroevolution, Pattern and Process," 332 pp. W. H. Freeman, San Francisco.
Stovall, J. W., Price, L. W. and Romer, A. S. (1966). The postcranial skeleton of the giant Permian pelycosaur *Cotylorhynchus romeri. Bull. Mus. comp. Zool. Harv.* **135**, 1–30.
Tatarinov, L. P. (1963). New late Permian therocephalian. *Palaeont. Zh. for 1963,* 76–94. (In Russian.)
Tatarinov, L. P. (1964). Anatomy of the therocephalian head. *Palaeont. Zh.* for 1964, 72–84. (In Russian.)
Tatarinov, L. P. (1967). The development of a system of blood vessels and nerves for the lips in the theriodonts. *Palaeont. Zh.* for 1967, 3–17 (In Russian.)
Tatarinov, L. P. (1968). Morphology and systematics of the Northern Dvina cynodonts (Reptilia, Therapsida; Upper Permian) *Postilla,* No. 126, 1–15.
Tatarinov, L. P. (1974). Theriodonts of U.S.S.R. *Trudi Palaeont. Inst. Acad. Sci. U.S.S.R.* **143**, 5–250. (In Russian.)
Van den Heever, J. A. (1980). On the validity of the therocephalian family Lycosuchidae (Reptilia, Therapsida). *Ann. S. Afr. Mus.* **81**, 111–125.
Van Heerden, J. (1976). The cranial anatomy of *Nanictosaurus rubidgei* Broom and the classification of the Cynodontia (Reptilia: Therapsida). *Navors. nas. Mus. Bloemfontein* **3**, 141–164.
Vaughn, P. P. (1958). On a new pelycosaur from the Lower Permian of Oklahoma, and on the origin of the family Caseidae. *J. Paleont.* **32**, 981–991.
Watson, D. M. S. (1916). The monotreme skull: a contribution to mammalian morphogenesis. *Phil. Trans. R. Soc.* **B207**, 311–374.
Watson, D. M. S. (1921). The bases of the classification of the Theriodonta. *Proc. zool. Soc. Lond.* for 1921, 35–98.
Watson, D. M. S. (1931). On the skeleton of a bauriamorph reptile. *Proc. zool. Soc. Lond.* for 1931, 1163–1205.
Watson, D. M. S. (1948). *Dicynodon* and its allies. *Proc. zool. Soc. Lond.* **118**, 823–877.
Watson,. D. M. S. (1960). The anomodont skeleton. *Trans. zool. Soc. Lond.* **29**, 131–208.
Watson, D. M. S. and Romer, A. S. (1956). A classification of therapsid reptiles. *Bull. Mus. comp. Zool. Harv.* **111**, 37–89.
Webster, D. B. (1966). Ear structure and function in modern mammals. *Am. Zool.* **6**, 451–466.
Westoll, T. S. (1945). The mammalian middle ear. *Nature, Lond.* **155**, 114–115.
Woodburne, M. O. and Tedford, R. H. (1975). The first Tertiary monotreme from Australia. *Am. Mus. Novitates* No. 2588, 1–11.

# *Appendix: Classification of the Mammal-like Reptiles*

The classification presented below is not cladistic (see page 14). Paraphyletic taxa, which are taxa not containing all the descendants of the hypothetical ancestor of the taxon, have been retained. Some suspected or even well-established relationships have not been expressed by the creation of appropriate higher taxa. Therefore this classification remains readily comparable to existing classifications such as that of Romer (1966), and the creation of unfamiliar new taxa has been kept to a minimum. The most important change concerns the recognition of the relationship between the Therocephalia and the Cynodontia, which is expressed by the new taxon Therosauria. This has been removed from the Therapsida and elevated to the status of an Order, alongside the Orders Pelycosauria and Therapsida. All these three main divisions of the mammal-like reptiles are paraphyletic; the Pelycosauria contains the ancestor of the Therapsida; the Therapsida contains the ancestor of the Therosauria; the Therosauria contains the ancestor of the mammals. Nevertheless, this tripartite division has the value of reflecting the three main stages in the evolution of the mammal-like reptiles.

At a lower taxonomic level, several new taxa are introduced to reflect some of the more important proposed new relationships. Within the Pelycosauria, a Suborder Caseamorpha is created on the assumption that the eothyridids are the sister-group of the caseids. The Suborder Eupelycosauria is similarly new, and contains the family Ophiacodontidae along with its undoubted relatives the sphenacodontids.

Within the now restricted Order Therapsida, the Dinocephalia have been removed from the Anomodontia, since there is no satisfactory basis for a belief in the close relationship of these two groups. The gorgonopsians have been combined with the other primitive carnivorous therapsids as the Suborder Eotheriodontia, which is probably paraphyletic.

The classification of the Suborder Cynodontia, within the new Order Therosauria, has been modified from present schemes. A paraphyletic Infraorder Procynosuchia has been created for reception of the primitive cynodonts, including the semi-advanced family Galesauridae. The belief that all the advanced cynodonts arose monophyletically is indicated by the creation of the Infraorder Eucynodontia for their reception. No attempt has been made to recognise formally the interrelationships between the various families of eucynodonts, and both the Tritylodontidae and the Tritheledontidae have been left in this group, notwithstanding the possibility that one or both of them may prove to be better classified with the mammals in the future.

## ORDER PELYCOSAURIA

| | | |
|---|---|---|
| Suborder Caseamorpha | Family Eothyrididae | |
| | Family Caseidae | |
| Suborder Edaphosauria | Family Edaphosauridae | |
| Suborder Eupelycosauria | Family Ophiacodontidae | |
| | Family Varanopidae | |
| | Family Sphenacodontidae | Subfamily Haptodontinae |
| | | Subfamily Secodontosaurin |
| | | Subfamily Sphenacodontina |

## ORDER THERAPSIDA

| | | |
|---|---|---|
| Suborder Dinocephalia | Infraorder Estemmenosuchia | Family Estemmenosuchidae |
| | Infraorder Brithopia | Family Brithopodidae |
| | | Family Anteosauridae |
| | Infraorder Titanosuchia | Family Titanosuchidae |
| | | Family Tapinocephalidae |
| Suborder Eotheriodontia | Infraorder Eotitanosuchia | Family Phthinosuchidae |
| | | Family Eotitanosuchidae |
| | Infraorder Gorgonopsia | Family Ictidorhinidae |
| | | Family Gorgonopsidae |
| Suborder Anomodontia | Infraorder Venjukovoidea | Family Otsheriidae |
| | | Family Venjukoviidae |
| | | ? Family Dromasauridae |
| | Infraorder Dicynodontia | Family Eodicynodontidae |
| | | Family Endothiodontidae |
| | | Family Kingoridae |
| | | Family Diictodontidae |
| | | Family Dicynodontidae |
| | | Family Lystrosauridae |
| | | Family Kannemeyeriidae |

ORDER THEROSAURIA

| | | |
|---|---|---|
| Suborder Therocephalia | Family Crapartinellidae | |
| | Family Pristerognathidae | |
| | Family Moschorhinidae | |
| | Family Whaitsiidae | |
| | Family Ictidosuchidae | |
| | Family Scaloposauridae | |
| | Family Ericiolacertidae | |
| | Family Bauriidae | |
| Suborder Cynodontia | Infraorder Procynosuchia | Family Procynosuchidae |
| | | Family Dviniidae |
| | | Family Galesauridae |
| | Infraorder Eucynodontia | Family Cynognathidae |
| | | Family Diademodontidae |
| | | Family Traversodontidae |
| | | Family Chiniquodontidae |
| | | Family Tritylodontidae |
| | | Family Tritheledontidae |

# Notes Added in Proof

1. (p. 33). Langston and Reisz (1981) have recently described an almost complete skeleton of a varanopid sphenacodont from the Arroyo Formation in New Mexico as a new species of *Aerosaurus*.

2. (p. 138). *Myosaurus* is now known to occur in the Fremouw Formation of Antarctica, in beds of *Lystrosaurus*-zone age (Hammer and Cosgriff, 1981).

3. (p. 216). Bonaparte (1980) recently described another South American tritheledontid. It consists of a poorly preserved skull, named *Chaliminia*, and has a dentition similar to that of *Pachygenelus* (Fig. 86I). A specialisation unique to these two genera is an edentulous gap between the three upper incisors of each side, into which the lower incisors fitted when the jaws closed. *Chaliminia* also has an interpterygoid vacuity like that of *Diarthognathus*, indicating that this is a true feature of the family. However, Bonaparte also mentions the presence of a similar vacuity in the chiniquodontid *Probelesodon* (p. 208). The significance of such a structure thus remains rather obscure.

4. (p. 263). Gill (1974) quotes the dental formula of *Kuehneotherium* as $I_4$ $C_1$ $PM_6$ $M_{4-5}$. She also shows that resorbtion rather than replacement of the anterior premolars occurs, as is the case in morganucodontids. This seems to imply that the two types of animal had similar patterns of tooth replacement.

### *References*

BONAPARTE, J. F. (1980). El primer ictidosaurio (Reptilia-Therapsida) de America del sur, *Chaliminia musteloides*, del Triasico Superior de la Rioja, Republica Argentina. *Actas II* Congreso Argentino de Paleontologia y Biostratigrafia y I Congreso Latinamericano de Paleontologia. Buenos Aires 1978. 123–133.

GILL, P. G. (1974). Resorption of premolars in the early mammal *Kuehneotherium praecursoris*. *Archs oral Biol.* **19**, 327–328.

HAMMER, W. R. and COSGRIFF, J. W. (1981). *Myosaurus gracilis*, an anomodont reptile from the Lower Triassic of Antarctica and South Africa. *J. Paleont.* **55**, 410–424.

Langston, W. Jr and Reisz, R. R. (1981). *Aerosaurus wellesi*, new species, a varanopseid mammal-like reptile (Synapsida: Pelycosauria) from the Lower Permian of New Mexico. *J. Vert. Paleont.* **1**, 73–96.

# Subject Index

Numbers in italics indicate pages on which entries relate to figure legends.